Stochastic Processes

(Photo by Walter L. Smith)

Gopinath Kallianpur, Chapel Hill, 1980

Stamatis Cambanis   Jayanta K. Ghosh
Rajeeva L. Karandikar   Pranab K. Sen
Editors

Stochastic Processes

A Festschrift in Honour of Gopinath Kallianpur

Springer-Verlag

New York Berlin Heidelberg London Paris
Tokyo Hong Kong Barcelona Budapest

MATH-STAT.

Stamatis Cambanis
Department of Statistics
University of North Carolina
Chapel Hill, NC 27599-3260 USA

Jayanta K. Ghosh
Indian Statistical Institute
203, B.T. Road
Calcutta, 700035 India

Rajeeva L. Karandikar
Indian Statistical Institute
7, S.J. S. Sansanwal Marg
New Delhi, 110016 India

Pranab K. Sen
Department of Statistics
University of North Carolina
Chapel Hill, NC 27599-3260 USA

With eight illustrations.

Mathematics Subject Classifications (1991): 60G, 60H

Library of Congress Cataloging-in-Publication Data
Stochastic processes : a festschrift in honour of Gopinath Kallianpur /
    Stamatis Cambanis . . . [et al.].
        p.   cm.
    Includes bibliographical references
    ISBN 0-387-97921-2
    1. Stochastic processes.   I. Cambanis, S. (Stamatis), 1943–   .
II. Kallianpur, G.
QA274.S822   1992
519.2 – dc20   92-31111

Printed on acid-free paper.

Production managed by Henry Krell; manufacturing supervised by Genieve Shaw.
Camera-ready copy prepared by the editors.
Printed and bound by Edwards Brothers, Inc., Ann Arbor, MI.
Printed in the United States of America.

9 8 7 6 5 4 3 2 1

ISBN 0-387-97921-2 Springer-Verlag New York Berlin Heidelberg
ISBN 3-540-97921-2 Springer-Verlag Berlin Heidelberg New York

Preface

*On behalf of those of us who in various ways have con-
tributed to this volume, and on behalf of all of his colleagues,
students and friends throughout the world-wide scientific com-
munity, we dedicate this volume to* Gopinath Kallianpur *as a
tribute to his work and in appreciation for the insights which he
has so graciously and generously offered, and continues to offer,
to all of us.*

Stochastic Processes contains 41 articles related to and frequently influenced by Kallianpur's work. We regret that space considerations prevented
us from including contributions from his numerous colleagues (at North
Carolina, ISI, Minnesota, Michigan), former students, co–authors and other
eminent scientists whose work is akin to Kallianpur's. This would have
taken several more volumes!

All articles have been refereed, and for their valuable assistance in this
we thank many of the contributing authors, as well as: R. Bradley, M.H.A.
Davis, R. Davis, J. Hawkins, J. Horowitz, C. Houdré, N.C. Jain, C. Ji,
P. Kokoszka, T. Kurtz, K.S. Lau, W. Linde, D. Monrad, D. Stroook, D.
Surgailis and S. Yakowitz.

We also thank June Maxwell for editorial assistance, Peggy Ravitch for
help with the production of the volume, and Lisa Brooks for secretarial
assistance. Finally, we are indebted to Dr. Martin Gilchrist, the Statistics
editor, and the Springer editorial board for their excellent cooperation and
enthusiastic support throughout this project.

Stamatis Cambanis
Jayanta K. Ghosh
Rajeeva L. Karandikar
Pranab K. Sen
July, 1992

Contents

Gopinath Kallianpur

Early education in India. Gopinath Kallianpur was born on April 16, 1925 in Mangalore, India. After attending schools in various towns in South Kanara District (now a part of the state of Karnataka), he spent his first two undergraduate years at St. Aloysius College, Mangalore. He received his B.A. (Honours) in 1945 and M.A. in 1946 from Madras University at St. Joseph's College, Trichy, in south India. Both degrees were in Mathematics and both were with First Class Honours. In those days, the Indian Civil Service was considered to be the most lucrative career for the country's best students, but Kallianpur's academic bent of mind and fascination with mathematics led him to pursue a career in mathematical research. He went to Bombay immediately after he received his M.A., found that research scholarships in mathematics were very scarce indeed, and so he took a position as lecturer in Mathematics at a local college.

It was during his stay in Bombay that Kallianpur first became aware that probability and statistics were important branches of the mathematical sciences. He attended a course of lectures by Professor D.D. Kosambi at the Royal Institute of Science which were based on Kolmogorov's monograph, *Foundations of Probability* which at that time had not been translated into English. His introduction to statistics was through Cramér's classical book *Mathematical Methods of Statistics*, which had been published a year earlier and copies of which had only just reached India.

Studies in the U.S. Though reluctant at first to leave his homeland, he was strongly encouraged by his colleagues to go abroad for higher studies. In the summer of 1949, Kallianpur came to Chapel Hill and initially enrolled in the Master's program. The Department of Mathematical Statistics, as it was then called, was small but with a star-studded faculty. At the end of the summer session, the chairman, Harold Hotelling, convinced Kallianpur to switch to the Ph.D. program and provided him with a research assistantship. Besides Hotelling, Kallianpur's teachers were Raj Chandra Bose, Samarendra Nath Roy, and two young assistant professors, Wassily Hoeffding and Herbert Robbins. Kallianpur worked with Robbins in the newly emerging field of stochastic processes and one summer he went to the University of California at Berkeley to attend a course on stochastic processes taught by Paul Lévy.

After finishing his Ph.D. in 1951, Kallianpur went for a year to the statistical laboratory at Berkeley, the forerunner of the present Department of Statistics. He spent 1952–53 as a Member of the Institute for Advanced

Study (IAS) at Princeton where he continued working with Robbins, who was visiting there. They proved certain types of limit theorems for sums of independent random variables and obtained weak ergodic theorems for one- and two-dimensional Brownian motion. Though the methods have now been superseded by more powerful techniques, there has been a revival of interest in these results which have been considerably generalized.

At the Indian Statistical Institute (ISI). In 1953, Kallianpur returned to India and joined the faculty at the ISI in Calcutta. He had some misgivings, for at that time the ISI was primarily involved in such applied work as sample surveys, crop experiments, economic planning, etc., and he expected to find that the theoretical work he had been doing might not exactly be encouraged. However, he soon found that he had complete freedom to work in any area of his choice, a tradition at the ISI that has continued to this day.

Many of Kallianpur's friends and colleagues who have known him in recent years would be surprised to learn that he took part in some of the applied activities. One was the problem of estimating areas using line grids. An empirical method of estimation had been devised by a colleague, J.M. Sengupta. Kallianpur was asked by Professor Mahalanobis, director of the ISI, to see if this was an unbiased estimate and if so, to prove it and obtain a formula for the variance. This he did, and the results he obtained were similar to the power of chords formulae associated with M.W. Crofton in integral geometry. This work was published in 1992. Also, the ISI was engaged in preparing a draft for the second five year plan for the Indian economy (subsequently implemented by the government), and at the urging of Mahalanobis, Kallianpur also took part in these planning activities.

Since 1938, R.A. Fisher had been a regular visitor to the ISI Calcutta University complex and, during his 1955 visit, gave seminars on asymptotic properties of maximum likelihood estimates (m.l.e.). Inspired by these lectures, Kallianpur, together with C.R. Rao, introduced the notion of Fisher consistency and they were able to derive Fisher's lower bound for the asymptotic variance for a class of Fisher consistent estimators which are smooth functionals of the empirical distribution function. Subsequently, Kallianpur showed that the m.l.e. is a member of this class. The work used techniques such as Fréchet derivatives on the space of estimators. At the time, the work did not receive much attention, but recently this approach has generated considerable interest.

Professor Norbert Wiener made two visits to Calcutta. The first was a short trip in 1954 and the second in 1955 lasted five months. Kallianpur and Wiener interacted extensively, working on problems of nonlinear prediction. At Wiener's initiative, Kallianpur took a leave of absence and assumed a faculty position at Michigan State University.

Michigan, Indiana, Minnesota. The main purpose of his leave was to continue his work with Wiener; however, Kallianpur became seriously ill soon after his arrival in 1956 and was hospitalized for a prolonged period. Though the planned collaboration did not materialize, Wiener's ideas had a profound influence on Kallianpur. When they later met at Indiana University, where Kallianpur spent 1959-1961, Wiener stressed the importance of developing a theory of nonlinear prediction based on his homogeneous chaos expansions.

In the early sixties, R.E. Kalman and R.S. Bucy, with engineering applications in mind, considered the problem of prediction and filtering in the following framework: The signal process is the output of a linear dynamical system driven by Gaussian white noise and the signal is observed corrupted by noise, assumed to be Gaussian white noise. In this approach, the stationarity assumption on the signal process (a basic requirement in the Kolmogorov–Wiener theory) is unnecessary and the observations need not be over an infinite time interval. Also, the prediction or filtering can be done recursively.

Kallianpur returned to Michigan State University in 1961, and in 1963 he moved to the University of Minnesota where he stayed until 1976. He began working on a nonlinear analogue of this theory with Professor C. Striebel. They considered the special case when signal and observational noise are independent and obtained a function space version of the Bayes formula which was an important step in the derivation of the stochastic differential equation (SDE) for the optimal filter. This has since played a central role in the so-called theory of robust filtering.

In collaboration with M. Fujisaki and H. Kunita, Kallianpur also derived an SDE for the general case of correlated signal noise. The resulting paper contains an early example of a stochastic integral representation for square integrable martingales, which is more general then the standard result for Brownian motion.

Kallianpur worked extensively individually and with colleagues on Gaussian measures and Gaussian processes. He used the reproducing kernel Hilbert space to study linear and subsequently nonlinear problems in Gaussian processes, their equivalence and singularity, the nonanticipative representation of equivalent Gaussian processes, zero one laws and the support for Gaussian measures, the structure of abstract Wiener spaces, etc.

During this period Kallianpur's pioneering work on filtering theory and on Gaussian processes served as inspiration to an entire generation of mathematicians and mathematically inclined engineers.

At the ISI as Director. After thirteen years at the University of Minnesota, Kallianpur agreed to take over the directorship of the ISI in 1976, a post he held until 1979, on leave from the University of Minnesota.

The editors of this volume all began extensive contact with him during the late seventies. J.K. Ghosh was a professor and R.L. Karandikar a graduate student at ISI and, when Kallianpur moved to Chapel Hill in 1979, he became an inspiring colleague of S. Cambanis and P.K. Sen.

When Kallianpur became director of the ISI, its main campus was in Calcutta, and a center had been established in Delhi. At his initiative and as a result of his persistent efforts, another center was established in Bangalore. Since then, the Bangalore Center has continued to grow and now has an active research faculty and graduate program in statistics and probability. His efforts to improve the research atmosphere knew no bounds. He arranged for many visitors to come to the ISI, and he cut through red tape to set up laboratories without which, for example, J.K. Ghosh's long term collaboration with one of the Institutes leading geologists, Supriya Sengupta, would not have borne fruit. He arranged with the Ford Foundation to provide funds to send ISI scholars abroad and this is how Kesar Singh went to Stanford where he wrote his now classic paper on the bootstrap.

Kallianpur also took a keen interest in improving the living conditions of the students as well as their access to library facilities. Steps initiated by him have had a long term impact on the students at ISI Calcutta.

During the ISI years Kallianpur wrote a book on *Stochastic Filtering Theory* (published in 1980), which included such topics as stochastic differential equations with memory, homogeneous chaos expansions, reproducing kernel Hilbert spaces, etc. He also joined the consortium of editors of Sankhyā, the Indian Journal of Statistics, a post he still holds.

At North Carolina. In 1979, Kallianpur returned to the Department of Statistics of the University of North Carolina at Chapel Hill as Alumni Distinguished Professor. Soon after his arrival he established, along with his colleagues S. Cambanis and M.R. Leadbetter, the Center for Stochastic Processes within the Statistics Department. Since the early eighties, the Center has provided the framework for substantial research in the area of stochastic processes, and for significant interaction among faculty, visitors and graduate students. His contributions to this enterprise have been both intensive and enjoyable and his tireless efforts are reflected in the Center's technical report series.

Meanwhile, the developments in nonlinear filtering theory had to meet the criticism that the results obtained were not useful in practice since they could not be implemented. Another objection concerned modelling noise via Brownian motion. Professor A.V. Balakrishnan advocated the use of white noise defined over a finitely additive probability space to model the observational noise. In another context, these ideas had originated with I.E. Segal.

In 1982, in collaboration with Karandikar, Kallianpur began a systematic study of cylinder measures on Hilbert spaces and Gaussian white noise. Using this theory, equations for the optimal filter were derived. For the case of finite dimensional signal processes, these equations turned out to be partial differential equations in which the observation path appears as a parameter. This approach does not require enlarging the natural sample space of observations and yields a pathwise filter. Thus began a long collaboration on white noise calculus which culminated with the publication in 1988 of *White Noise Theory of Prediction, Filtering and Smoothing.* Kallianpur had suggested writing this book just as Karandikar was leaving Chapel Hill in 1984 after a two year visit. Thus the writing was done with Kallianpur in Chapel Hill and Karandikar in Delhi (without e-mail at that time!) and with Karandikar visiting Chapel Hill for two months each year until the monograph was completed. Since then, their collaboration has continued in the broad area of Stochastic Analysis.

The Feynman integral is a theme which has intrigued Kallianpur, and he initiated the use of the finitely additive Gauss measure on the Hilbert space of paths in its study. This showed that the two approaches, one via analytical continuation and the other using polygonal approximations, actually lead to the same answer when one compared them using the notion of liftings.

Recently, Kallianpur has been interested in infinite dimensional problems which include stochastic differential equations (SDE's) with values in nuclear spaces, approximation of such an SDE driven by a Poisson random measure by an SDE driven by a Wiener process, and propagation of chaos for a system of interacting SDE's. Kallianpur was drawn to these infinite dimensional SDE's as a stochastic model for the behaviour of neurons. While editing contributions to a volume in honour of his colleague Norman L. Johnson, P.K. Sen was attracted to the novelty of Kallianpur's approach to modelling neuronal behavior as developed in his article in that volume and, indeed, the last ten years have witnessed a phenomenal growth of research in this area.

Some of Kallianpur's reflections are contained in *Glimpses of India's Statistical Heritage.*

Kallianpur has found the time and energy to present inspiring lectures all over the world; to organize numerous international conferences and publish their proceedings; to edit journals, including *Applied Mathematics and Optimization,* of which he has been the editor since 1985; and to direct doctoral dissertations. At this writing he is pursuing with youthful enthusiasm a multitude of research projects and has plans to write two monographs, one on Random Fields and one on Infinite Dimensional Stochastic Differential Equations.

Publications

Books

Stochastic Filtering Theory, Springer–Verlag, 1980.
White Noise Theory of Prediction, Filtering and Smoothing (with R.L. Karandikar), Gordon & Breach, 1988.

Books Edited

Measure Theory Applications to Stochastic Analysis (with D. Kölzow), Lecture Notes in Mathematics **695**, Springer-Verlag, 1977.
Statistics and Probability, Essays in Honor of C.R. Rao (with P.R. Krishnaiah and J.K. Ghosh), North Holland, 1982.
Theory and Application of Random Fields, Lecture Notes in Control and Information Sciences **49**, Springer-Verlag, 1983.
Stochastic Methods in Biology (with M. Kimura and T. Hida), Lecture Notes in Biomathematics **70**, Springer-Verlag, 1987.

Papers

Intégrale de Stieltjes stochastique et un théorème sur les fonctions aléatoires d'ensembles, *C.R. Acad. Sci. Paris* **232** (1951), 922-923.

Ergodic property of the Brownian motion process (with H. Robbins), *Proc. Nat. Acad. Sci. USA* **39** (1953), 525-533.

The sequence of sums of independent random variables (with H. Robbins), *Duke Math. J.* **21** (1954), 285-308.

A note on the Robbins-Monro stochastic approximation method, *Ann. Math. Statist.* **25** (1954), 386-388.

On a limit theorem for dependent random variables (in Russian), *Dokl. Akad. Nauk. SSSR (NS)* **101** (1955), 13-16.

On an ergodic property of a certain class of Markov processes, *Proc. Amer. Math. Soc.* **6** (1955), 159-169.

On Fisher's lower bound asymptotic variance of a consistent estimate (with C.R. Rao), *Sankhyā* **15** (1955), 331-342.

A note on perfect probability, *Ann. Math. Statist.* **30** (1959), 169-172.

A problem in optimum filtering with finite data, *Ann. Math. Statist.* **30** (1959), 659-669.

On the amount of information contained in a σ–field, *Contributions to Probability and Statistics*, I. Olkin et al. eds., Stanford Univ. Press, (1960), 265-273.

The topology of weak convergence of probability measures, *J. Math. Mech.* **10** (1961), 947-969.

On the equivalence and singularity of Gaussian measures (with H. Oodaira), *Time Series Analysis*, M. Rosenblatt ed., Wiley, (1962), 279-291.

Von Mises functionals and maximum likelihood estimation, *Sankhyā* A **25** (1963), 149-158.

On the connection between multiplicity theory and O. Hanner's time domain analysis of weakly stationary processes (with V. Mandrekar), *Essays in Probability and Statistics*, R.C. Bose et al. eds., Univ. of North Carolina Press, (1964), 1-13.

Multiplicity and representation theory of weakly stationary processes (with V. Mandrekar), *Theory Probab. Appl.* **10** (1965), 553-581.

Semi-groups of isometries and the representation and multiplicity of weakly stationary stochastic processes (with V. Mandrekar), *Ark. Mat.* **6** (1966), 319-335.

Estimation of stochastic processes: arbitrary system process with additive white noise observation errors (with C. Striebel), *Ann. Math. Statist.* **39** (1968), 785-801.

Stochastic differential equations occurring in the estimation of continuous parameter stochastic processes (with C. Striebel), *Theory Probab. Appl.* **14** (1969), 567-594.

Stochastic differential equations in statistical estimation problems (with C. Striebel), *Multivariate Analysis* II, P.R. Krishnaiah ed., Academic Press, (1969), 367-388.

A zero-one law for Gaussian processes, *Trans. Amer. Math. Soc.* **149** (1970), 199-211.

Uniform convergence of stochastic processes (with N.C. Jain), *Ann. Math. Statist.* **41** (1970), 1360-1362.

The role of reproducing kernel Hilbert spaces in Gaussian stochastic processes, *Advances in Probability II*, P. Ney ed., Dekker, (1970), 49-83.

Norm convergent expansions for Gaussian processes (with N.C. Jain), *Proc. Amer. Math. Soc.* **25** (1970), 890-895.

Supports of Gaussian measures (with M.G. Nadkarni), *Proc. Sixth Berkeley Symp. Probab. Math. Statist.*, Univ. of California, Berkeley, **2**, (1970), 375-387.

The Bernstein-von-Mises theorem and Bayes estimation in Markov processes (with J. Borwanker and B.L.S. Prakasa Rao), *Ann. Math. Statist.* **42** (1971), 1241-1253.

A stochastic differential equation of Fisk type for estimation and nonlinear filtering problems (with C. Striebel), *SIAM J. Appl. Math.* **21** (1971), 61-72.

Abstract Wiener processes and their reproducing kernel Hilbert spaces, *Z. Wahr. verw. Geb.* **17** (1971), 113-123.

Spectral theory for H-valued stationary processes (with V. Mandrekar), *J. Multivariate Anal.* **1** (1971), 1-16.

Stochastic differential equations for the nonlinear filtering problem (with M. Fujisaki and H. Kunita), *Osaka J. Math.* **9** (1972), 19-40.

Oscillation function of a multiparameter Gaussian process (with N.C. Jain), *Nagoya Math. J.* **47** (1972), 15-28.

Homogeneous chaos expansions, *Statistical Models and Turbulence*, M. Rosenblatt et al. eds, Lecture Notes in Physics **12**, Springer-Verlag, (1972), 230-254.

Nonlinear filtering, *Optimizing Methods in Statistics*, J.S. Rustagi ed., Academic Press, (1972), 211-232.

Non-anticipative representations of equivalent Gaussian processes (with H. Oodaira), *Ann. Probab.* **1** (1973), 104-122.

Non-anticipative canonical representations of equivalent Gaussian processes, *Multivariate Analysis* III, P.R. Krishnaiah ed., Academic Press, (1973), 31-44.

Canonical representations of equivalent Gaussian processes, *Sankhyā* A **35** (1973), 405-416.

The Markov property for generalized Gaussian random fields, *Ann. Inst. Fourier* **24** no. 2 (1974), 143-167.

Canonical representations of equivalent Gaussian processes, *Stochastic Processes and Related Topics*, M.L. Puri ed., Academic Press, (1975), 195-221.

The square of a Gaussian Markov process and non-linear prediction (with T. Hida), *J. Multivariate Anal.* **5** (1975), 451-461.

A general stochastic equation for the non-linear filtering problem, *Optimization Techniques IFIP Technical Conference*, G.I. Marchuk ed., Lecture Notes in Computer Science **27**, Springer-Verlag, (1975), 198-204.

A stochastic equation for the optimal non-linear filter, *Multivariate Analysis* IV, P.R. Krishnaiah ed., North Holland, (1977), 267-281.

Non-anticipative transformations of the two parameter Wiener process and a Girsanov theorem (with N. Etemadi), *J. Multivariate Anal.* **7** (1977), 28-49.

A linear stochastic system with discontinuous control, *Stochastic Differential Equations*, K. Itô ed., Wiley, (1978), 127-140.

Freidlin-Wentzell estimates for abstract Wiener processes (with H. Oodaira), *Sankhyā A* **40** (1978), 116-137.

Representation of Gaussian random fields (with C. Bromley), *Stochastic Differential Systems*, B. Grigelionis ed., Lecture Notes in Control and Information Sciences **25**, Springer-Verlag, (1980), 129-142.

Gaussian random fields (with C. Bromley), *Appl. Math. Optimization* **6** (1980), 361-376.

A stochastic equation for the conditional density in a filtering problem, *Multivariate Analysis* V, P.R. Krishnaiah ed., North Holland, (1980), 137-150.

Some ramifications of Wiener's ideas on nonlinear prediction, *Norbert Wiener Collected Works*, III, P. Masani ed., MIT Press, (1981), 402-424.

Some remarks on the purely nondeterministic property of second order random fields, *Stochastic Differential Systems*, M. Arato et al. eds., Lecture Notes in Control and Information Sciences **36**, Springer-Verlag, (1981), 98-109.

A generalized Cameron-Feynman integral, *Statistics and Probability*, G. Kallianpur, P.R. Krishnaiah and J.K. Ghosh eds., North Holland, (1982), 369-374.

On the diffusion approximation to a discontinuous model for a single neuron, *Contributions to Statistics*, P.K. Sen ed., North Holland, (1983), 247-258.

Nondeterministic random fields and Wold and Halmos decompositions for commuting isometries (with V. Mandrekar), *Prediction Theory and Harmonic Analysis*, V. Mandrekar and H. Salehi eds., North Holland, (1983), 165-190.

Commuting semigroups of isometries and Karhunen representation of stationary random fields (with V. Mandrekar), *Theory and Application of Random Fields*, G. Kallianpur ed., Lecture Notes in Control and Information Sciences **49**, Springer–Verlag, (1983), 126-145.

A finitely additive white noise approach to nonlinear filtering (with R.L. Karandikar), *Appl. Math. Optimization* **10** (1983), 159-185.

On the splicing of measures (with D. Ramachandran), *Ann. Probab.* **11** (1983), 819-822

Some recent developments in nonlinear filtering theory (with R.L. Karandikar), *Acta Appl. Math.* **1** (1983), 399-434.

Generalized Feynman integrals using analytic continuation in several complex variables (with C. Bromley), *Stochastic Analysis*, M. Pinsky ed., Dekker, (1984), 217-267.

Regularity property of Donsker's delta function (with H.H. Kuo), *Appl. Math. Optimization* **12** (1984), 89-95.

Measure valued equations for the optimum filter in the finitely additive nonlinear filtering theory (with R.L. Karandikar), *Z. Wahr. verw. Geb.* **66** (1984), 1-17.

Infinite dimensional stochastic differential equation models for spatially distributed neurons (with R. Wolpert), *Appl. Math. Optimization* **12** (1984), 125-172.

The nonlinear filtering problem for the unbounded case (with R.L. Karandikar), *Stochastic Proc. Appl.* **18** (1984), 57-66.

Markov property of the filter in the finitely additive white noise approach to nonlinear filtering (with R.L. Karandikar), *Stochastics* **13** (1984), 177-198.

The finitely additive approach to nonlinear filtering: a brief survey (with R.L. Karandikar), *Multivariate Analysis* **VI**, P.R. Krishnaiah ed., North Holland, (1985), 335-344.

White noise theory of filtering: some robustness and consistency results, *Stochastic Differential Systems*, M. Metivier and E. Pardoux eds., Lecture Notes in Control and Information Sciences **69**, Springer–Verlag, (1985), 217-223.

Analytic and sequential Feynman integrals on abstract Wiener and Hilbert spaces and a Cameron-Martin formula (with D. Kannan and R.L. Karandikar), *Ann. Inst. H. Poincaré Probab. Statist.* **21** (1985), 323-361.

White noise calculus and nonlinear filtering (with R.L. Karandikar). *Ann. Probab.* **13** (1985), 1033-1107.

White noise calculus for two-parameter filtering (with A.H. Korezlioglu), *Stochastic Differential Systems*, H. Engelberg and W. Schmidt eds., Lecture Notes in Control and Information Sciences **96**, Springer-Verlag, (1986), 61-69.

Stochastic differential equations in duals of nuclear spaces with some applications, *IMA Technical Report No. 244*, Univ. of Minnesota, (1986).

Weak convergence of stochastic neuronal models (with R. Wolpert), *Stochastic Methods in Biology*, Kimura et al. eds., Lecture Notes in Biomathematics **70**, Springer–Verlag, (1987), 116-145.

The filtering problem for infinite dimensional stochastic processes (with R.L. Karandikar), *Stochastic Differential Systems, Stochastic Control, Theory and Applications*, W. Fleming and P.L. Lions eds., Springer–Verlag, (1987), 215-223.

Stochastic differential equations for neuronal behavior (with S.K. Christensen), *Adaptive Statistical Procedures and Related Topics*, J. Van Ryzin ed., IMS Lecture Notes Monograph Series **8**, (1987), 400-416.

Stochastic evolution equations driven by nuclear space-valued martingales (with V. Perez-Abreu), *Appl. Math. Optimization* **17** (1988), 237-272.

Smoothness properties of the conditional expectation in finitely additive white noise filtering (with H. Hucke and R.L. Karandikar), *J. Multivariate Anal.* **27** (1988), 261-269.

Weak convergence of solutions of stochastic evolution equations in nuclear spaces (with V. Perez-Abreu), *Stochastic Partial Differential Equations and Applications*, G. Da Prato and L. Tubaro eds., Lecture Notes in Mathematics **1390**, Springer–Verlag, (1989), 133-139.

Some remarks on Hu and Meyer's paper and infinite dimensional calculus on finitely additive canonical Hilbert space (with G.W. Johnson), *Theory Probab. Appl.* **34** (1989), 742-752.

Diffusion equations in duals of nuclear spaces (with I. Mitoma and R. Wolpert), *Stochastics* **20** (1990), 285-329.

Infinite dimensional stochastic differential equations with applications, *Stochastic Methods in Experimental Sciences*, W. Kasprzak and A. Weron eds., World Scientific, (1990), 227-238.

On the prediction theory of two parameter stationary random fields (with A.G. Miamee and H. Niemi), *J. Multivariate Anal.* **32** (1990), 120-149.

Multiple Wiener integrals on abstract Wiener spaces and liftings of p-linear forms (with G.W. Johnson), *White Noise Analysis*, T. Hida et al. eds., World Scientific, (1990), 208-219.

Propagation of chaos and the McKean-Vlasov equation in duals of nuclear spaces, (with T.S. Chiang and P. Sundar), *Appl. Math. Optimization* **24** (1991), 55-83.

A skeletal theory of filtering, *Stochastic Analysis*, E. Mayer-Wolf et al. eds., Academic Press, (1991), 213-243.

Traces, natural extensions and Feynman distributions, *Gaussian Random Fields*, K. Itô and T. Hida eds., World Scientific, (1991), 14-27.

A line grid method in areal sampling and its connection with some early work of H. Robbins, *Amer. J. Math. Manag. Sci.* **11** (1991), 40-53.

Parameter estimation in linear filtering (with R. Selukar), *J. Multivariate Anal.* **39** (1991), 284-304.

The Skorohod integral and the derivative operator of functionals of a cylindrical Brownian motion (with V. Perez-Abreu), *Appl. Math. Optimization* **25** (1992), 11-29.

The analytic Feynman integral of the natural extension of p-th homogeneous chaos (with G.W. Johnson), *Rend. Circ. Mat. Palermo* (1992), to appear.

Propagation of chaos for systems of interacting neurons (with T.S. Chiang and P. Sundar), *Proc. Trento Conf. on Stochastic Partial Differential Equations 1990*, G. Da Prato et al. eds., (1992), to appear.

Homogeneous chaos, p-forms, scaling and the Feynman integral (with G.W. Johnson), *Trans. Amer. Math. Soc.*, to appear.

Estimation of Hilbert space valued parameters by the method of sieves (with R. Selukar), *Current Issues in Statistics and Probability*, J.K. Ghosh et al. eds., Wiley Eastern, to appear.

A nuclear-space-valued stochastic differential equation driven by Poisson random measures (with J. Xiong), *Stochastic PDE's and their Applications*, B.L. Rozovskii and R.B. Sowers eds., Lectures Notes in Control and Information Sciences **176**, Springer–Verlag, (1992), 135–143.

Periodically correlated and periodically unitary processes and their relationship to $L^2[0, T]$–valued stationary sequences (with H. Hurd), *Nonstationary Stochastic Processes and their Applications*, J.C. Hardin and A.G.

Miamee eds., World Scientific, (1992) to appear.

A Segal-Langevin type stochastic differential equation on a space of generalized functionals (with I. Mitoma), *Canadian J. Math.*, to appear.

Stochastic differential equation models for spatially distributed neurons and propagation of chaos for interacting systems, *J. Math. Biol.*, to appear.

Other Articles

On the Indian Statistical Institute, *Encyclopedia of Statistics Sciences*, S. Kotz and N.L. Johnson eds., Wiley, **4** (1983).

On P.C. Mahalanobis, ibid., **5** (1985).

On prediction and filtering, ibid, **7** (1986).

Review of stationary sequences and random fields, by M. Rosenblatt, *Bull. Amer. Math. Soc.* **21** (1989), 133-139.

Random Reflections, *Glimpses of India's Statistical Heritage*, J.K. Ghosh et al. eds., Wiley Eastern, (1992), 47–66.

A remark on the support of cadlag processes

S. Albeverio* Z. M. Ma** M. Röckner***

* Fakultät für Mathematik, Ruhr-Universität Bochum, D-4630 Bochum 1, Germany,
   SFB 237 - Essen - Bochum - Düsseldorf, BiBoS - Bielefeld - Bochum, CERFIM - Locarno;
** Institute for Applied Mathematics, Academia Sinica, P.O.Box 2734, Beijing 100080
   China;
*** Institut für Angewandte Mathematik, Universität Bonn, Wegelerstr. 6, 5300 Bonn,
   Germany.

Abstract

We show that every cadlag process on a metrizable co-Souslin space is supported by a K_σ-set. The result is central for the analytic characterization of Dirichlet forms associated with right-continuous strong Markov processes.

1. Introduction

In [LR 90] one main step to prove that the capacity of a Dirichlet form having associated to it a conservative diffusion $(X_t)_{t \geq 0}$ with state space E was to show that $(X_t)_{t \geq 0}$ is supported by a K_σ-set in E, i.e., a countable union of compact subsets of E. In this paper we generalize this result to arbitrary cadlag (non-conservative) stochastic processes. Though we do not use Dirichlet forms below we emphasize that this generalization is central for the (analytic) characterization of those Dirichlet forms having an asssociated right continuous strong Markov process (cf. [MR 91] and [AMR 91a,b]).

2. Every cadlag process is supported by a K_σ-set

Let E be a metrizable co-Souslin space, i.e., it is topologically isomorphic to the complement of a $(\mathcal{K}(\overline{E})$-)analytic subset of a polish space $\overline{E}$ (cf. [DM 75]). Here $\mathcal{K}(\overline{E})$ denotes the system of all compact subsets of $\overline{E}$. We denote the Borel σ-algebra of $\overline{E}$ by $\mathcal{B}(\overline{E})$.

Remark 2.1. By [DM 75, III 13], $\mathcal{B}(\overline{E})$-analyticity is equivalent with

$\mathcal{K}(\overline{E})$-analyticity, hence by [DM 75, III 8] each $B \in \mathcal{B}(\overline{E})$ is $\mathcal{K}(\overline{E})$-analytic. In particular, if E is a metrizable Lusin space (i.e., topologically isomorphic to a Borel subset of a polish space), then E is co-Souslin. For more details we refer to [DM 75] whose terminology we shall adopt below.

Theorem 2.2. Let $(X_t)_{t \geq 0}$ be a stochastic process with state space E and life time ζ on a measurable space $(\Omega, \mathcal{F})$. Assume that P is a probability measure on $(\Omega, \mathcal{F})$ such that P-a.s., $t \mapsto X_t$ is right continuous on $[0, \infty[$ and has left limits X_{t-} for all $t \in]0, \zeta[$. Then there exists an increasing sequence $(K_n)_{n \in \mathbb{N}}$ of compact subsets of E such that

$$P[\lim_{n \to \infty} \sigma_{E \setminus K_n} < \zeta] = 0$$

where for $U \subset E, U$ open, $\sigma_U := \inf\{t > 0 | X_t \in U\}$ is the *first hitting time* of U.

Before we prove 2.2 we need some preparations concerning the Skorohod topology which we shortly describe below following [Mai 72].

Let ρ denote the metric on $\overline{E}$. We may assume that ρ is bounded. As usual a function a from $[0, \infty]$ to $\overline{E}$ is called *cadlag* if $t \mapsto a(t)$ is right continuous on $[0, \infty[$ and has left limits $a(t-)$ for all $t \in]0, \infty[$. Let $\mathcal{D}$ be the set of all cadlag functions from $[0, \infty[$ to $\overline{E}$. Let Λ denote the set of all real-valued increasing functions on $[0, \infty[$ such that $\lambda(0) = 0$. For $\lambda \in \Lambda$ define

$$\|\lambda\| := \sup_{s,t \geq 0, s \neq t} |\log \frac{\lambda(t) - \lambda(s)}{t - s}| + \sup_{t \geq 0} |\lambda(t) - t|.$$

Note that if $\|\lambda\| < \infty$, then λ is continuous, strictly increasing and onto. In this case we have for its inverse λ^{-1} that $\|\lambda\| = \|\lambda^{-1}\|$. For $a, b \in \mathcal{D}$ define

(2.1) $d(a, b) := \inf\{\|\lambda\| + \sup_{t \geq 0} e^{-t} \rho(a(t), b(\lambda(t))) | \lambda \in \Lambda\}.$

For $t \geq 0$ define maps π_t, π_{t-} from $\mathcal{D}$ to $\overline{E}$ by

$$\pi_t(a) := a(t) \ , \ \pi_{t-}(a) := a(t-) \ , \ a \in \mathcal{D}.$$

Let Π, Π_- denote the maps from $[0, \infty[\times \mathcal{D},]0, \infty[\times \mathcal{D}$ respectively to $\overline{E}$ given by

$$\Pi(t, a) := \pi_t(a) \ , \Pi_-(t, a) := \pi_{t-}(a).$$

Proposition 2.3. (i) $(\mathcal{D}, d)$ is a complete separable metric space.
(ii) If $\mathcal{B}(\mathcal{D})$ denotes the Borel σ-algebra of $(\mathcal{D}, d)$, then $\mathcal{B}(\mathcal{D}) = \sigma\{\pi_t | t \geq 0\}$

Proof. [Mai 72]. □

Lemma 2.4. Let $\Gamma \subset \mathcal{D}$, Γ compact, and $M > 0$. Then $\Pi([0, M] \times \Gamma)$ is a relatively compact subset of $\overline{E}$ and its closure is contained in $\Pi([0, M] \times \Gamma) \cup \Pi_-(]0, M] \times \Gamma)$

Proof. Let $a_n(t_n) \in \Pi([0, M] \times \Gamma), n \in I\!N$. Selecting a subsequence if necessary we may assume that $t_n \to t$ and $a_n \to a$ in $(\mathcal{D}, d)$ as $n \to \infty$ for some $t \in [0, M], a \in \mathcal{D}$. By definition (2.1) there exist $\lambda_n \in \Lambda, n \in I\!N$, such that $\|\lambda_n\| \xrightarrow[n \to \infty]{} 0$ and

$$\lim_{n \to \infty} \sup_{t \geq 0} e^{-t} \rho(a_n(t), a(\lambda_n(t))) = 0.$$

Hence $\lim_{n \to \infty} \lambda_n(t_n) = t$ and again selecting a subsequence we may assume that either $\lambda_n(t_n) \downarrow t$ or $\lambda_n(t_n) \uparrow t$ as $n \to \infty$. Consequently,

$$\lim_{n \to \infty} a_n(t_n) = a(t) \in \overline{E} \quad \text{or} \quad \lim_{n \to \infty} a_n(t_n) = a(t-) \in \overline{E}$$

(since e.g. $\rho(a(t), a_n(t_n)) \leq \rho(a(t), a(\lambda_n(t_n))) + \rho(a(\lambda_n(t_n)), a_n(t_n))$ for all $n \in I\!N$). □

Proof of 2.2. We may assume that $E \subset \overline{E}$ and that for every $\omega \in \Omega, t \mapsto X_t(\omega)$ is right continuous on $[0, \infty[$ and has left limits $X_{t-}(\omega)$ for all $t \in]0, \zeta(\omega)[$. Define for $\omega \in \Omega$

$$Y_t(\omega) := \begin{cases} X_t(\omega) & \text{if } \zeta(\omega) = \infty \\ X_{t\zeta(1+t)^{-1}}(\omega) & \text{if } \zeta(\omega) < \infty. \end{cases}$$

Then $Y := (Y_t)_{t \geq 0}$ defines a map from Ω to $\mathcal{D}$ such that $\pi_t \circ Y (= Y_t)$ is $\mathcal{F}$-measurable for all $t \geq 0$. Hence by 2.3(ii), Y is $\mathcal{F}/\mathcal{B}(\mathcal{D})$-measurable. Let $Q := P \circ Y^{-1}$ be the image measure of P under Y defined on $(\mathcal{D}, \mathcal{B}(\mathcal{D}))$. Let $\mathcal{D}_0 \subset \mathcal{D}$ denote the set of all $a \in \mathcal{D}$ such that $a(t)$, $a(t-) \in E$ for all $t \geq 0, t > 0$ respectively, and set

$$S := \mathcal{D} \setminus \mathcal{D}_0.$$

Then

$$S = \{a \in \mathcal{D} | (t, a) \in G \text{ for some } t \geq 0\}$$

where

$$G := \{(t, a) \in [0, \infty[\times \mathcal{D} | a(t) \in \overline{E} \setminus E \text{ or } a(t-) \in \overline{E} \setminus E\}.$$

Since both Π and Π_- are $\mathcal{B}([0, \infty[) \otimes \mathcal{B}(\mathcal{D})/\mathcal{B}(\overline{E})$-measurable and $\overline{E} \setminus E$ is $\mathcal{K}(\overline{E})$-analytic, by [DM 75, III 11] we see that G is an analytic subset of

$[0, \infty[\times \mathcal{D}$. Hence by [DM 75, III 13] S is an analytic subset of $\mathcal{D}$ which in turn implies that $S \in \mathcal{B}(\mathcal{D})^Q$ ($:=$ completion of $\mathcal{B}(\mathcal{D})$ w.r.t. Q) by [DM 75, III 33]. Consequently, $\mathcal{D}_0 \in \mathcal{B}(\mathcal{D})^Q$ and since $Y(\Omega) \subset \mathcal{D}_0$ we can find $\mathcal{D}_1 \in \mathcal{B}(\mathcal{D})$ such that $\mathcal{D}_1 \subset \mathcal{D}_0$ and $Q(\mathcal{D}_1) = 1$. Since by 2.3(i), Q is inner regular on $(\mathcal{D}, \mathcal{B}(\mathcal{D}))$ we can find an increasing sequence $(\Gamma_n)_{n \in I\!\!N}$ of compact subsets of $\mathcal{D}_1$ such that

$$Q(\Gamma_n) \geq Q(\mathcal{D}_1) - \frac{1}{n} = 1 - \frac{1}{n} \text{ for all } n \in I\!\!N.$$

Fix $n \in I\!\!N$ and let K_n denote the closure of $\Pi([0, n] \times \Gamma_n)$ in $\overline{E}$, then K_n is a compact subset of E by 2.4. Furthermore, if $\omega \in Y^{-1}(\Gamma_n)$ then for all $t \in [0, \frac{n\zeta(\omega)}{1+n} \wedge n]$

$$X_t(\omega) = \begin{cases} Y_t(\omega) \in K_n & \text{if } \zeta(\omega) = \infty \\ Y_{\frac{t}{\zeta - t}}(\omega) \in K_n & \text{if } \zeta(\omega) < \infty \end{cases}$$

which implies that

$$\sigma_{E \setminus K_n}(\omega) \geq \frac{n\zeta(\omega)}{1+n} \wedge n.$$

Consequently,

$$P[\lim_{m \to \infty} \sigma_{E \setminus K_m} \geq \frac{n\zeta(\omega)}{1+n} \wedge n] \geq P[\sigma_{E \setminus K_n} \geq \frac{n\zeta(\omega)}{1+n} \wedge n]$$
$$\geq P[Y^{-1}(\Gamma_n)] = Q[\Gamma_n]$$
$$\geq 1 - \frac{1}{n}$$

and letting $n \to \infty$ we obtain the assertion. $\square$

Acknowledgement

We would to thank P. Fitzsimmons for discussions about this paper. Financial support of the Chinese National Natural Science Foundation, the research project BiBoS and the Sonderforschungsbereich 256 are gratefully acknowledged.

References

[AMR 91a]   Albeverio, S., Ma, Z.M., Röckner, M., Non-symmetric Dirichlet forms and Markov processes on general state space.

Preprint (1991). To appear in C.R. Acad. Sci. Paris, Série I.

[AMR 91b]   Albeverio, S., Ma, Z.M., Röckner, M., Quasi-regular Dirichlet forms and Markov processes. Preprint (1991).

[DM 75]   Dellacherie, C., Meyer, P.A., Probabilités et potentiel, Chapitres I-IV. Paris: Hermann 1975.

[LR 90]   Lyons, T., Röckner, M., A note on tightness of capacities associated with Dirichlet forms. Preprint Edinburgh 1990. To appear in Bull. London Math. Soc.

[MR 91]   Ma, Z. M., Röckner, M., An introduction to the theory of (non-symmetric) Dirichlet forms. Book in preparation (1991).

[Mai 72]   Maisonneuve, B., Topologies du type de Skorohod. In: Séminaire de Probabilités VI, 113-117. Berlin-Heidelberg-New York: Springer 1972.

LARGE DEVIATION RESULTS
FOR BRANCHING PROCESSES

K.B. Athreya and A. Vidyashankar

ABSTRACT. Let Z_n be a supercritical simple branching process with a finite offspring mean m and zero extinction probability. Then $Z_{n+1}Z_n^{-1}$ converges to m with probability one. This note deals with the large deviation aspects of this convergence and shows that under exponential moment hypothesis the rate is geometric. Extensions to multitype case are stated without proof. Some open problems are indicated

1. Introduction.

Let $\{Z_n\}_0^\infty$ be a Galton-Watson branching process with offspring probability distribution $\{p_j : j = 0_11, 2 \ldots \}$. Assume $p_0 = 0, 1 < m \equiv \sum jp_j < \infty$ and $\sigma^2 \equiv \sum j^2 p_j - m^2 > 0$. It is known that (see Athreya & Ney [2] pp. 24) if $P(Z_0 > 0) = 1$ then $P(Z_n \to \infty) = 1$ and $Z_{n+1}Z_n^{-1} \to m$ w.p.1. Also by the central limit theorem for random sums $(Z_{n+1}Z_n^{-1} - m)\sqrt{m^n}$ converges in distribution to a mixture of normal distributions provided $0 < \sigma^2 < \infty$. There is a law of the iterated logarithm as well. The problems of large deviations of $Z_{n+1}Z_n^{-1} - m$ have not been treated in the literature and this note fills this gap. The next section treats the one dimensional case. Extension to the multitype case is outlined (without proof) in section 3.

2. The Main Result.

In this section we prove the following:

Theorem 1. *Let the offspring distribution $\{p_j\}$ satisfy $p_0 = 0$ and $p_1 \neq 0$. Let A be a Borel set in R such that for random variables $\{X_i\}_1^\infty$ that are independent with distribution $\{p_j\}$ there exists a λ in (0,1) such that*

$$(1) \qquad G(n, A) \equiv P(\overline{X}_n - m \in A) = 0(\lambda^n) \text{ as } n \to \infty$$

1991 *Mathematics Subject Classification.* 60J, 60F.

Key words and phrases. Branching processes, Large deviations.

† Research supported in part by NSF Grant DMS 907182

where $\overline{X}_n = n^{-1} \sum_1^n X_i$ and $m = \sum j p_j$. Then, if $P(Z_0 = 1) = 1$

(2) $\qquad p_1^{-n} P((Z_{n+1} Z_n^{-1} - m) \in A) \rightarrow \sum_{j=1}^{\infty} G((j, A)q_j < \infty$

where $\{q_j\}$ is defined by its generating function $Q(s) \equiv \sum_1^{\infty} q_j s^j$, as the unique solution of the functional equation

(3) $\qquad \begin{cases} Q(f(s)) = p_1 Q(s) \text{ for } 0 \le s < 1 \\ Q(0) = 0, \end{cases}$

where $f(s) = \sum s^j p_j$

Proof. Let $f_n(s)$ be the probability generating function of Z_n with $P(Z_0 = 1) = 1$. Then it is known that (see Athreya & Ney [2] pp. 38)

(4) $\qquad \lim_n p_1^{-n} f_n(s) \equiv Q(s) < \infty \text{ exists for } 0 \le s < 1$

and $Q(\cdot)$ is the unique solution of the functional equation (3).

Now, by conditioning on $Z_0, Z_1 \ldots Z_n$ and using the branching property, we see that

(5) $\qquad P((Z_{n+1} Z_n^{-1} - m) \in A) = \sum_j G(j, A) P(Z_n = j)$

Let $h_n(j) = G(j, A) P(Z_n = j) p_1^{-n}$. By hypothesis there exists constants $C \in (0, \infty)$ and $\lambda \in (0, 1)$ such that $G(j, A) \le C\lambda^j$ for all $j \ge 1$.

Let $r_n(j) = C\lambda^j P(Z_n = j) p_1^{-n}$

Then by (4)

$$\sum_j r_n(j) = C p_1^{-n} f_n(\lambda) \rightarrow CQ(\lambda) < \infty$$

Also we have, for $j = 1, 2, \ldots$.

$$0 \le h_n(j) \le r_n(j)$$

and

$$h_n(j) \rightarrow G(j, A) q_j.$$

(This last assertion follows from (4)).

So by a slight generalization of the dominated convergence theorem (see Royden [5] pp. 92, pp. 270) applied to the counting measure space on the positive integers we get from (5).

$$p_1^{-n} P((Z_{n+1} Z_n^{-1} - m) \in A) \rightarrow \sum_j G(j, A) q_j < \infty. \quad \square$$

Corollary 1. *Let $\{p_j\}$ satisfy $f(\theta_0) < \infty$ for some $1 < \theta_0 < \infty$ and $p_0 = 0, 0 < p_1 < 1$. Let A be any Borel set $\subset R - (-\epsilon, \epsilon)$ for some $\epsilon > 0$. Then (2) holds.*

Proof.

$$G(n, A) \leq P(|\overline{X}_n - m| > \epsilon)$$
$$\leq P(\overline{X}_n > m + \epsilon) + P(\overline{X}_n < m - \epsilon)$$
$$\leq P(\alpha^{S_n} > \alpha^{n(m+\epsilon)}) + P(\beta^{S_n} > \beta^{n(m-\epsilon)})$$

where $S_n = \sum_1^n X_i$, α and β are arbitrary constants in $(1, \theta_0)$ and $(0,1)$ respectively. Thus, by Markov's inequality

$$G(n, A) \leq (f(\alpha)\alpha^{-(m+\epsilon)})^n + (f(\beta)\beta^{-(m-\epsilon)})^n.$$

It can be verified that for every $0 < \epsilon < 1$ there exists α_0 in $(1, \theta_0)$ and β_0 in $(0,1)$ such that

$$0 < (f(\alpha_0)\alpha_0^{-(m+\epsilon)}) < 1 \text{ and } 0 < f(\beta_0)\beta_0^{-(m-\epsilon)} < 1.$$

This yields (1) and so (2) follows.

Corollary 2. *Let $\{p_j\}$ satisfy $f(\theta_0) < \infty$ for some $1 < \theta_0 < \infty$. Let $p_0 > 0, p_1 > 0$ and $1 < m = \sum jp_j < \infty$. Let $P(Z_0 = 1) = 1$. Then for every Borel set $A \subset R - (-\epsilon, \epsilon)$ for some $\epsilon > 0$*

$$(6) \qquad \gamma^{-n}P((Z_{n+1}Z_n^{-1} - m) \in A | Z_n > 0) \to \sum_j G(j, A)\left(\frac{q_j}{1-q}\right)$$

where $\gamma = f'(q), q$ the smallest root of $s = f(s)$ in $[0,1]$, $Q(s) \equiv \sum_0^\infty q_j s^j$, the unique solution of the equation

$$\begin{cases} Q(f(s)) &= \gamma Q(s) \qquad 0 \leq s < 1 \\ Q(0) &= 0 \end{cases}$$

Proof. As in the proof of Theorem 1

$$P((Z_{n+1}Z_n^{-1} - m) \in A | Z_n > 0)$$
$$= \sum_{j=1}^\infty G(j, A)P(Z_n = j | Z_n > 0).$$

It is known that (see Athreya and Ney [2] pp. 40) if $P(Z_0 = 1) = 1, p_0 > 0, 1 < m < \infty$ then $\gamma^{-n} P(Z_n = j | Z_n > 0) \to \frac{q_j}{(1-q)}$. The rest of the argument is the same as in theorem 1 and Corollary 1.

Remark 1. The conclusion (2) is a stronger form of the usual large deviation type result involving rate functions for the exponential decay. Indeed if (1) holds for a Borel set $A \subset R$ then

(7) $$\frac{1}{n} \log P((Z_{n+1} Z_n^{-1} - m) \in A) \to \log p_1.$$

In particular, if $f(\theta_0) < \infty$ for some $1 < \theta_0 < \infty$ then by Corollary 1, (7) above holds for all Borel sets $A \subset R - (-\epsilon, \epsilon)$ for some $\epsilon > 0$. Thus if a rate function $I(\cdot)$ were to exist that satisfies the usual conditions for being called a *good rate function*, (see Deuschel & Stroock [3]) then by considering an open interval $(x_0 - h, x_0 + h)$ and a closed interval $[x_0 - h, x_0 + h]$ in $(0, \infty) \cup (-\infty, 0)$ we see that

$$- \liminf_{x_0 - h < x < x_0 + h} I(x) \leq \log p_1$$

and

$$- \liminf_{x_0 - h \leq x \leq x_0 + h} I(x) \geq \log p_1.$$

Since $I(\cdot)$ is lower semicontinuous on R it must follow that $I(x) = -\log p_1$ for all $x \neq 0$. On the otherhand if $A = (0, \infty)$ and if $\sum j^2 p_j < \infty$ (and in particular if $f(\theta_0) < \infty$ for some $1 < \theta_0 < \infty$) then by the central limit theorem (see Athreya and Ney [2] pp. 55)

$$P((Z_{n+1} Z_n^{-1} - m) \in A) \to \frac{1}{2}$$

and hence

$$\frac{1}{n} \log P((Z_{n+1} Z_n^{-1} - m) \in A) \to 0.$$

This is also true for $A = [0, \infty)$. Thus, the rate function must satisfy

$$0 \leq \inf_{x \geq 0} I(x) \leq \inf_{x > 0} I(x) \leq 0.$$

But $\inf_{x > 0} I(x) = -\log p_1$ and so we have a contradiction. This shows that there is no good rate function for the large deviation problem on hand. Nevertheless (7) does hold for all A satisfying (1) and in particular, for all Borel sets $A \subset R - (-\epsilon, \epsilon)$ for some $\epsilon > 0$.

Remark 2. Regarding complete convergence, that is, convergence of the series $\sum_n P(Z_{n+1}Z_n^{-1} - m \in A)$ we note that if $\sum_j j^2 p_j < \infty$ and $A \subset R - (-\epsilon, \epsilon)$ for some $\epsilon > 0$ then by the Erdos-Hsu Robbins theorem (see [4])

$$\sum G(j, A) < \infty$$

where G is as in (1).

Thus

$$\sum_n P(Z_{n+1}Z_n^{-1} - m \in A) = \sum_n EG(Z_n, A) = E\left(\sum_n G(Z_n, A)\right)$$
$$\leq \sum_j G(j, A) < \infty$$

We conjecture that $EG(Z_n, A) < \infty$ just with $1 < m < \infty$. (See also Asmussen and Kurtz [1]).

Remark 3. (Open problem) The assumption that (1) holds could be too strong for (2). From (5) we see that

(8) $$\varliminf p_1^{-n} P(Z_{n+1}Z_n^{-1} - m \in A) \geq \sum_j G(j, A)q_j.$$

At the moment not much is known about $\{q_j\}$. An interesting open problem is to investigate the growth rate of $\{q_j\}$ and relate that to the convergence of $\sum_j G(j, A)q_j$ as well as improving (8) to a full convergence result.

3.   Extensions to multitype case.

Let $\{Z_n\}_0^\infty$ be a p-type $(p > 1)$, postively regular, supercritical branching process with offspring generating functions $f^{(i)}(s)$ for $i = 1, 2 \ldots p$. Assume $f^{(i)}(0) = 0$ for all i. Let $m_{ij} = \frac{\partial f^{(i)}}{\partial s_j}$ (1) where $\mathbf{1} = (1, 1, \ldots, 1)$ and $a_{ij} = \frac{\partial f^{(i)}}{\partial s_j}(0)$. Let γ be the Perron-Frobenius root of the matrix $A = ((a_{ij}))$ which is assumed to be positively regular. Let ρ be the maximal eigenvalue of M with normalized eigenvectors u and v such that $u'M = \rho u', Mu = \rho u, u \cdot \mathbf{1} = 1, u \cdot v = 1$. Let $f_1(s) = (f^{(1)}(s), f^{(2)}(s), \ldots, f^{(p)}(s))$ be a map of the unit cube $C = \{s : s = (s_1, s_2, \ldots s_p), 0 \leq s_i \leq 1\}$ onto itself and $f_n(\cdot)$ be its nth iterate. Then the following results hold.

Theorem 2. *There exists a map Q of the open unit cube C to R_+^p such that*

$$\frac{f_n(s)}{\gamma^n} \to Q(s)$$

and Q is the unique solution of

$$Q(f(s)) = \gamma Q(s)$$
$$Q(0) = 0$$

Theorem 3. *Assume $f^{(i)}(\theta_0 1) < \infty$ for some $1 < \theta_0 < \infty$ and for all $1 \le i \le p$. Let $P(1 \cdot Z_0 = 1) = 1$. Then for any vector ℓ and $\epsilon > 0$ $\lim_n \frac{1}{\gamma^n} P(\frac{\ell \cdot Z_{n+1}}{1 \cdot Z_n} - \frac{\ell \cdot Z_n M}{1 \cdot Z_n} > \epsilon)$ and $\lim_n \frac{1}{\gamma^n} P(\frac{\ell \cdot Z_n}{1 \cdot Z_n} > \ell \cdot v + \epsilon)$ exist and are finite and positive.*

Theorem 2 which is a p-type extension of Theorem 1 (pp. 40 of Athreya and Ney [2]) is new. The proof of Theorem 3 uses Theorem 2 as in the proof of Theorem 1 although the arguments are more involved. The proofs of both these results will be given elsewhere. The problem of obtaining the conclusion of Theorem 3 without the exponential moment hypothesis is an interesting open problem.

REFERENCES

1. Asmussen, S. and Kurtz, T.G., *Necessary and Sufficient Conditions for Complete Convergence in the Law of Large Number*, Annals of Probability (1980), 176-182.
2. Athreya, K.B. and Ney, P.E., *Branching Processes*, Springer-Verlag, Berlin, 1972.
3. Deuschel, J.-D. and Stroock, D.W., *Large Deviations*, Academic Press, N.Y., 1989.
4. Hsu, P.L. and Robbins, H., *Complete Convergence and the law of large numbers*, Proc. Natl. Acad. Sci. U.S.A. 33 (1947), 25-31.
5. Royden, H.L., *Real Analysis*, Macmillan Publishing Company, N.Y., 1987, Third Edition.

IOWA STATE UNIVERSITY
AMES, IOWA 50011 U.S.A.

Random Iterations of Two Quadratic Maps

Rabi N. Bhattacharya* and B.V. Rao

Abstract. We study invariant measures of Markov processes obtained by the action of successive independent iterations of a map chosen at random from a set of two quadratic maps.

1. Introduction.

Markov processes may be viewed as random perturbations of dynamical systems. Indeed, if the state space S is a Borel subset of a Polish space one may represent a Markov process with any prescribed transition probability and an arbitrary initial distribution as $\alpha_n \alpha_{n-1} \cdots \alpha_1 X_0$, where $\{\alpha_n : n \geq 1\}$ is an i.i.d. sequence of random maps on S into itself and X_0 is independent of $\{\alpha_n : n \geq 1\}$ (see Kifer [7], p.8). In the case of a dynamical system α_n is degenerate, i.e., $P(\alpha_n = f) = 1$ for a given single map f on S. This point of view of a Markov process is useful for the study of Markov processes as well as dynamical systems. Often a chaotic dynamical system admits uncountably many ergodic invariant probabilities only one of which, the so-called Kolmogorov measure, is physically relevant. This measure is the limit of the invariant probabilities of Markov processes obtained as appropriate random perturbations of the dynamical system, as the distribution of α_1 approaches the Dirac measure at f (see Kifer [8], Ruelle [9], and Katok and Kifer [6]). This, however, is not the focus of the present article.

Consider the quadratic family of functions $\{F_\mu : 0 \leq \mu \leq 4\}$, where F_μ is the map on [0,1] defined by

$$(1.1) \qquad F_\mu(x) := \mu x(1 - x), \qquad 0 \leq x \leq 1.$$

Dynamical systems with $f = F_\mu$, and similar ones, have been extensively studied in the literature (see, e.g., Devaney [4] and Collet and Eckman [3]). Given a pair of parameter values $\mu < \lambda$ and a number $\gamma \in (0,1)$ we consider an i.i.d. sequence of maps $\{\alpha_n : n \geq 1\}$ with $P(\alpha_1 = F_\mu) = \gamma$, $P(\alpha_1 = F_\lambda) = 1 - \gamma$. For certain choices of μ, λ, we study the uniqueness and other properties of invariant probabilities of the resulting Markov processes. It turns out that even for those F_μ (and F_λ) which are simple as dynamical systems, the above randomization often leads to Markov processes with interesting invariant probabilities some times supported on Cantor sets of Lebesgue measure zero.

* Research was supported in part by NSF Grant DMS 9206937

Section 2 on iterated random monotone maps on $[0, 1]$ is based largely on Dubins and Freedman [5], and provides the basic tool for deriving the main results. In Section 3 we review certain aspects of the quadratic maps $F_\mu (0 \leq \mu \leq 1 + \sqrt{5})$, such as attracting and repelling fixed points and period–two orbits, and identify pairs F_μ, F_λ which have a common invariant interval on which they are both monotone. Section 4 contains the main results, summarized in Theorem 4.1. It will be clear from the proofs that some of the results extend to more general classes of maps than the family (1.1), but we do not pursue such extensions in this article.

2. Iterations of i.i.d monotone maps.

Let $a < b$ be given reals, and $(\Omega, \mathcal{F}, P)$ a probability space on which is defined a sequence of i.i.d. continuous maps $\alpha_n (n \geq 1)$ on $[a, b]$ into $[a, b]$. This means (i) for each $\omega \in \Omega, x \to \alpha_n(\omega)x$ is continuous (for all $n \geq 1$), (ii) for each B belonging to the Borel sigmafield $\mathcal{B}$ on $[a, b]$, $\{(\omega, x) : \alpha_n(\omega)x \in B\} \in \mathcal{F} \otimes \mathcal{B}$, and (iii) for every finite set $\{x_1, x_2, \ldots, x_k\} \subset [a, b]$, the sequence of random vectors $(\alpha_n x_1, \alpha_n x_2, \ldots, \alpha_n x_k)$, $n \geq 1$, are i.i.d. If X_0 is a random variable (with values in $[a, b]$) independent of $\{\alpha_n : n \geq 1\}$ (i.e., of $\sigma\{\alpha_n x : x \in [a, b], n \geq 1\}$), then $X_0, X_n \equiv \alpha_n \ldots \alpha_1 X_0 (n \geq 1)$, is a Markov process on $[a, b]$ having transition probability $p(x, B) := P(\alpha_1 x \in B)$ and initial distribution $\mu(B) := P(X_0 \in B), B \in \mathcal{B}$. In particular, if $X_0 \equiv x$ then we write $X_n(x)$ for this Markov process. The n–step transition probability may then be expressed as $p^n(x, B) = P(X_n(x) \in B)$. Note that the continuity of $\alpha_1(\omega)$ implies $x \to p(x, dy)$ is weakly continuous. For weakly continuous transition probabilities (on some metric space) a well known elementary criterion for the existence of an invariant probability for p is the following: *If for some x and some sequence of integers $n_1 < n_2 < \ldots < n_k < \ldots$, there exists a probability measure π such that*

$$(2.1) \qquad \frac{1}{n_k} \sum_{m=1}^{n_k} p^m(x, dy) \xrightarrow{weakly} \pi(dy),$$

then π is invariant. If for some x, the sequence $\frac{1}{n} \sum_{m=1}^n p^m(x, dy)$ is tight, then (2.1) holds for some sequence $n_k(k \geq 1)$ and some probability measure π.

We now state a basic result due to Dubins and Freedman [5] for monotone maps on $[a, b]$. For this case the *splitting condition* is said to hold if there exist x_0 and a positive integer m such that

$$(2.2) \qquad P(X_m(x) \leq x_0 \forall x) > 0, \quad P(X_m(x) \geq x_0 \forall x) > 0.$$

Let p^* denote the adjoint operator on the space of all finite signed measures on $[a, b]$,

$$(2.3) \qquad (p^* \nu)(B) := \int p(x, B) \nu(dx), \quad B \in \mathcal{B},$$

with norm

(2.4)
$$\|\nu\| := \sup\{|\nu([a,x])| : a \le x \le b\},$$
$$\|p^*\nu\| = \sup\{|\int p(y,[a,x])\nu(dy)| : a \le x \le b\}.$$

Write $p^{*n}\nu = p^*(p^{*(n-1)}\nu)(n \ge 2), p^{*1} = p^*$.

PROPOSITION 2.1. (Dubins and Freedman [5]). *Suppose $\alpha_n(n \ge 1)$ are i.i.d. monotone continuous maps on $[a,b]$ into $[a,b]$. (a) If the splitting condition holds then $\|p^{*n}\nu_1 - p^{*n}\nu_2\| \equiv \|p^{*n}(\nu_1 - \nu_2)\|$ goes to zero exponentially fast as $n \to \infty$, uniformly for every pair of probability measures ν_1, ν_2; and there exists a unique invariant probability π which is the limit of $p^{*n}\nu$ for every probability ν. (b) If α_1 is strictly increasing a.s., and there is no c such that $P(\alpha_1(\omega)c = c) = 1$, then splitting is also necessary for the conclusion in (a) to hold.*

REMARK 2.1.1. Under the hypothesis of part (a) the invariant probability is nonatomic, i.e., its distribution function is continuous. For this take ν_1 nonatomic and $\nu_2 = \pi$ in the statement and note that the continuous distribution functions of $p^{*n}\nu_1$ converge uniformly to that of π.

REMARK 2.1.2. Part (a) of the theorem holds if the state space is an arbitrary interval not necessarily compact. Indeed, this result can be extended to appropriate subsets of $I\!R^k$ and coordinatewise monotone maps (see Bhattacharya and Lee [2]).

For our purposes a different version of this result will be useful. To state it define $Y_n(x) := \alpha_1 \cdots \alpha_n x$. If α_1 is increasing on $[a,b]$ then $Y_n(a) \uparrow$ and $Y_n(b) \downarrow$ as $n \uparrow$. Let $\underline{Y}, \bar{Y}$ denote the respective limits. Note that $X_n(x)$ and $Y_n(x)$ have the same distribution, namely, $p^n(x, dy)$. A proof of part (b) of Proposition 2.1 is included in the proof of the following result.

PROPOSITION 2.2. *Let α_1 be a.s. continuous and increasing on $[a,b]$. Consider the following statements: (i) $\underline{Y} = \bar{Y}$ a.s. (ii) There exists a unique invariant probability. (iii) Splitting holds. (iv) $\underline{Y} = \bar{Y}$ a.s. and $\underline{Y}$ is not constant a.s. (v) There exists a unique invariant probability and it is nonatomic. (a) The following implications hold: (v) $\Longrightarrow$ (iv) $\Longrightarrow$ (iii) $\Longrightarrow$ (ii) $\Leftrightarrow$ (i). (b) If α_1 is strictly increasing a.s. then (iii) $\Leftrightarrow$ (iv) $\Leftrightarrow$ (v).*

Proof. (a) By criterion (2.1), the distributions of $\underline{Y}$ and $\bar{Y}$ are both invariant. Since $\underline{Y} \le \bar{Y}$, these invariant probabilities are the same if and only if $\underline{Y} = \bar{Y}$ a.s. Also, $Y_n(a) \le Y_n(x) \le Y_n(b)$ for all x. Therefore, $\underline{Y} = \bar{Y}$ a.s. implies $Y_n(x) \to \underline{Y}$ a.s., so that $p^n(x, dy)$ converges weakly to the common distribution of $\underline{Y}$ and $\bar{Y}$, *for all* x. This easily implies uniqueness of the invariant probability. Hence (i) $\Leftrightarrow$ (ii). Also, (iii) $\Rightarrow$ (ii) by Proposition 2.1(a).

Now assume (iv) holds. Then there exists x_0 such that $P(\underline{Y} < x_0)(= P(\bar{Y} < x_0)) > 0$ and $P(\underline{Y} > x_0) > 0$. This implies (2.2) for all sufficiently large m. Thus (iv) $\Rightarrow$ (iii). Since (ii) $\Rightarrow$ (i), clearly (v) $\Rightarrow$ (iv).

(b) Assume (iii) holds. Then (i) holds. If $\underline{Y} = \bar{Y} = c$ a.s. for some constant c then the Dirac measure δ_c is invariant, which implies $P(\alpha_1 c = c) = 1$. This in turn implies $P(X_n(c) = c) = 1$ for all $n \geq 1$. Since α_1 is strictly increasing, so is $x \to X_n(x)$ (for every $n \geq 1$). Therefore, if $a \leq x_0 \leq c < b$ then $P(X_n(b) \leq x_0) = 0$, and if $a < c \leq x_0 \leq b$ then $P(X_n(a) \geq x_0) = 0$. Hence splitting does not occur. Thus (iii) $\Rightarrow$ (iv), so that (by (a)) (iii) $\Rightarrow$ (iv). By Remark 2.1.1, (iii) $\Rightarrow$ (v), so that (iii) $\Leftrightarrow$ (v). $\square$

Suppose now that α_1 is decreasing a.s. Then $\alpha_1 \alpha_2$ is increasing a.s. and $\underline{Z} := \lim Y_{2n}(a)$, $\bar{Z} := \lim Y_{2n}(b)$ (as $n \to \infty$) exist. Proposition 2.2 then holds for the two–step transition probability p^2 (in place of p), if $\underline{Y}$ and $\bar{Y}$ are replaced by $\underline{Z}$ and $\bar{Z}$, respectively. Since every invariant probability for p is invariant for p^2, the following corollary is immediate.

COROLLARY 2.2. *Suppose α_1 is continuous and either strictly increasing a.s. or strictly decreasing a.s. on $[a, b]$. In addition assume that there does not exist a c such that $P(\alpha_1 c = c) = 1$. Then splitting is a necessary and sufficient condition for the existence of a unique invariant probability. This probability is nonatomic.* 3. **Quadratic maps.**

We will henceforth confine our attention to the family of maps $F_\mu (0 \leq \mu \leq 4)$ defined by (1.1). If $\mu \neq 0$, F_μ is strictly increasing on $[0, \frac{1}{2}]$ and strictly decreasing on $[\frac{1}{2}, 1]$ attaining its maximum value $\mu/4$ at $x = \frac{1}{2}$.

If $0 \leq \mu \leq 1$, then $F_\mu(x) < x$ for $x \in (0, 1]$. Hence $F_\mu^n(x) \downarrow$ as $n \uparrow$. The limit must be a fixed point. But the only fixed point is $x = 0$. Hence 0 is an *attracting fixed point*: $F_\mu^n(x) \to 0$ as $n \to \infty$, for all $x \in [0, 1]$.

If $1 < \mu \leq 4$, then F_μ has a second fixed point $p_\mu = 1 - \frac{1}{\mu}$. Suppose $1 < \mu \leq 2$. Then $F_\mu(x) > x$ for $x \in (0, p_\mu)$ and F_μ is increasing on $(0, p_\mu)$. Hence $F_\mu^n(x)$ increases to the fixed point p_μ as n increases. For $x \in (p_\mu, 1)$, $F_\mu(x) < x$. Thus either $F_\mu^n(x)$ decreases to p_μ as n increases, or there exists n_0 such that $F_\mu^{n_0}(x) \in (0, p_\mu]$ and $F_\mu^n(x) F_\mu^{n_0}(x) \uparrow p_\mu$ as $n \uparrow$. Therefore, $F_\mu^n(x) \to p_\mu$ for all $x \in (0, 1)$, so that p_μ is an attracting fixed point.

For $\mu > 2$ one has $p_\mu > 1/2$, and the above approach fails. Let us try to find an interval $[a, b]$ on which F_μ is monotone, and which is left invariant by $F_\mu : F_\mu([a, b]) \subset [a, b]$. One must have $\frac{1}{2} \leq a \leq b \leq 1$. It is simple to check that $[\frac{1}{2}, \mu/4]$ is such an interval provided $F_\mu(\mu/4) \equiv F_\mu^2(\frac{1}{2}) \geq 1/2$.

This holds iff $2 \leq \mu \leq 1 + \sqrt{5}$. For such a μ, F_μ is strictly decreasing, and F_μ^2 strictly increasing, on $[\frac{1}{2}, \mu/4]$. Hence $F_\mu^{2n}(1/2) \uparrow$ and $F_\mu^{2n}(\mu/4) \downarrow$ as $n \uparrow$. Let $\alpha \equiv \alpha(\mu), \beta \equiv \beta(\mu)$ be the respective limits. Then $F_\mu^{2n}(x) \to \alpha$ for $x \in [\frac{1}{2}, \alpha], F_\mu^{2n}(x) \to \beta$ for $x \in [\beta, \mu/4]$. In particular, $\alpha \leq \beta$ are fixed points of F_μ^2. Since, for $2 < \mu \leq 3, F_\mu^2$ has no fixed points other than $0, p_\mu$, it follows that in this case $\alpha = \beta = p_\mu$, so that $F_\mu^n(x) \to p_\mu$ for all $x \in (0, 1)$.

Consider $3 < \mu \leq 1 + \sqrt{5}$. In this case $|F_\mu'(p_\mu)| = \mu - 2 > 1$. Therefore, p_μ is a repelling fixed point, so that $\alpha < p_\mu < \beta$. This implies that $\{\alpha, \beta\}$ is an *attracting period–two orbit* of F_μ. Since F_μ^2 is a fourth degree polynomial, $\{0, \alpha, p_\mu, \beta\}$ are the only fixed points for it. Since $0, p_\mu$ are repelling, it follows that $F_\mu^{2n}(x) \to \alpha$ or β for all $x \neq 0, 1$ or a preimage of p_μ. Note that $F_\mu^2(x) - x$ does not change sign on (α, p_μ) or on (p_μ, β). This analysis does not extend beyond $1 + \sqrt{5}$. We conjecture that for $\mu > 1 + \sqrt{5}$ a stable period–four orbit appears, while the period–two orbit (as well as the fixed points) becomes unstable.

For later purposes we consider intervals $[a, b]$ contained in $[0, \frac{1}{2}]$ or $[\frac{1}{2}, 1]$ and the set $I(a, b) := \{\mu \in [0, 4] : F_\mu([a, b]) \subset [a, b]\}$. Straight forward calculations show

$$(3.1a) \qquad I(0, b) = \left[0, \frac{1}{1 - b}\right] \text{ if } 0 \leq b \leq \frac{1}{2},$$

$$(3.1b) \qquad I(a, b) = \left[\frac{1}{1 - a}, \frac{1}{1 - b}\right] \text{ if } 0 < a \leq b \leq \frac{1}{2},$$

$$(3.1c) \qquad I(a, b) = \left[\frac{a}{b(1 - b)}, \frac{b}{a(1 - a)}\right] \text{ if } \frac{1}{2} \leq a \leq b \leq 1 \text{and}$$
$$a^2(1 - a) \leq b^2(1 - b).$$

The second requirement in (3.1c) may be expressed as

$$(3.1c)' \qquad a \leq b \leq b^*(a) := \frac{1 - a}{2} + \frac{1}{2}\sqrt{(1 - a)(1 + 3a)},$$

which implies the further restriction

$$(3.1c)'' \qquad \frac{1}{2} \leq a \leq \frac{2}{3}.$$

The maximum value of $\mu \in [0, 4]$ in the union of the sets $I(a, b)$ in (3.1c) (subject to (3.1c)', (3.1c)'') is $\mu = 1 + \sqrt{5}$. In particular,

$$(3.2) \qquad I\left(\frac{1}{2}, \frac{\mu}{4}\right) = \left[\frac{8}{\mu(4 - \mu)}, \mu\right], \qquad 2 < \mu \leq 1 + \sqrt{5}.$$

4. Main results: random iterations of two quadratic maps.

In this section we consider the Markov process X_n as defined in section 2, with $P(\alpha_1 = F_\mu) = \gamma$ and $P(\alpha_1 = F_\lambda) = 1 - \gamma$ for appropriate pairs $\mu < \lambda$ and $\gamma \in (0, 1)$. If $0 \leq \mu, \lambda \leq 1$, then it follows from Section 3 that $Y_n(x) \to 0$ a.s. for all $x \in [0, 1]$, so that the Dirac measure δ_0 is the unique invariant probability. Now take $1 < \mu < \lambda \leq 2$, and let $a = p_\mu \equiv 1 - 1/\mu$ and $b = p_\lambda$ in (3.1b). Then F_μ, F_λ are both strictly increasing on $[p_\mu, p_\lambda]$ and leave this interval invariant. Since p_μ, p_λ are attracting for F_μ, F_λ, respectively, (2.2) holds for any $x_0 \in (p_\mu, p_\lambda)$ if m is large enough. It follows from Proposition 2.1 (or, Proposition 2.2) that there exists a unique invariant probability π on $[p_\mu, p_\lambda]$. Since $P(X_n(x) \in [p_\mu, p_\lambda]$ for some $n) = 1$ for all $x \in (0, p_\mu) \cup (p_\lambda, 1)$ it follows that the Markov process has a unique invariant probability, namely, π on the state space $(0, 1)$ and that π is nonatomic. It is clear that both p_μ and p_λ belong to the *support* $S(\pi)$ *of* π as do the set of all points of the form

$$(4.1) \qquad F_{\varepsilon_1 \varepsilon_2 \cdots \varepsilon_k} p_\mu \equiv f_{\varepsilon_1} f_{\varepsilon_2} \cdots f_{\varepsilon_k} p_\mu \quad (f_0 := F_\mu, f_1 := F_\lambda),$$
$$F_{\varepsilon_1 \varepsilon_2 \cdots \varepsilon_k} p_\lambda \quad (k \geq 1),$$

for all k–tuples $(\varepsilon_1, \varepsilon_2, \ldots, \varepsilon_k)$ of 0's and 1's and for all $k \geq 1$. Write $\mathrm{Orb}(x; \mu, \lambda) = \{F_{\varepsilon_1 \varepsilon_2 \cdots \varepsilon_k} x : k \geq 0, \varepsilon_i = 0$ or $1 \forall i\}$ ($k = 0$ corresponds to x). It is easy to see that if $x \in S(\pi)$ then $S(\pi) = \overline{\mathrm{Orb}(x; \mu, \lambda)}$. This support, however, need not be $[p_\mu, p_\lambda]$. Indeed, if $F_\lambda(p_\mu) > F_\mu(p_\lambda)$, i.e.,

$$(4.2) \qquad \frac{1}{\lambda^2} - \frac{1}{\lambda^3} < \frac{1}{\mu^2} - \frac{1}{\mu^3} \qquad (1 < \mu < \lambda \leq 2),$$

then $S(\pi)$ is a *Cantor subset* of $[p_\mu, p_\lambda]$. Before proving this assertion we identify pairs μ, λ satisfying (4.2). On the interval $[1, 2]$ the function $g(x) := x^{-2} - x^{-3}$ is strictly increasing on $[1, 3/2]$ and strictly decreasing on $[3/2, 2]$, and $g(1) = 0, g(3/2) = 4/27, g(2) = 1/8$. Therefore, (4.2) holds iff

$$(4.3) \qquad \lambda \in (3/2, 2] \text{ and } \mu \in [\hat{\lambda}, \lambda),$$

where $\hat{\lambda} \leq 3/2$ is uniquely defined for a given $\lambda \in (3/2, 2]$ by $g(\hat{\lambda}) = g(\lambda)$. Since the smallest value of $\hat{\lambda}$ as λ varies over $(3/2, 2]$ is $\sqrt{5} - 1$ which occurs when $\lambda = 2$ ($g(2) = 1/8$), it follows that μ can not be smaller than $\sqrt{5} - 1$ if (4.2) (or, (4.3)) holds.

To show that $S(\pi)$ is a Cantor set (i.e., a closed, no where dense set having no isolated point) for μ, λ satisfying (4.2), or (4.3), write $I = [p_\mu, p_\lambda], I_0 = F_\mu(I), I_1 = F_\lambda(I), I_{\varepsilon_1 \varepsilon_2 \cdots \varepsilon_k} = F_{\varepsilon_1 \varepsilon_2 \cdots \varepsilon_k}(I)$ for $k \geq 1$ and k–tuples $(\varepsilon_1 \varepsilon_2 \cdots \varepsilon_k)$ of 0's and 1's. Here $F_{\varepsilon_1 \varepsilon_2 \cdots \varepsilon_k} = f_{\varepsilon_1} f_{\varepsilon_2} \cdots f_{\varepsilon_k}$ as defined in (4.1). Under the present hypothesis $F_\lambda(p_\mu) > F_\mu(p_\lambda)$ (or, (4.2)), the 2^k intervals $I_{\varepsilon_1\varepsilon_2\cdots\varepsilon_k}$ are disjoint, as may be easily shown by induction, using the fact that F_μ, F_λ are strictly increasing on $[p_\mu, p_\lambda]$. Let $J_k = \cup I_{\varepsilon_1\varepsilon_2\cdots\varepsilon_k}$ where the union is over the 2^k k-tuples $(\varepsilon_1, \varepsilon_2, \cdots, \varepsilon_k)$. Since $X_k(x) \in J_k$ for all $x \in I$, and $J_k \downarrow$ as $k \uparrow, S(\pi) \subset J_k$ for all k, so that $S(\pi) \subset J :=$ $\cap_{k=1}^\infty J_k$. Further, F_μ is a *strict contraction* on $[p_\mu, p_\lambda]$ while $F_\lambda^n(I) \downarrow \{p_\lambda\}$ as $n \uparrow \infty$. Hence the lengths of $I_{\varepsilon_1\varepsilon_2\cdots\varepsilon_k}$ go to zero as $k \to \infty$, for a sequence $(\varepsilon_1, \varepsilon_2, \cdots)$ which has only finitely many 0's or finitely many 1's. If there are infinitely many 0's and infinitely many 1's in $(\varepsilon_1, \varepsilon_2, \cdots)$ then for large k with $\varepsilon_k = 0$ one may express $f_{\varepsilon_1} f_{\varepsilon_2} \cdots f_{\varepsilon_k}$ as a large number of compositions of functions of the type F_μ^n or $F_\lambda^n F_\mu(n \geq 1)$. Since for all μ, λ satisfying (4.2) the derivatives of these functions on $[p_\mu, p_\lambda]$ are bounded by $F_\lambda'(p_\mu)F_\mu'(p_\mu) < 1$ (use induction on n and the estimate $F_\lambda'(F_\lambda p_\mu) < 1$) it follows that the lengths of the nested intervals $I_{\varepsilon_1\varepsilon_2\cdots\varepsilon_k}$ go to zero as $k \to \infty$ (for every sequence $(\varepsilon_1, \varepsilon_2, \cdots)$). Thus, J does not contain any (nonempty) open interval. Also, $J \subset \overline{Orb(p_\mu; \mu, \lambda)} = \overline{Orb(p_\lambda; \mu, \lambda)}$ by the same reasoning, so that $J = S(\pi)$. Since π is nonatomic, $S(\pi)$ does not include any isolated point, completing the proof that $S(\pi)$ is a Cantor set.

Write $|A|$ for the Lebesgue measure of A. If, in addition to (4.2),

$$(4.4) \qquad \lambda < \left(\frac{\mu - 1}{2 - \mu}\right)\mu,$$

then $|J| = 0$. Indeed, for any subinterval I' of I one has $|F_\lambda(I')| \leq F_\lambda'(p_\mu)|I'|$, $|F_\mu(I')| \leq F_\mu'(p_\mu)|I'|$, from which it follows that $|J_{k+1}| \leq (2 - \mu)(1 + \lambda/\mu)|J_k|$. If (4.4) holds then $c \equiv (2 - \mu)(1 + \lambda/\mu) < 1$, so that $|J| = 0$ if (4.2), (4.4) hold.

Note that the proof that $|I_{\varepsilon_1\varepsilon_2\cdots\varepsilon_k}| \to 0$ depends only on the facts that on $[p_\mu, p_\lambda]$, (i) F_μ is a contraction, (ii) $F_\lambda^n(I) \downarrow \{p_\lambda\}$, and (iii) $F_\lambda'(F_\lambda p_\mu) < 1, F_\lambda'(p_\mu)F_\mu'(p_\mu) < 1$. The last condition (iii) may be expressed as

$$(4.5) \qquad \lambda - 2\lambda^2(\mu - 1)/\mu^2 < 1, \ \lambda < \mu/(2 - \mu)^2 \ \ (1 < \mu < \lambda \leq 2).$$

If (4.5) holds, but (4.2) does not, then the 2^k intervals $I_{\varepsilon_1\varepsilon_2\cdots\varepsilon_k}$ cover $I = [p_\mu, p_\lambda]$. Since the endpoints of $I_{\varepsilon_1\varepsilon_2\cdots\varepsilon_k}$ are in $S(\pi)$, it follows that $S(\pi) = [p_\mu, p_\lambda]$.

A point of additional interest is that if (4.2) (or (4.3)) holds then the Markov process X_n restricted to the invariant set $J = S(\pi)$ is isomorphic to one on $\{0,1\}^N$ having the transition probability

$$(4.6) \qquad (\varepsilon_1, \varepsilon_2, \cdots) \to \begin{cases} (0, \varepsilon_1, \varepsilon_2, \cdots) & \text{with probability } \gamma, \\ \\ (1, \varepsilon_1, \varepsilon_2, \cdots) & \text{with probability } 1 - \gamma. \end{cases}$$

The isomorphism is defined by $y \to (\varepsilon_1, \varepsilon_2, \cdots)$ where

$$y = \lim_{k \to \infty} f_{\varepsilon_1} f_{\varepsilon_2} \cdots f_{\varepsilon_k} p_\mu$$

In this representation for a fixed $\gamma \in (0,1)$ the Markov processes on $J = J_{\mu,\lambda}$ are the same for all μ, λ satisfying (4.2).

Next consider the case $2 < \mu < \lambda \leq 3$. If $\mu \in I(\frac{1}{2}, \lambda/4) \equiv [8/\lambda(4 - \lambda), \lambda)$ (see (3.2)) then F_μ, F_λ may be restricted to $[1/2, \lambda/4]$ and are strictly decreasing on it. Since p_μ, p_λ are attracting for F_μ, F_λ, respectively, (2.2) holds for any $x_0 \in (p_\mu, p_\lambda)$ if m is sufficiently large. It follows from Section 2 that there exists a unique invariant probability in $[p_\mu, p_\lambda]$ and it is nonatomic.

Finally, if $2 < \mu \leq 3 < \lambda < 1 + \sqrt{5}$ and $\mu \in I(1/2, \lambda/4) \equiv [8/\lambda(4 - \lambda), \lambda)$, then p_μ is attracting for F_μ and $\beta \equiv \beta(\lambda)$ (see Section 3) is an attracting fixed point for F_λ^2. It follows that (2.2) holds on $[p_\mu, p_\lambda]$ in this case also if $x_0 \in (p_\lambda, \beta)$ and m is even and sufficiently large, so that the invariant probability on $[p_\mu, p_\lambda]$ (and also on $(0,1)$) is unique and nonatomic.

We state the main results proved above as a theorem. Note that δ_0 is invariant on $[0,1]$ for all $0 \leq \mu < \lambda \leq 4$.

THEOREM 4.1. (a) *If $0 \leq \mu < \lambda \leq 1$, then δ_0 is the unique invariant probability on $[0,1]$.* (b) *If $1 < \mu < \lambda \leq 2$ then there exists a unique invariant probability π on $(0,1)$. This probability is nonatomic. If μ, λ satisfy (4.2) (or (4.3)) then the support $S(\pi)$ of π is a Cantor subset $J \equiv J_{\mu,\lambda}$ of $[p_\mu, p_\lambda]$. If, (4.2) and (4.4) both hold, then $|J| = 0$.* (c) *If the inequality (4.2) does not hold, but (4.5) holds, then $S(\pi) = [p_\mu, p_\lambda]$.* (d) *If $2 < \mu < \lambda < 1 + \sqrt{5}$ and $\mu \in [8/\lambda(4 - \lambda), \lambda)$ then there exists a unique invariant probability on $(0,1)$, which is nonatomic and has its support contained in $[1/2, \lambda/4]$.*

EXAMPLES.
1. If $\mu = \sqrt{5} - 1 = 1.232 \cdots, \lambda = 2$, then (4.2) holds, but (4.4) does not, and $S(\pi) = J$ is a Cantor set.

2. If $\lambda = 2, -\frac{1}{2} + \frac{1}{2}\sqrt{17} < \mu < 2$, then (4.2), (4.4) both hold, and $S(\pi)$ is a Cantor set of Lebesgue measure zero.

3. If $\lambda = 3/2, 6/(3 + \sqrt{5}) < \mu < 3/2$, then (4.2) does not hold, but (4.5) does, and $S(\pi) = [p_\mu, p_\lambda]$.

4. Suppose $0 < \mu < 1, 1 < \lambda < 2$. Theorem 4.1 does not apply. But if $\gamma \in (0,1)$ is such that $\mu^\gamma \lambda^{1-\gamma} < 1$, then δ_0 is the unique invariant probability (see Barnsley and Elton [1]).

We conclude with two remarks.

Remark 4.1.1 If $[a, b]$ is an invariant interval under F_μ and $F_\lambda (\mu < \lambda)$, then $[a, b]$ is invariant under F_γ for all $\gamma \in [\mu, \lambda]$. In particular, if $1 \leq \mu < \lambda \leq 2$, then the maps $F_\gamma (\gamma \in [\mu, \lambda])$ are all increasing on the invariant interval $[p_\mu, p_\lambda]$, and the splitting condition is satisfied by the Markov process on $[p_\mu, p_\lambda]$ corresponding to every randomization of F_γ's, $\gamma \in [\mu, \lambda]$. Thus there exists a unique invariant probability in this case on $[p_\mu, p_\lambda]$ (and on $(0, 1)$). If the support of the distribution of the random parameter γ is $[\mu, \lambda]$, then the support of the invariant probability is $[p_\mu, p_\lambda]$. A similar consideration applies to $2 < \mu < \lambda < 1 + \sqrt{5}$ if $\mu \in [8/\lambda(4 - \lambda), \lambda]$.

Remark 4.1.2. Let $1 < \mu < 2 < \lambda < 4$ be arbitrary. The interval $[1 - \frac{1}{\mu}, \frac{\lambda}{4}]$ is invariant under F_γ for all $\gamma \in [\mu, \lambda]$. Let F_γ be chosen at random such that the distribution of γ has a positive density with respect to Lebesgue measure m on $[\mu, \lambda]$. One may then show that the coresponding Markov process is m–irreducible on $[1 - \frac{1}{\mu}, \frac{\lambda}{4}]$. It follows from standard Markov process theory that in this case there exists a unique invariant probability. If, moreover, $\lambda > \frac{16}{\mu^3} - \frac{16}{\mu^2} + 4$, then one can show that the transition probability density $p(x, y)$ is no smaller than a nonzero, non–negative function $f(y)$ for all x in $[1 - \frac{1}{\mu}, \frac{\lambda}{4}]$. It is then easy to check that the n–step transition probability density $p^{(n)}(x, y)$ converges in L^1 to the invariant probability uniformly in x.

Acknowledgment. Remark 4.1.2 is in response to a question raised by the referee. We wish to thank the referee for his comments.

REFERENCES

1. Barnsley, M.F. and Elton, J.H. (1988). A new class of Markov processes for image encoding. *Adv. Appl. Prob.* **20** 14–32.

2. Bhattacharya, R.N. and Lee, O. (1988). Asymptotics of a class of Markov processes which are not in general irreducible. *Ann. Probab.* **16** 1333–1347, Correction, ibid (1992).

3. Collet, P. and Eckman, J-P. (1980). *Iterated Maps on the Interval as Dynamical Systems.* Birkhauser, Boston.

4. Devaney, R.L. (1989). *An Introduction to Chaotic Dynamical Systems*, Second Ed., Addison–Wesley, New York.

5. Dubins, L. E. and Freedman, D.A. (1966). Invariant probabilities for certain Markov processes. *Ann. Math. Statist.* **37** 837–847.

6. Katok, A. and Kifer, Y. (1986). Random perturbations of transformations of an interval. *J. D'Analyse Math.* **47** 193–237.

7. Kifer, Y. (1986). *Ergodic Theory of Random Transformations.* Birkhauser, Boston.

8. Kifer, Y. (1988). *Random Perturbations of Dynamical Systems.* Birkhauser, Boston.

9. Ruelle, D. (1989). *Chaotic Evolution and Strange Attractors.* Cambridge Univ. press, Cambridge.

Indiana University
Department of Mathematics
Bloomington, IN 47405

and

Indian Statistical Institute
203 B.T. Road
Calcutta 700 035, India

ZERO-ONE LAW FOR SEMIGROUPS
OF MEASURES ON GROUPS

TOMASZ BYCZKOWSKI[1]

AND

BALRAM S. RAJPUT[2]

ABSTRACT. Let $(\mu_t)_{t>0}$ be a convolution semigroup of probability measures of Poisson type on a complete separable metric abelian group. The purpose of this note is to provide a short and elementary proof of the zero-one law for $(\mu_t)_{t>0}$.

In 1951, Cameron and Graves proved that every measurable rational subspace of $C[0,1]$ has Wiener measure zero or one [8]. Up to that time and until Kallianpur's fundamental and pioneering paper [15] appeared in 1970, this result seemed to exemplify one more of a special feature of the Wiener process. In [15], Kallianpur not only showed that such a zero-one law is valid for a large class of Gaussian processes, but he also applied successfully reproducing kernel Hilbert space methods in investigating properties of subgroups and linear sub-spaces of sample paths of stochastic processes. Kallianpur's results and methods were supplemented and augmented by a number of authors (see, e.g., [1], [7] and [12]).

A more geometric method of proof for zero-one laws for Gaussian processes was initiated by Fernique in [10]. A somewhat similar approach was applied later by Dudley and Kanter [9] to prove a zero-one law for stable measures, which was later generalized, using different methods of proof, for semi-stable measures in [16].

A group-theoretic approach for zero-one laws for Gaussian elements in groups was presented, for the first time, in [3] and a complete solution, in the abelian case, was given in [4]. The main idea was that the image

[1] The research of this author is supported by KBN Grant, the University of Tennessee, and the University of Tennessee Science Alliance, a State of Tennessee Center of Excellence.

[2] The research of this author is supported by AFSOR Grant # 90–016 8, and the University of Tennessee Science Alliance, a State of Tennessee Center of Excellence.

1991 *Mathematics Subject Classification*. Primary 60B15, 60F20, 60E07.

Key words and phrases. Convolution semigroups, infinitely divisible probability measures, zero-one laws.

of a Gaussian measure under measurable homomorphism from one Polish abelian group into another cannot have idempotent factors. Gaussian random elements (or measures) were defined in the sense of Fernique : A random element X is Gaussian if for any independent copies X_1, X_2 of X the random elements $X_1 + X_2$ and $X_1 - X_2$ are also independent. Unfortunately, it turned out that this definition and the techniques based on it are not suitable in the non-abelian case [5].

A more general aproach for zero-one laws for Gaussian and other infinitely divisible measures, based on semigroup techniques, was used in [13] and [5]. In these papers zero-one laws were proved for normal subgroups. The latter work [5] turned out to be more general, and a complete solution, based on the methods of [5], appeared in [2] and [18]. These proofs, however, are complicated and based on Trotter's approximation theorem, which is a rather sophisticated tool. As far as we know, even the abelian case (treated in [13]) requires complicated algebraic and measure-theoretic concepts related to the purity laws (see [11]). A reader interested in a more complete survey is referred to [14].

The purpose of our note is to present a relatively short and simple proof of the zero-one law for convolution semigroups of Poisson type on complete separable abelian groups. We point out that our proof relies on a kind of density argument for generators of semigroups involved and uses only very elementary facts concerning semigroups of operators.

We also mention a recent paper [17], where a version of our Theorem was proved in the setting of Banach spaces. The proof is based on Le Page type series representation of infinitely divisible distributions and a generalization of a very interesting theorem of Lévy.

Throughout the paper $(G, \mathcal{B})$ will denote a complete metric separable abelian group with its Borel σ-field $\mathcal{B}$. We will employ an additive notation for the group operation. Before formulating our theorem, we state first a version of Lévy-Khintchine formula for convolution semigroups on G (see Section 3 in [6]). A (convolution) semigroup $(\mu_t)_{t>0}$ is called continuous if $\mu_t \Rightarrow \delta_0$ weakly, as $t \downarrow 0$; it is called symmetric if all μ_t's are symmetric.

If m is a finite probability measure, then

$$\exp tm = e^{-tm(G)} \sum_{k=0}^{\infty} \left(\frac{t^k}{k!}\right) m^{*k},$$

with $m^{*0} = \delta_0$, is a continuous semigroup.

Suppose now that $(\mu_t)_{t>0}$ is a symmetric continuous semigroup. Then there exists a nonnegative measure ν such that for every open neighborhood U of 0 the restriction $\nu|_{U^c}$ is finite and $(1/t)\mu_t|_{U^c}$ converges weakly to $\nu|_{U^c}$, as $t \to 0$, whenever $\nu(\partial U) = 0$. Moreover $\mu_t = \chi_t * \gamma_t$, where $(\chi_t)_{t>0}$ and $(\gamma_t)_{t>0}$ are symmetric and continuous convolution semigroups,

$\gamma_t = \lim \exp t\nu|_{F_n}$, for every increasing sequence $\{F_n\}$ of symmetric Borel subsets such that $\nu|_{F_n}$ is finite and $\bigcap_n F_n^c = \{0\}$ and

$$\lim_{t \to 0+} (1/t)\chi_t(U^c) = 0,$$

for every open neighborhood of 0. The measure ν is called the Lévy measure of $(\mu_t)_{t>0}$ and the semigroup γ_t will be denoted by $\exp t\nu$. For every $\eta > 0$, we have $\mu_t = \exp(t\nu|_{q>\eta}) * \exp(t\nu|_{q\leq\eta})$ with $\nu|_{q>\eta}$ being finite.

Now, we recall some basic facts concerning the theory of semigroups of operators. By $C_u \equiv C_u(G)$, we denote the space of all uniformly continuous real bounded functions on G. For a convolution semigroup $(\mu_t)_{t>0}$, we define a semigroup of probability operators (T_t) by

$$T_t f(x) = \int_G f(x + y)\mu_t(dy), \quad f \in C_u(G).$$

It is clear that $T_t f \in C_u$, $T_t T_s = T_{t+s}$ and $T_t f \to f$ uniformly, for $f \in C_u(G)$

Now, suppose that H is Borel subgroup and that $\mu_t(H) > 0$, for all $t > 0$. Define

$$\mu = \int_0^\infty e^{-u}\mu_u \, du.$$

The main tool employed for the proof of the 0–1 law consists of the $L^1(\mu)$ method, developed in the paper [5]. To make our paper self-contained we collect in the Lemma some elementary facts from this paper.

Lemma. *Suppose that $(\mu_t)_{t>0}$ is a continuous semigroup on G and that $\mu_t(H) > 0$ for all $t > 0$. Then $(\mu_t)_{t>0}$ acts continuously on $L^1(\mu)$ so that $\mu_t(H) \to 1$ as $t \to 0$. Moreover, if $\pi : G \to G/H$ is the canonical homomorphism onto G/H with the induced σ-field, then $\pi \circ \mu_t = \exp t\gamma$, with γ finite on G/H.*

Proof. Let f be a nonnegative Borel function on G. Then

$$\|T_{\mu_t}f\|_{L^1(\mu)} = \int_G \int_G f(x + y)d\mu_t(y)d\mu(x)$$

$$= \int_0^\infty \int_G \int_G e^{-u}f(x + y)d\mu_t(x)d\mu_u(y)du$$

$$= \int_0^\infty e^{-u}\int_G f(z)d\mu_{t+u}(z)du$$

$$= e^t\int_t^\infty e^{-u}\int_G f(z)d\mu_u(z)du \leq e^t\|f\|_{L^1(\mu)},$$

so

$$\|T_{\mu_t}\| \le e^t.$$

Since $(\mu_t)_{t>0}$ acts continuously on C_u, which is dense in $L^1(\mu)$, $(\mu_t)_{t>0}$ is continuous on $L^1(\mu)$. In particular, $T_{\mu_t}\mathbf{I}_H \to \mathbf{I}_H$ in $L^1(\mu)$, as $t \to 0$. Hence

$$\mu(H)\mu_t(H) = \int\limits_H \int \mathbf{I}_H(x+y)\mu_t(dx)\mu(dy)$$

$$= \int\limits_H T_{\mu_t}\mathbf{I}_H\,du \to \int\limits_H \mathbf{I}_H\,d\mu = \mu(H),$$

so

$$\mu_t(H) \to 1, \text{ as } t \to 0.$$

Denote $\lambda_t = \pi \circ \mu_t$. Then (λ_t) is uniformly continuous on Borel bounded functions:

$$\|T_{\lambda_t}g - g\|_\infty = \sup_{x \in G/H} \left| \int g(x+y)(\lambda_t - \delta_H)(dy) \right|$$

$$\le \|g\|_\infty \|\lambda_t - \delta_H\| = \|g\|_\infty \left(\lambda_t|_{H^c} + (1 - \lambda_t|_H) \right) = 2\|g\|_\infty (1 - \mu_t(H)).$$

Hence

$$\|T_{\lambda_t} - I\|_\infty \le 2(1 - \mu_t(H)) \to 0.$$

Theorem. *Let $\mu_t = \exp t\nu$ be a symmetric convolution semigroup of Poisson type on G and let H be a Borel subgroup of G. Then $\nu(H^c) = \infty$ implies that $\mu_t(H - x) = 0$ for all $t > 0$ and all $x \in G$, while if $\nu(H^c) = 0$ then either $\mu_t(H - x) = 0$ for all $t > 0$ and all $x \in G$, or $\mu_t(H) = 1$, for all $t > 0$.*

Proof. 1. We first prove the last statement. Let q be a seminorm generating the topology of G. Observe first that if $\nu(G) < \infty$ then obviously $(\exp t\nu)(H) = 1$. So, assume that $\nu(G) = \infty$. Let $0 < \eta_n \downarrow 0$. Denote $\nu_n = \nu|_{\eta_n < q \le \eta_{n-1}}, \eta_0 = \infty$. Then

$$\mu_1 = \mathop{*}_{n=1}^{\infty} \exp \nu_n.$$

So, if X_n has the distribution of $\exp \nu_n$ and X_n's are independent, then $S = \sum X_n$ has the distribution of μ_1 (see Corollary 1, p. 227 and Theorem 3, p. 231 [19]). Write

$$S = S_n + R_n, \quad S_n = \sum_{i=1}^{n} X_i, \quad R_n = \sum_{i>n} X_i.$$

Since $\nu_n(A) = \nu_n(H \cap A)$, for a Borel set A, we have that $P\{X_i \in H\} = P\{S_n \in H\} = 1$, so

$$P\{S \in H\} = P\{R_n \in H\}.$$

However, $\limsup_{n \to \infty} \mathbf{I}_H(R_n) = 0$ or 1 a.s., by the Kolmogorov's $0 - 1$ law. Since $S_n \in H$, a.s., for $n = 1, 2, \ldots$, so if $R_n(\omega) \in H$ for some n then $R_n(\omega) \in H$ for $n = 0, 1, \ldots$, a.s., which means that $\lim_{n \to \infty} \mathbf{I}_H(R_n) = \mathbf{I}_H(S)$ a.e. so either $P\{S \in H\} = 0$ or 1.

2. Let $\nu(H^c) = \infty$. We show that $(\exp t\nu)(H - y) = 0$, for all $y \in G$. To do this, we use a kind of density argument for generators of semigroups involved. We begin with recalling one more fact from [5].

Suppose that N and $\mathcal{N}$ are generators of $(\mu_t)_{t>0}$ acting on C_u and $L^1(\mu)$, respectively. Then $\bar{N} = \mathcal{N}$, where $\bar{N}$ is the closure of N in $L^1(\mu)$. Indeed, it is clear that $N \subseteq \mathcal{N}$. Let now $f \in \mathcal{D}(\mathcal{N})$ = the domain of $\mathcal{N}$. Then there exists an element $g \in L^1(\mu)$ such that $f = \mathcal{R}(1, \mathcal{N})g$, where $\mathcal{R}(1, \mathcal{N}) = \int_0^\infty e^{-u} T_{\mu_u} \, du$ is the resolvent of $(\mu_t)_{t>0}$. There exists a sequence $g_n \in C_u$ such that $g_n \to g$ in $L^1(\mu)$. Then we have $f_n = \mathcal{R}(1, N)g_n \to \mathcal{R}(1, \mathcal{N})g = f$ in $L^1(\mu)$, with $f_n \in \mathcal{D}(N) \subseteq C_u$. Furthermore,

$$f_n - N f_n = (I - N)\mathcal{R}(1, N)g_n = g_n \to g \text{ in } L^1(\mu).$$

Hence

$$N f_n = f_n - g_n \to f - g = \mathcal{N}f \text{ in } L^1(\mu);$$

so

$$\bar{N} = \mathcal{N}.$$

Now, for $\eta > 0$ we denote $\nu_\eta = \nu|_{q>\eta}$. We then have

$$\mu_t = \exp t\nu_\eta * \chi_t^\eta,$$

for a convolution semigroup χ_t^η. Denoting by N_η and N^η the generators of $\exp t\nu_\eta$ and χ_t^η, respectively, on C_u, we have

$$N = N_\eta + N^\eta.$$

Of course

$$N_\eta f(x) = \int f(x + y)\nu_\eta(dy) - c_\eta f(x) = T_{\nu_\eta} f(x) - c_\nu f(x),$$

with $c_\eta = \nu\{q > \eta\}$. On the other hand, denoting

$$\chi^\eta = \int_0^\infty e^{-(1+c_\eta)u} \chi_u^\eta \, du, \quad \chi^{\eta,1} = \int_0^\infty e^{-(1+c_\eta)u} u \, \chi_u^\eta \, du,$$

we have, as before, that $(\chi_t^\eta)_{t>0}$ acts continuously on $L^1(\chi^\eta)$. Let $\mathcal{N}^\eta$ be the generator of $(\chi_t^\eta)_{t>0}$ on $L^1(\chi^\eta)$. Then as above

$$\bar{N}^\eta = \mathcal{N}^\eta \tag{1}$$

where the closure is in $L^1(\chi^\eta)$. Furthermore, for $f \geq 0$

$$\|T_{\nu_\eta} f\|_{L^1(\chi^{\eta,1})} = \int e^{-(1+c_\eta)u} u \int f(z) \, \nu_\eta * \chi_u^\eta(dz) du = \|f\|_{L^1(\nu_\eta * \chi^{\eta,1})}.$$

Observe that

$$\mu = \int_0^\infty e^{-u} \exp(u\nu_\eta) * \chi_u^\eta du$$

$$= \int_0^\infty e^{-u(1+c_\eta)} \chi_u^\eta du + \sum_{k=1}^\infty (\nu_\eta^{*k}/k!) * \int_0^\infty e^{-u(1+c_\eta)} u^k \chi_u^\eta du;$$

so, in particular, $\chi^\eta + \nu_\eta * \chi^{\eta,1} \leq \mu$ and

$$\|T_{\nu_\eta} f\|_{L^1(\chi^{\eta,1})} \leq \|f\|_{L^1(\mu)}.$$

If we now assume that $f_n \to f$ in $L^1(\mu)$ with $|f_n| \leq C$, then, since the measures $\chi^{\eta,1}$ and χ^η are equivalent, we also obtain that

$$T_{\nu_\eta} f_n \to T_{\nu_\eta} f \quad \text{in } L^1(\chi^\eta).$$

If now additionally $f_n \in \mathcal{D}(N)$ and $Nf_n \to \mathcal{N}f$ in $L^1(\mu)$ (so in $L^1(\chi^\eta)$, as well) then $N^\eta f_n$ converges in $L^1(\chi^\eta)$ also, with $\mathcal{D}(N^\eta) \ni f_n \to f$ in $L^1(\chi^\eta)$. By (1) we obtain that $f \in \mathcal{D}(\mathcal{N}^\eta)$ and

$$\mathcal{N}f = \mathcal{N}_\eta f + \mathcal{N}^\eta f \quad \text{in } L^1(\chi^\eta). \tag{2}$$

Finally, if $\mu_t(H) > 0$ then also $\chi_t^\eta(H) > 0$ for all $t > 0$, so $\chi^\eta(H) > 0$. By the Lemma $\pi \circ \mu_t = \exp t\gamma$, $\pi \circ \chi_t^\eta = \exp t\gamma^\eta$ with $\gamma(G/H) < \infty$, $\gamma^\eta(G/H) < \infty$. The proof thus will be finished if we show that $f = \mathbf{I}_{H^c}$ can be put in (2). Indeed, integrating (2) over H with respect to χ^η, we will obtain

$$\gamma(H^c)\chi^\eta(H) = (\nu(H^c \cap \{q > \eta\}) + \gamma^\eta(H^c))\chi^\eta(H);$$

so

$$\nu(H^c \cap \{q > \eta\}) \leq \gamma(H^c) < \infty,$$

which will show that $\nu(H^c) < \infty$, contrary to our assumption that $\nu(H^c) = \infty$.

Thus, we have to show that there exists a sequence $\mathcal{D}(N) \ni f_n \to \mathbf{I}_{H^c}$ in $L^1(\mu)$ and such that $|f_n| \leq C$. To do this, observe that, because $\pi \circ \mu_t = \exp t\gamma$, we have $\mathbf{I}_{H^c} \in \mathcal{D}(\mathcal{N})$, and it is easy to see that

$$\mathcal{N}\mathbf{I}_{H^c}(x) = \gamma(H^c)\mathbf{I}_H(x) - \sum_{y \notin H} \gamma(H - y)\,\mathbf{I}_{H-y}(x).$$

Since $f = \mathcal{R}(1,\mathcal{N})g \iff g = f - \mathcal{N}f$, so the corresponding function $g = -\gamma(H^c)\mathbf{I}_H(x) + \sum_{y \notin H}(1 + \gamma(H - y))\mathbf{I}_{H-y}(x)$

Approximating g in measure μ by $g_n \in C_u, |g_n| \leq 2$ we obtain $\mathcal{D}(N) \ni f_n = \mathcal{R}(1, N)g_n$ with $\|f_n\|_\infty \leq \|g_n\| \leq 2$ and $Nf_n \to \mathcal{N}f$ in $L^1(\mu)$. This ends the proof.

REFERENCES

1. C.R.Baker, *Zero-one laws for Gaussian measures on Banach spaces*, Trans. Amer. Math. Soc. **186** (1973), 291–308.
2. H.Byczkowska, T.Byczkowski, *Zero-one law for subgroups of paths of group-valued stochastic processes*, Studia Math. **89** (1988), 65–73.
3. T.Byczkowski, *Gaussian measures on L_p spaces $0 \leq p < \infty$*, Studia Math. **59** (1977), 249–261.
4. T.Byczkowski, *Zero-one laws for Gaussian measures on metric abelian groups*, Studia Math. **69** (1980), 159–189.
5. T.Byczkowski, A.Hulanicki, *Gaussian measure of normal subgroups*, Ann. Prob. **11** (1983), 685–691.
6. T.Byczkowski, K.Samotij, *Absolute continuity of stable seminorms*, Ann. Prob. **14** (1986), 299–312.
7. S.Cambanis, B.S.Rajput, *Some zero-one laws for Gaussian processes*, Ann. Prob. **14** (1973), 304–312.
8. R.H.Cameron, R.E.Graves, *Additive functionals on a space of continuous functions I*, Trans. Amer. Math. Soc. **70** (1951), 160–176.
9. R.M.Dudley, M.Kanter, *Zero-one laws for stable measures*, Proc. Amer. Math. Soc. **45** (1974), 245–252.
10. X.Fernique, *Certaines propriétés des éléments aléatoires gaussiens*, Instituto Nazionale di Alta Matématica **9** (1972), 37–42.
11. C.C.Graham, O.C.McGehee, *Essays in commutative harmonic analysis*, Springer Verlag, New York, 1979.
12. N.C.Jain, *A zero-one law for Gaussian processes*, Proc. Amer. Math. Soc. **29** (1971), 585–587.
13. A.Janssen, *Zero-one laws for infinitely divisible probability measures on groups*, Z. Wahr. verw. Geb. **60** (1982), 119–138.
14. A.Janssen, *A survey about zero-one laws for probability measures on linear spaces and locally compact groups*, Probability Measures on Groups VII, Lect. Notes in Math. **1064** (1984), 551–563.
15. G.Kallianpur, *Zero-one laws for Gaussian processes*, Trans. Amer. Math. Soc. **149** (1970), 199-211.

16. D.Louie, B.S.Rajput and A.Tortrat, *A zero-one dichotomy theorem for r-semistable laws on infinite dimensional linear spaces*, Sankhyā Series A **42** (1980), 9–18.
17. J.Rosinski,, *An application of series representations to zero-one laws for infinitely divisible random vectors*, Progress in Probability **21** (1990), 189–200; Birkhauser.
18. E.Siebert, *Decomposition of convolution semigroups on Polish groups and zero-one law*, Hokkaido Math. J. **16** (1987), 235–255.
19. A.Tortrat, *Lois de probabilité sur un espace topologique complètement régulier et produits infinis à termes indépendants dans un group topologique*, Ann. Inst. Henri Poincaré **1** (1965), 217–237.

INSTITUTE OF MATHEMATICS, WROCLAW TECHNICAL UNIVERSITY, 50 - 370 WROCLAW, AND THE UNIVERSITY OF TENNESSEE, KNOXVILLE, TN 37996-1300.

DEPARTMENT OF MATHEMATICS, THE UNIVERSITY OF TENNESSEE, KNOXVILLE, TN 37996-1300.

Multiplicity Properties of
Stationary Second Order Random Fields

CHIANG TSE-PEI

Abstract. In this paper, we give a necessary and sufficient spectral criterion by which the multiplicity M_0 can be completely determined. The equivalence relation between the weak commutation property and $M_0 = 1$ is proved.

1. Introduction

In this paper, we study multiplicity properties of stationary second order random fields(SSORF's) $\{X(m, n)\}$, $(m,n) \in Z^2$.

Recently, Kallianpur and Mandrekar [1] investigated the quarter plane prediction problem for SSORF's. They have undertaken a time domain analysis of SSORF's in which, taking quarter plane (in the south west corner) as the past, they introduce three different types of innovation subspaces:

$$I_m{}^1 = [L_1(x : m) \ominus L_1(x : m - 1)] \cap L_2(x : -\infty), \qquad (1.1)$$

$$I_n{}^2 = [L_2(x : n) \ominus L_2(x : n - 1)] \cap L_1(x : -\infty), \qquad (1.2)$$

$$I_{mn} = [L_1(x : m) \ominus L_1(x : m - 1)] \cap [L_2(x : n) \ominus L_2(x : n - 1)] \qquad (1.3)$$

AMS 1980 Subject Classification: 60G60, 60G25.

Key Words and Phrases. *stationary random fields, Wold decompositions, commutation property, innovation subspaces, multiplicity, spectral criterion.*

Research supported in part by the National Natural Sciences Fundation of China.

where

$$L_1(x:m) = \overline{SP}\{X(s,t): \quad s \leq m, \quad t \in Z\} \tag{1.4}$$

$$L_2(x:n) = \overline{SP}\{X(s,t): \quad s \in Z, \quad t \leq n\} \tag{1.5}$$

$$L_1(x:-\infty) = \cap_m L_1(x:m), \quad L_2(x:-\infty) = \cap_n L_2(x:n) \tag{1.6}$$

The innovation subspaces $I_m^j : \quad m \in Z$ have the same dimension M_j. As for the joint innovation subspaces I_{mn}, one can show that they have the same dimension M_0. M_0, M_1 and M_2 are three multiplicities associated with the SSORF $\{X(m, n)\}$. The multiplicity problems were first investigated in [2, 3] by Kallianpur, Miamee and Niemi. They obtain necessary and sufficient criteria (cf. Theorem VI.3 in [3]), by which the multiplicities M_1 and M_2 can be completely determined for j-purely nondetermininstic random fields, under the assumption that the random field has weak commutation property. They also show that if the SSORF $\{X(m, n)\}$ has the strong commutation property, then M_0 is either 0 or 1, and furthermore, $M_0 = 1$, iff $\xi_{mn} \neq 0$ (cf. Theorem VI.1 in [3]).

Korezlioglu and Loubaton in [4] have studied M_0. They show that if the SSORF $\{X(m,n)\}$ has the weak commutation property, then $\{X(m,n)\}$ admits a four-fold decomposition which coincides with the one given in [1], and moreover (i) $M_0 \geq 1$ iff $\int_{-\pi}^{\pi} \log \frac{F_x(\lambda,\mu)}{d\lambda d\mu} d\mu > -\infty$, $a.e.(d\lambda)$ and $\int_{-\pi}^{\pi} \log \frac{F_x(\lambda,\mu)}{d\lambda d\mu} d\lambda > -\infty$, $a.e.(d\mu)$ (ii) $M_0 = 0$ iff the augmented half-plane (AMHP) purely nondeterministic component field $\{X^{(R)}(m,n)\}$ is horizontal HP deterministic or vertical HP deterministic (cf. Proposition V.4 in [4]). They also prove that if $\{X(m,n)\}$ has the weak commutation property with $M_0 \geq 1$ and $dF_x(\lambda,\mu) \ll d\lambda d\mu$, then $M_0 = 1$ (cf. Proposition V.8 in [4]).

More recently, Chiang [5] obtained necessary and sufficient criteria (cf. Theorem 3.1 in [5]), by which the multiplicities M_1 and M_2 can be completely determined for any AMHP nondeterministic random fields. For AMHP deterministic random fields, spectral criteria for determining the multiplicities M_1 and M_2 has also been given in [5], under the weak commutation assumption. It has been proved in [5] that, for any SSORF $\{X(m,n)\}$, we have $M_0 \leq 1$, and moreover, if $\{X(m,n)\}$ has weak commutation property, then $\{X(m,n)\}$ is AMHP nondeterministic iff $M_0 = 1$ (cf. Theorem 3.3 in [5]).

The purpose of this paper is to give a necessary and sufficient criterion by which the multiplicity M_0 can be completely determined.

Suppose that $\{X(m,n)\}$ is either an AMHP nondeterministic field or is a non-vanishing horizontal and vertical HP purely nondeterministic field. We will prove that the following three conditions are equivalent:

    (i) $M_0 = 1$,

    (ii) X(m,n) has the weak commutation property,

(iii) there exist Borel functions M_1 and M_2 with unit modulus such that

$$M_1(\mu)\Gamma_1^*(\lambda, \mu) = M_2(\lambda)\Gamma_2^*(\lambda, \mu), \quad a.e.(d\lambda d\mu) \tag{1.7}$$

where

$$\Gamma_1^*(\lambda, \mu) = \lim_{r \to 1^-} \Gamma_1(re^{-i\lambda}, \mu), \tag{1.8}$$

$$\Gamma_2^*(\lambda, \mu) = \lim_{r \to 1^-} \Gamma_2(\lambda, re^{-i\mu}), \tag{1.9}$$

$$\Gamma_1(z, \mu) = (2\pi)exp\left\{\frac{1}{4\pi}\int_{-\pi}^{\pi}\frac{e^{-i\lambda} + z}{e^{-i\lambda} - z}\log\frac{dF_x(\lambda, \mu)}{d\lambda d\mu}d\lambda\right\}, \tag{1.10}$$

$$\Gamma_2(\lambda, z) = (2\pi)exp\left\{\frac{1}{4\pi}\int_{-\pi}^{\pi}\frac{e^{-i\mu} + z}{e^{-i\mu} - z}\log\frac{dF_x(\lambda, \mu)}{d\lambda d\mu}d\mu\right\}, \tag{1.11}$$

and where $dF_x(\lambda, \mu)$ is the spectral measure of the SSORF $\{X(m,n)\}$, and $\frac{dF_x(\lambda,\mu)}{d\lambda d\mu}$ is the Randon-Nikodym derivative of $dF_x(\lambda, \mu)$ w.r.t. $d\lambda d\mu$.

From the equivalence of conditions (i)-(iii) and the fact that $M_0 \leq 1$ for any SSORF, we obtain the following spectral characterization of the multiplicity M_0: M_0 is either 0 or 1, and $M_0 = 1$ iff the spectral condition (1.7) holds true.

Remark. Korezlioglu and Loubaton in [6] have proved that a AMHP nondeterministic random field has the weak commutation property with $M_0 = 1$ iff the spectral condition (1.7) holds true.

2. Proof of the Equivalence Relation Between the Weak Commutation Property and $M_0 = 1$

Theorem 2.1. Let the SSORF $\{X(m,n)\}$ be a non-vanishing horizontal and vertical HP purely nondeterministic field. Then $M_0 = 1$ iff $\{X(m,n)\}$ has the weak commutation property.

Proof: If $\{X(m,n)\}$ has the weak commutation property, then from Theorem 2.4 and Theorem 3.3 in [5], it follows immediately that $M_0 = 1$.

Conversely, if $M_0 = 1$, then we choose an element η_{00} in I_{00} such that $E\{|\eta_{00}|^2\} = 1$. Let L(x) be the closed linear manifold spanned by all X(m,n): (m,n)$\in Z^2$. V_1 and V_2 will denote the respective horizontal and vertical shift operator of the SSORF $\{X(m,n)\}$. Let

$$\eta_{mn} = V_1^m V_2^n\{\eta_{00}\}. \tag{2.1}$$

Then $\{\eta_{mn}\} : (m, n) \in Z^2$ is a two-dimensional white noise. There exists a square summable function $\phi(\lambda, \mu)$ w.r.t. $dF_x(\lambda, \mu)$ such that

$$\eta_{mn} = \int_{-\pi}^{\pi}\int_{-\pi}^{\pi} exp\{i(m\lambda + n\mu)\}\phi(\lambda, \mu)dZ_x(\lambda, \mu), \quad (m, n) \in Z^2, \tag{2.2}$$

where $dZ_x(\lambda, \mu)$ is the corresponding random spectral measure of the SSORF $\{X(m,n)\}$. We have

$$E\{\eta_{mn}\eta_{00}\} = \int_{-\pi}^{\pi}\int_{-\pi}^{\pi} exp\{i(m\lambda + n\mu)\}|\phi(\lambda, \mu)|^2 dF_x(\lambda, \mu), \quad (m,n) \in Z^2.$$

(2.3)

Since $\{\eta_{mn}\}$ is a two-dimensional white noise, we conclude from (2.3) that

$$|\phi(\lambda, \mu)|^2 dF_x(\lambda, \mu) = \frac{1}{4\pi^2}d\lambda d\mu.$$

(2.4)

We shall prove that

$$L(x) = \sum_{m=-\infty}^{\infty}\sum_{-\infty}^{\infty} I_{mn},$$

(2.5)

Suppose β is an element in $L(x) \ominus [\sum_{m=-\infty}^{\infty}\sum_{-\infty}^{\infty} I_{mn}]$. Then

$$\beta \perp \eta_{mn}, \quad (m,n) \in Z^2$$

(2.6)

and there exists a square summable function $b(\lambda, \mu)$ w.r.t. $dF_x(\lambda, \mu)$ such that

$$\beta = \int_{-\pi}^{\pi}\int_{-\pi}^{\pi} b(\lambda, \mu)dZ_x(\lambda, \mu),$$

(2.7)

which, together with (2.2) and (2.6), implies

$$\int_{-\pi}^{\pi}\int_{-\pi}^{\pi} exp\{-i(m\lambda + n\mu)\}b(\lambda, \mu)\overline{\phi(\lambda, \mu)}dF_x(\lambda, \mu) = 0, \quad (m,n) \in Z^2$$

(2.8)

whence

$$b(\lambda, \mu)\overline{\phi(\lambda, \mu)}dF_x(\lambda, \mu) = 0.$$

(2.9)

A comparison of (2.9) with (2.4) shows that

$$b(\lambda, \mu) = 0 \quad a.e.(d\lambda d\mu).$$

(2.10)

We need the following lemma whose proof offers no difficulty, and is thus omitted.

Lemma. *A SSORF $\{X(m,n)\}$ is a non-vanishing horizontal and vertical HP purely nondeterministic iff $dF_x(\lambda, \mu) = f_x(\lambda, \mu)d\lambda d\mu$, $\int_{-\pi}^{\pi} \log f_x(\lambda, \mu)d\lambda > -\infty$, a.e.($d\mu$) and $\int_{-\pi}^{\pi} \log f_x(\lambda, \mu)d\mu > -\infty$, a.e.($d\lambda$).*

Using (2.7), (2.10) and the above lemma, we get

$$E|\beta|^2 = \int_{-\pi}^{\pi}\int_{-\pi}^{\pi} |b(\lambda, \mu)|^2 dF_x(\lambda, \mu) = \int_{-\pi}^{\pi}\int_{-\pi}^{\pi} |b(\lambda, \mu)|^2 f_x(\lambda, \mu)d\lambda d\mu = 0$$

(2.11)

which gives (2.5).

We are now in a position to prove that the SSORF $\{X(m,n)\}$ has the weak commutation property. Since $\{X(m,n)\}$ is horizontal and vertical HP purely nondeterministic, we have

$$L(x) = \sum_{m=-\infty}^{\infty} \oplus \left[L_1(x:m) \ominus L_1(x:m-1) \right], \qquad (2.12)$$

and

$$L(x) = \sum_{n=-\infty}^{\infty} \oplus \left[L_2(x:n) \ominus L_2(x:n-1) \right], \qquad (2.13)$$

which, together with (2.5), implies

$$L_2(x:n) \ominus L_2(x:n-1) = \sum_{m=-\infty}^{\infty} \oplus \left(\left[L_1(x:m) \ominus L_1(x:m-1) \right] \right.$$
$$\left. \cap \left[L_2(x:n) \ominus L_2(x:n-1) \right] \right). \qquad (2.14)$$

From (2.12), (2.14) and the Lemma 3 in [5], it follows that

$$P_{[L_2(x:n) \ominus L_2(x:n-1)]} \quad \text{commutes with} \quad P_{[L_1(x:m) \ominus L_1(x:m-1)]},$$

and hence, by Lemma 4 in [5],

$$P_{L_1(x:m)} \quad \text{commutes with} \quad P_{L_2(x:n)},$$

i.e. $\{X(m,n)\}$ has the weak commutation property. The proof of the theorem is now complete. ∎

Theorem 2.2. *Let the SSORF $\{X(m,n)\}$ be a AMHP nondeterministic random field. Then $M_0 = 1$ iff $\{X(m,n)\}$ has the weak commutation property.*

Proof: It has been shown in [6, p.39] that a AMHP nondeterministic random field $\{X(m,n)\}$ has the weak commutation property iff its purely nondeterministic component field $\{X^{(R)}(m,n)\}$ has the weak commutation property. On the other hand, since $\{X^{(R)}(m,n)\}$ is a non-vanishing horizontal and vertical HP purely nondeterministic field, we then conclude from Theorem 2.1 that $\{X(m,n)\}$ has the weak commutation property iff $\{X^{(R)}(m,n)\}$ has one-dimensional joint innovation spaces $I_{mn}^{(R)}$. Hence for the proof of our theorem, it is sufficient to show that

$$I_{mn} = I_{mn}^{(R)}. \qquad (2.15)$$

By formulas (2.16) and (2.17) in [5], we have

$$I_{mn} = [\overline{SP}\{\eta_1(m,t) : t \in Z\} \oplus \overline{SP}\{\xi_1(m,t) : t \in Z\}]$$
$$\cap \, [\overline{SP}\{\eta_2(s,n) : s \in Z\} \oplus \overline{SP}\{\xi_2(s,n) : s \in Z\}].$$

$$(2.16)$$

On the other hand, using (2.35) in [5], we get

$$I_{mn} \perp L(\xi_k), \, (k = 1, 2) \qquad (2.17)$$

which, together with (2.16), implies

$$I_{mn} = \overline{SP}\{\eta_1(m,t) : t \in Z\} \cap \overline{SP}\{\eta_2(s,n) : s \in Z\} = I_{mn}^{(R)} \qquad (2.18)$$

which gives (2.15). ∎

3. Spectral Characterization of
The Weak Commutation Property.

Theorem 3.1. *Let the SSORF $\{X(m,n)\}$ be a non-vanishing horizontal and vertical HP purely nondeterministic random field. Then $\{X(m,n)\}$ has the weak commutation property iff the spectral condition (1.7) holds true.*

Proof of The Necessity Part of The Theorem: Since $\{X(m,n)\}$ has the commutation property, we have, by Theorem 2.1, $M_0 = 1$. Then, just as we did in the proof of Theorem 2.1 we obtain a two-dimensional white noise $\{\eta_{mn}\} : (m,n) \in Z^2$ such that

$$\eta_{mn} = V_1^m V_2^n \{\eta_{00}\} \in I_{mn}. \qquad (3.1)$$

By Theorem 8 in [7], we have the following vertical HP moving average representation:

$$X(m,n) = \sum_{t=0}^{\infty} \sum_{s=-\infty}^{\infty} c_{st} u(m-s, n-t), \qquad (3.2)$$

where $\{u(s,t)\}$ is a two-dimensional white noise such that

$$L_2(x : t) = L_2(u : t), \quad t \in Z, \qquad (3.3)$$

and

$$\sum_{t=0}^{\infty} \sum_{s=-\infty}^{\infty} c_{st} exp\{-i(s\lambda + t\mu)\} = \Gamma_2^*(\lambda, \mu), \qquad (3.4)$$

in which, $\Gamma_2^*(\lambda, \mu)$ is determined by (1.8) and (1.10).

From (3.3), we have

$$L_2(x:n) \ominus L_2(x:n-1) = \overline{SP}\{u(s,n) : s \in Z\}. \tag{3.5}$$

There exists Borel function $M_2(\lambda)$ with unit modulus such that

$$\eta_{mn} = \int_{-\pi}^{\pi} \int_{-\pi}^{\pi} exp\{i(m\lambda + n\mu)\}(M_2(\lambda))^{-1}dZ_u(\lambda, \mu), \tag{3.6}$$

whence, by (3.2) and (3.4),

$$\eta_{mn} = \int_{-\pi}^{\pi} \int_{-\pi}^{\pi} exp\{i(m\lambda + n\mu)\}(M_2(\lambda)\Gamma_2^*(\lambda, \mu))^{-1}dZ_x(\lambda, \mu). \tag{3.7}$$

Again, by Theorem 8 in [7] we get the following horizontal HP moving average representation:

$$X(m,n) = \sum_{s=0}^{\infty} \sum_{t=-\infty}^{\infty} d_{st}v(m-s, n-t), \tag{3.2'}$$

where $\{v(s,t)\}$ is a two-dimensional white noise such that

$$L_1(x:s) = L_1(v:s), \quad s \in Z, \tag{3.3'}$$

and

$$\sum_{s=0}^{\infty} \sum_{t=-\infty}^{\infty} d_{st}exp\{-i(s\lambda + t\mu)\} = \Gamma_1^*(\lambda, \mu), \tag{3.4'}$$

in which, $\Gamma_1^*(\lambda, \mu)$ is determined by (1.7) and (1.9).

Arguing with (3.2') as we did with (3.2) we conclude that

$$L_1(x:m) \ominus L_1(x:m-1) = \overline{SP}\{v(m,t) : t \in Z\} \tag{3.5'}$$

and there exists Borel function $M_1(\mu)$ with unit modulus such that

$$\eta_{mn} == \int_{-\pi}^{\pi} \int_{-\pi}^{\pi} exp\{i(m\lambda + n\mu)\}(M_1(\mu)\Gamma_1^*(\lambda, \mu))^{-1}dZ_x(\lambda, \mu). \tag{3.7'}$$

A comparison of the above expression with (3.7) shows that the spectral condition (1.7) holds true. Hence the necessity part of the theorem is proved.

Proof of The Sufficiency Part of The Theorem: By the assumption of the theorem, we still can apply Theorem 8 in [7]. We then also have the vertical HP moving average representation (3.2) and (3.2'), and moreover, (3.5) and (3.5') also hold true. Let us consider

$$\rho_{mn} = \int_{-\pi}^{\pi} \int_{-\pi}^{\pi} exp\{i(m\lambda + n\mu)\}(M_2(\lambda))^{-1} dZ_u(\lambda, \mu) \qquad (3.8)$$

$$\rho'_{mn} = \int_{-\pi}^{\pi} \int_{-\pi}^{\pi} exp\{i(m\lambda + n\mu)\}(M_1(\mu))^{-1} dZ_v(\lambda, \mu) \qquad (3.8')$$

whence $\rho_{mn} \in L_2(x:n) \ominus L_2(x:n-1)$ and $\rho'_{mn} \in L_1(x:m) \ominus L_1(x:m-1)$ on account of (3.5) and (3.5').

According to Theorem 2.1, it suffices to prove $M_0 = 1$. Thus it suffices to prove:

$$\rho_{mn} = \rho'_{mn}. \qquad (3.9)$$

Combing (3.2), (3.4), and (3.8), we obtain

$$\rho_{mn} = \int_{-\pi}^{\pi} \int_{-\pi}^{\pi} exp\{i(m\lambda + n\mu)\}(M_2(\lambda)\Gamma_2^*(\lambda, \mu))^{-1} dZ_x(\lambda, \mu), \qquad (3.10)$$

Correspondingly, we have

$$\rho_{mn} = \int_{-\pi}^{\pi} \int_{-\pi}^{\pi} exp\{i(m\lambda + n\mu)\}(M_1(\mu)\Gamma_1^*(\lambda, \mu))^{-1} dZ_x(\lambda, \mu), \qquad (3.10')$$

which, together with (3.10) and the spectral condition (1.7), gives (3.9). The sufficiency part of the theorem is proved. ∎

Theorem 3.2. Let SSORF $\{X(m,n)\}$ be an AMHP nondeterministic random field. Then $\{X(m,n)\}$ has the weak commutation property iff the spectral condition (1.7) holds true.

Proof: There is no restriction in assuming that $\{X(m,n)\}$ is AMHP purely nondeterministic random field. Since an AMHP purely nondeterministic field is non-vanishing horizontal and vertical purely nondeterministic field, the desired result follows from Theorem 3.1. ∎

4. Examples

We shall give two illustrative examples.

Example 1. Let $\{X(m,n)\}$ be a SSORF having the spectral density:

$$f_x(\lambda, \mu) = exp\{-\frac{1}{(|\lambda| + |\mu|)^2}\}. \qquad (4.1)$$

It has been shown in [5, p.55] that $\{X(m,n)\}$ is horizontal and vertical HP purely nondeterministic but is AMHP determinnistic, and moreover the SSORF $\{X(m,n)\}$ does not possess the weak commutation property. Therefore $M_0 = 0$, by Theorem 3.1.

Example 2. In [5, p.61], we have pointed out that there exists an AMHP purely nondeterministic random field not having the weak commutation property. This is demonstrated by the fact that an example has been given in [8] showing that there exists an AMHP purely nondeterministic random field which is AMHP strongly deterministic (for the definition of strong determinism, we refer to [9]). The example given in [8] is the following: Let

$$f(\lambda) = \begin{cases} 1, & -\pi \le \lambda \le 0, \\ (\lambda - \frac{\pi}{2^n}) / \frac{\pi}{2^{n+1}}, & \frac{\pi}{2^n} \le \lambda \le \frac{3\pi}{2^{n+1}}, \\ (\frac{\pi}{2^{n-1}} - \lambda) / \frac{\pi}{2^{n+1}}, & \frac{3\pi}{2^{n+1}} \le \lambda \le \frac{\pi}{2^{n-1}}, \quad n = 1, 2, \dots \end{cases} \tag{4.2}$$

Then there exists an H_2 function $A(z) = \sum_{n=0}^{\infty} a_n z^n$ such that $|A(e^{i\lambda})|^2 = f(\lambda)$. Suppose that $\{u(m, n)\}_{(m,n) \in Z^2}$ is a two-dimensional white noise. Our required SSORF is

$$X(m, n) = \sum_{k=0}^{\infty} a_k u(m + k, n - k), \quad (m, n) \in Z^2 \tag{4.3}$$

Since $\{X(m,n)\}$ does not possess the weak commutation property, we have $M_0 = 0$, by Theorem 3.2.

References

[1]. Kallianpur, G., and Mandrekar, V. (1983), *Nondeterministic random fields and Wold and Halmos decompositions for commuting isometries*, Prediction theory and Harmonic Analysis. The Pesi Masani Volume, pp. 165-190, North-Holland, Amsterdam.

[2]. Kallianpur, G., Miamee, A. G., and Niemi, H. (1987), *On the prediction theory of two parameter stationary random fields*, Technical Report 178, Center for Stochastic Processes, Department of Statistics, University of North Carolina.

[3]. Kallianpur, G., Miamee, A.G., and Niemi, H. (1990), *On the prediction theory of two parameter stationary random fields*, J. Multivariate Anal. , **32**, 120-149.

[4]. Korezlioglu, H., and Loubaton, P. H. (1987), *Prediction and spectral decomposition of wide sense stationary processes on Z^2*, In Spatial Processes and Spatial Time Series Analysis. Proceedings, 6th Franco-Belgian Meeting of Statisticians, 1985 (F. Droesbeke, Ed.). Publ. Facult. Univ. Saint-Louis, Brussels.

[5]. Chiang, T.P.(1991), *The prediction theory of stationary random fields. III. Fourfold Wold decompositions*, J. Multivariate Anal. , **37**, 46-65.

[6]. Korezlioglu, H., and Loubaton, P.H. (1986), *Spectral factorization of wide sense stationary processes on Z^2*, J. Multivariate Anal. , **19**, 24-47.

[7]. Chiang, T.P.(1989), *The prediction theory of stationary random fields (I): Half-plane prediction*, Acta Sci. Nat. Univ. Peking. , **25**, 25-50.

[8]. Chiang, T.P. (1985), *On the strong regularity of stationary random fields*, Chinese J. Appl. Probab. Statist. , **1**, 125-126.

[9]. Soltani, A.R. (1984), *Extrapolation and moving average representation for stationary random fields and Beurling's theorem*, Ann. Probab. , **12**, 102-132.

Department of Probability and Statistics
Peking University
Beijing, 100871, China

Multiple Time Scale Analysis
of Hierarchically Interacting Systems

DONALD A. DAWSON[1] and ANDREAS GREVEN

Abstract. A hierarchically interacting stochastic model is introduced and its long time behavior is identified by multiple time scale analysis and an associated interaction chain.

1. Introduction.

We consider a class of stochastic models which are systems with *infinitely* many interacting components whose interactions are organized in a *hierarchical* manner. The kth level of the hierarchy is comprised of N_k objects of the $(k-1)st$ level and the strength of the interaction between two individuals decreases as their hierarchical distance (the first level at which they are members of the same class) increases. Although a wide range of hierarchical models of this type arise in different fields, from mathematical physics to mathematical biology, we will restrict our attention to one typical model which was originally introduced in population biology by Sawyer and Felsenstein [5].

This is a model of a population in which N_1 individuals form a site, N_2 sites form a group, N_3 groups form a clan, N_4 clans form a village and so on. Each individual is of one of two types and we consider the proportions of type I individuals at different levels when the basic population dynamics is neutral genetic sampling, that is, at a given rate individuals die and are replaced by one of that type with probability equal to the empirical frequency of that type at the site (*resampling*). In the limit as the number of individuals, N_1, goes to infinity and in time scale $\sqrt{N_1}t$ the proportion of type I individuals at a site follows the *Fisher-Wright diffusion*, associated with the stochastic differential equation

$$dy(t) = \sqrt{2y(t)(1 - y(t))}dw(t), \quad y(0) \in [0, 1],$$

where $w(\cdot)$ is a Brownian motion.

If we now consider N_2 sites of this type and allow *migration* of individuals between sites with migration rate c_0 and uniform one-step distribution we obtain for $i = 1, \ldots, N_2$,

$$dy_i^{N_2}(t) = c\Big(\frac{1}{N_2} \sum_{k=1}^{N_2} y_k^{N_2}(t) - y_i^{N_2}(t)\Big)dt + \sqrt{2y_i^{N_2}(t)(1 - y_i^{N_2}(t))}dw_i(t)$$

where $y_i^{N_2}$ denotes the type I proportion at site i, and $\{w_i(\cdot) : i = 1, \ldots, N_2\}$ are independent Brownian motions. This finite and one level system in which the N_2 sites interact in a symmetric way is an example of a *mean-field interaction*. In order to construct an *infinite* system and incorporate more

---

[1] Research supported by NSERC.

levels of the population hierarchy mentioned above, a general hierarchical framework will be described in section 2. We shall focus on the case $N_2 = N_3 = \ldots$ for notational convenience. The remaining sections 3 and 4 will analyse the behavior of the system for $t \to \infty$; first in 3 we develop the multiple time scale analysis and in 4 we apply this to the study of the long time behavior.

2. Multilevel interactions and the hierarchical group

In this section we first formulate the hierarchical structure and then introduce the basic hierarchical stochastic model.

The *hierarchical group* Ω_N, with N a natural number or infinity, has as elements the (countable) set of all sequences $\xi = [\xi_1, \xi_2, \ldots]$ with coordinates $\xi_i \in \{0, \ldots, N-1\}$, and $\xi_i = 0$ for all but finitely many i. Then ξ denotes the ξ_1-st site of a group which is the ξ_2-nd group of a clan which is the ξ_3-rd clan of a village ... which is the ξ_k-th member of a k-level set, $k \geq 1$. Ω_N is an Abelian group with addition defined componentwise modulo N and with zero element $\mathbf{0} = (0, 0, 0, \ldots)$.

We define a hierarchical distance d on Ω_N by

$$d(\xi, \zeta) = d(\xi - \zeta, 0) := \begin{cases} k & \text{if } \xi_k \neq \zeta_k \text{ and } \xi_j = \zeta_j \ \forall j > k \\ 0 & \text{if } \xi = \zeta. \end{cases}$$

Two individuals, ξ and ζ, are said to be *relatives of degree k* if $d(\xi, \zeta) = k$. The *k-block* associated with ξ is defined as $\xi_{[k]} := \{\zeta : d(\xi, \zeta) \leq k\}$, and the *$k$-block average* is given by

$$x_{\xi, k} = N^{-k} \sum_{\zeta \in \xi_{[k]}} x_{\zeta, 0}.$$

The full *hierarchical model* is a process on $[0, 1]^{\Omega_N}$ prescribed by the infinite system of Itô stochastic differential equations:
(1) for $\xi \in \Omega_N$,

$$dx_{\xi, 0}(t) = \Big(\sum_{k=1}^{\infty} \frac{c_{k-1}}{N^{k-1}} (x_{\xi, k}(t) - x_{\xi, 0}(t)) \Big) dt + \sqrt{2g(x_{\xi, 0}(t))} dw_\xi(t).$$

Here $g : [0, 1] \to \mathbf{R}^+$ is Lipschitz continuous, $g(x) > 0$ for $x \in (0, 1)$, and $g(0) = g(1) = 0$. This includes the Fisher-Wright case, that is $g(x) = x(1 - x)$, which was described above in the population genetics model. The $\{c_k\}$ are nonnegative numbers with $\sum c_k N^{-k} < \infty$. In (1) $x_{\xi, 0}(t)$ denotes the proportion of type I individuals at ξ. As initial distributions of the process we shall allow *homogeneous* measures μ on $[0, 1]^{\Omega_N}$ which are *ergodic* and we denote $\theta = E^\mu(x_{\xi, 0})$.

The infinite system of stochastic differential equations has a unique solution (cf. Shiga and Shimizu [7]) yielding a time homogeneous Markov diffusion process $\mathcal{X}$ with state space $[0, 1]^{\Omega_N}$ equipped with its product topology.

The significance of the interaction term in (1) is that the proportion of type I individuals at a site is influenced by k-block averages with interaction strengths c_{k-1}/N^{k-1}. It can also be rewritten

$$\left(\sum_{k=1}^{\infty} \frac{c_{k-1}}{N^{k-1}} (x_{\xi,k}(t) - x_{\xi,0}(t)) \right) = \left(\sum_{\zeta \in \Omega_N} q_{\xi,\zeta} (x_{\zeta,0}(t) - x_{\xi,0}(t)) \right)$$

where $q_{\xi,\xi} := - \sum_{\xi \neq \zeta} q_{\xi,\zeta}$, and

$$q_{\xi,\zeta} = q_{\zeta,\xi} := \sum_{k=1}^{\infty} c_k N^{-2k} \mathbf{1}\{\zeta \in \xi_{[k]}\} \quad \text{if} \ \xi \neq \zeta.$$

Consequently the interaction can be interpreted as a *migration* of individuals in which the migration rate between sites depends only on their hierarchical distance. In particular, the migration of a single individual can be viewed as a *symmetric random walk* on the hierarchical group with generator $Qf(\xi) = \sum_{\zeta} q_{\xi,\zeta} f(\zeta)$. As will demonstrated below, the qualitative nature of the long time behavior of the system (1) depends on the rate of decay of interaction strength with hierarchical distance and in particular whether the random walk on Ω_N with generator Q is *transient* or *recurrent*.

3. Multiple time scale behavior.

(i) The behavior of the system (1) for a fixed value of the parameter N as $t \to \infty$ can be determined by exploiting the duality relations with systems of finitely many coalescing random walks on the hierarchical group Ω_N, or using coupling methods (compare Shiga [6], Cox and Greven [1]). In particular we can determine the structure of the set of extremal equilibrium measures and their domain of attraction, as a function of the random walk with generator Q. If the random walk is transient then for every $\theta \in [0,1]$ we can find a spatially mixing extremal invariant measure with $E[x_{\xi,0}] = \theta$, whereas if the random walk is recurrent then the only extremal invariant measures are concentrated on one of the traps of the system, that is, either $\{x_{\xi,0} \equiv 0\}$ or $\{x_{\xi,0} \equiv 1\}$.

The *invariant measures* cannot be described in a simple fashion in closed form, since all components are correlated. However a study of the SDE (1), indicates that at the successive time scales of size $sN^k, k = 1, 2, 3 \ldots$, the evolution of a component is mainly influenced by the components within hierarchical distance k and the strength of the influence is measured by the parameter c_k. Hence even though an equilibrium will eventually develop, in which everything contributes to the collective behavior, we observe almost stationary situations up to the k-block level over time intervals of the form sN^k to $sN^k + tN^{k-1}$ $(s > 0)$. We refer to this collection of time scales and approximately stationary states as the *multiple time scale behavior*. Furthermore we expect that this sequence of almost stationary states, so-called *quasiequilibria*, will provide a closer and closer approximation to the *global equilibrium* and the behavior as $t \to \infty$.

Hence for a deeper understanding of the collective behavior of the infinite system we should identify the dynamics of the mechanism producing this sequence of quasiequilibria. For this purpose we shall introduce the notion of *interaction chain* in Section 4.

In what way can the above picture be made precise? The main obstacle is that although the time points s, sN, sN^2 are widely spread out for large N the time scales do not completely separate. Moreover, it is not possible to provide a condensed description of the state of the system at time sN^k in a block of size k which would be sufficient to fully describe the evolution of the state of a component over the time interval from sN^k to $sN^k + t$ (short of giving full information on the state of the system). Therefore it is useful to consider the limit $N \to \infty$ in which case the epochs $s, sN, sN^2, \ldots$ will completely separate for any fixed s. In this *hierarchical mean-field limit* block averages, $x_{\xi,k}$ alone will provide sufficient information on the state of the system in the block to fully describe the behavior at the level of the $(k-1)$-subblock, $x_{\xi,k-1}$ in the time interval sN^k to $sN^k + tN^{k-1}$. The point of this procedure is of course that the long time behavior of the infinite system for large but fixed N is then reasonably well approximated.

We carry out this program as follows. In subsection (iii) we shall state two precise theorems on the multiple time scale behavior, but first in section (ii) we shall give some heuristic explanation and present the basic results necessary to prove the theorem.

(ii) We proceed in two steps. First we consider the original time scale t and consider the special case of a system with $c_0 > 0, c_1 = c_2 = \ldots = 0$. Then

$$\mathcal{L}\left((\{x_{\xi,0}(t)\})\right) \underset{N \to \infty}{\Longrightarrow} \mathcal{L}\left((y_\xi(t))_{\xi \in \Omega_\infty}\right)_{t \in \mathbf{R}+},$$

where $\mathcal{L}\left(\{y_\xi\}_{\xi \in \Omega_\infty}\right)$ denotes the law of a system in which every component evolves independently according to the equation

(2) $dy(t) = c_0(\theta - y(t))dt + \sqrt{2g(y(t))}dw(t); \quad \theta = E(x_{\xi,0}).$

This is the *McKean-Vlasov* (or *mean-field*) limit. It is not hard to show by coupling arguments that in this time scale the same behavior also occurs for arbitrary sequences $\{c_k\}$, since the higher order terms are $O(N^{-1})$ and hence negligible in the limit $N \to \infty$ for $t \in [0, T]$.

The second step is to consider a two level system with $c_0, c_1 > 0, \ c_2 = c_3 = c_4, \ldots = 0$, but now in the *faster time scale tN*. By letting $N \to \infty$ we shall simplify the equations (1) in two steps by simplifying the drift term. We first note that $(x_{\xi,2}(t))_{t \in \mathbf{R}+}$ is a martingale with mean 0 and increasing process

$$\langle x_{\xi,2}(t) \rangle = \int_0^t \frac{1}{N^4} \left(\sum_{\xi' \in \xi_{[2]}} g\left(x_{\xi',0}(s)\right) \right) ds$$

$$= \frac{1}{N^2} \int_0^t \left(\frac{1}{N^2} \sum_{\xi' \in \xi_{[2]}} g\left(x_{\xi',0}(s)\right) \right) ds.$$

Therefore, $\langle x_{\xi,2}(Nt)\rangle \leq N^{-1}\|g\|_\infty \xrightarrow[N\to\infty]{} 0$ and $\mathcal{L}\left(x_{\xi,2}(Nt)_{t\in\mathbf{R}}\right) \Longrightarrow \delta_{\{Y_t\equiv\theta\}}$.
This means that instead of studying the original system we can consider the simplified equation

$$(3) \qquad dy_{\xi,0}(t) = c_0\left(y_{\xi,1}(t) - y_{\xi,0}(t)\right)dt + c_1 N^{-1}\left(\theta - y_{\xi,0}(t)\right)dt$$
$$+ \sqrt{2g\left(y_{\xi,0}(t)\right)}dw(t).$$

To understand the behavior of this equation as $N \to \infty$ we have to study $(y_{\xi,1}(tN))_{t\in\mathbf{R}+}$. This process is a semimartingale with increasing process A_t^N given by $A_t^N = \int_0^t \left(\frac{1}{N}\sum g(y_{\xi,0}(sN))\right)ds$. Since this is clearly a tight family, we can select a subsequence N_j such that

$$\mathcal{L}\left((y_{\xi,1}(tN_j))_{t\in\mathbf{R}+}\right) \underset{j\to\infty}{\Longrightarrow} \mathcal{L}\left((\theta_t)_{t\in\mathbf{R}+}\right) := Q_t(\theta,.).$$

By identifying $(\theta_t)_{t\in\mathbf{R}+}$ and observing that it does not depend on N_j we will later on conclude that the convergence holds for $N \to \infty$. To do this we must determine the behavior of $\{y_{\xi,0}(tN)\}_{\xi\in\Omega_N}$. It turns out (see below) that

$$(4) \qquad \mathcal{L}\left((y_{\xi,0}(tN))_{\xi\in\Omega_N}\right) \underset{N\to\infty}{\Longrightarrow} \int Q_t(\theta,d\tilde{\theta})\nu_{\tilde{\theta}},$$

where ν_θ denotes the unique equilibrium of (2). From this we conclude that in fact $A_t^N \Longrightarrow \int_0^t u_g(\theta_s)ds$ with $u_g(\theta) := \int g(y)\nu_\theta(dy)$. Hence the continuous process θ_t is characterized by the property that $M_t = \theta_t - c_0\int_0^t(\theta - \theta_s)ds$ is a martingale and in addition that $M_t^2 - \int_0^t u_g(\theta_s)ds$ is a martingale, so that (θ_t) must be the diffusion with generator $c_0(\theta - x)\frac{\partial}{\partial x} + \frac{1}{2}u_g(x)\left(\frac{\partial}{\partial x}\right)^2$.

The basic idea behind relation (4) is the following. Choose a sequence $T(N) \uparrow \infty$ with $T(N) = o(N)$. Then uniformly in $t \leq T(N)$, the process $\{y_{\xi,0}(t)\}_{\xi\in\Omega_N}$ will converge to a system with independent components evolving as described by (2). Now pick a subsequence N_k such that for t fixed $\mathcal{L}\left((y_{\xi,0}(tN_k - T(N_k))))\right)_{\xi\in\Omega_N}$ converges to a measure ν and the laws of the processes $((y_{\xi,1}(tN_k))_{t\in\mathbf{R}+})$ converge. It can be shown that under $\nu, X_{\xi,1}$ converges to $Q_t(\theta,.)$. Now consider a system with initial distribution ν in which the components evolve independently according to (2) and θ is a random variable with law $Q_t(\theta,.)$. It will then of course tend weakly to $\int Q_t(\theta,d\tilde{\theta})\nu_{\tilde{\theta}}$ and therefore $\mathcal{L}\left(y_{\xi,0}(tN_k)\right)$ will do the same. To justify these steps considerable work needs to be done. Details can be found in Dawson and Greven [3].

(iii) The facts presented so far are the building blocks needed to develop an understanding of the behavior of the infinite hierarchy in time scales $tN^k, k = 0, 1, 2, \ldots$. In order to describe the behavior we need the following ingredients:

DEFINITION 1.

(I) *time scales:* $\beta_k(N) = N^k$

(II) *quasi-equilibria:* $\Gamma^k_\theta(\cdot)$, the *associated diffusion* $\left(Z^{\theta,k}_t\right)_{t \in \mathbf{R}+}$ and the (state-dependent) *fluctuation coefficient* on the kth level $F_k(\cdot)$.

$$\Gamma^k_\theta(\cdot) \text{ is the unique equilibrium of } \left(Z^{\theta,k}_t\right)_{t \in \mathbf{R}+}$$

$$dZ^{\theta,k}_t = c_k\left(\theta - Z^{\theta,k}_t\right)dt + \sqrt{2F_k(Z^{\theta,k}_t)}dw(t)$$

$$F_0(x) = g(x), \quad F_{k+1}(x) = \int g(y)\Gamma^k_x(dy).$$

(III) *Level $k+1$ - Level j interaction kernel:* $\mu^{k,j}_\theta(\cdot)$.

$$\mu^{k,j}_\theta(\cdot) = \int \ldots \int \Gamma^k_\theta(d\theta_1)\Gamma^{k-1}_{\theta_1}(d\theta_2)\ldots\Gamma^{j+1}_{\theta_{k-j}}(\cdot).\blacksquare$$

Now we are ready to state our first theorem. Recall that $\theta = E^\mu(x_{\xi,0})$ and $x_{\xi,k}$ is the average over components within distance k of the point ξ.

THEOREM 1. *The multiple time scale behavior of the k-block $X_{\xi,k}$ is given by:*

$$k > j: \quad \mathcal{L}\left((x_{\xi,k}(tN^j))_{t \in \mathbf{R}+}\right) \underset{N \to \infty}{\Longrightarrow} \delta_{\{Y_t \equiv \theta\}}$$

$$k = j: \quad \mathcal{L}\left((x_{\xi,k}(tN^k))_{t \in \mathbf{R}+}\right) \underset{N \to \infty}{\Longrightarrow} \mathcal{L}\left(\left(Z^{\theta,k}_t\right)_{t \in \mathbf{R}+}\right)$$

$$k < j: \quad \mathcal{L}\left((x_{\xi,k}(sN^j + tN^k))_{t \in \mathbf{R}+}\right) \underset{N \to \infty}{\Longrightarrow} \mathcal{L}\left(\left(Z^{\theta^*,k}_t\right)_{t \in \mathbf{R}+}\right)$$

with $\mathcal{L}(\theta^*) = \mu^{j,k}_\theta(\cdot)$. $\blacksquare$

In other words block averages over blocks of size N^k remain constant and equal to the initial value in time scale N^j with $j < k$, fluctuate like a diffusion in an external field with force directed towards θ^* and strength c_k if $j \leq k$ where $\theta^* = \theta$ if $j = k$, and is a random variable if $j > k$.

This leaves open the question of how the entire collection $\left\{x_{\xi,0}(sN^k + t)\right\}_{\xi \in \Omega_N}$ as a process in t evolves. This is an *infinite interacting* system, viewed in a certain time scale with parameter N tending to infinity. Applying the same results as in Theorem 1 one can prove the following.

THEOREM 2.

$$\mathcal{L}\left((X^N(sN^k + t))_{t \in \mathbf{R}+}\right) \underset{N \to \infty}{\Longrightarrow} \mathcal{L}((X(t))_{t \in \mathbf{R}+})$$

where $X(t) = \{x_\xi(t)\}_{\xi \in \Omega_\infty}$ *is the stationary solution of:*

$$dx_\xi(t) = c_k(\theta_\xi - x_\xi(t))dt + \sqrt{F_k(x_\xi(t))}dw_\xi(t),$$

$\{w_\xi(t)\}_{\xi \in \Omega_\infty}$ are i.i.d. standard Wiener processes, and

$(\theta_\xi)_{\xi \in \Omega_\infty}$ is a $[0,1]$-valued random field specified as follows:
$\{\theta_{\xi^1}, \theta_{\xi^2}, \ldots, \theta_{\xi^m}\}$ are independent if $d(\xi^i, \xi^j) > k$ for $i \neq j$, otherwise the joint distribution is given by setting $\mathcal{L}\left(\{\theta_{(\eta_1, \eta_2, \ldots, \eta_k, \xi_{k+1}, \ldots)}\}_{\eta_1, \ldots, \eta_k \in \mathbf{N}}\right) = \mathcal{J}$, for every $\xi = (\xi_1, \xi_2, \ldots) \in \Omega_\infty$, and $\mathcal{J}$ will be defined below in (5). In addition we define $\nu_\theta^{(k)} = \mathcal{L}(X(t)) = \mathcal{L}(X(0))$. ∎

In order to construct $\mathcal{J}$ we must define a probability measure on $[0,1]^{\mathbf{N}^k}$. View this state space as a tree starting at a single root where at each root a number of edges labelled by $\mathbf{N}$ splits off. The depth of the tree is $k+1$. Assign to the root of the tree the value θ. At the next level assign independently random variables $(\theta_k^{i_1})_{i_1 \in \mathbf{N}}$ distributed according to $\Gamma_\theta^k(\cdot)$. At the next level assign independent variables $(\theta_k^{i_1, i_2})_{i_2 \in \mathbf{N}}$ distributed according to $\Gamma_{\theta^{i_1}}^{k-1}$. Iteration leads to a collection $(\theta i_1, \ldots, i_k)_{i_1, \ldots, i_k \in \mathbf{N}}$. Then define

$$(5) \qquad \mathcal{J} = \mathcal{L}\left((\theta^{i_1, \ldots, i_k})_{i_1, \ldots, i_k \in \mathbf{N}}^{\cdot}\right).$$

4. The longterm behavior of the system and the interaction chain.

(i) Theorem 1 tells us that the behavior of the system on the level of a single component and in time scale N^R is determined by $\mu_\theta^{R,0}(\cdot), R = 1, 2, \ldots$, the marginal distribution of ν_θ^R. In order to investigate the mechanics which produces this sequence of *quasi-equilibria* and to interpret $(\theta_k^{i_1, \ldots, i_j})$ of equation (5) we introduce the notion of *interaction chain*.

Define a double array of transition kernels on $[0,1] \times [0,1]$ by setting

$$K_n^j(x, A) = \Gamma_x^{n-j}(A), \quad x \in [0,1], \ A \in \mathcal{B}([0,1]), \ j \leq n, j, n \in \mathbf{N}.$$

Note that for the time scale tN^R the row $(K_R^j)_{j=1,\ldots,R}$ is relevant.

DEFINITION 2.
The *interaction chain of order* R is the Markov chain on $[0,1]$ which starts at time $-R$ in θ and has transition kernel K_R^j at time $-R+j$ for $j = 1, \ldots, R$. The chain is denoted $(Z_j^R)_{j \in \{-R, -R+1, \ldots, 0\}}$.

Sometimes it is useful to work with the reversed process $(\hat{Z}_j^R)_{j \in \{0, \ldots, R\}}$ defined by $\hat{Z}_j^R = Z_{-j}^R$. ∎

Recall that (cf. (5)) $\mu_\theta^{R,j}(\cdot) = \mathcal{L}(Z_{-j}^R) = \mathcal{L}(\hat{Z}_j^R)$, $\mathcal{L}(\theta_R^{i_1, \ldots, i_j}) = \mathcal{L}(\hat{Z}_j^R)$ and that $(\hat{Z}_j^R)_{j=0,1,\ldots,R}$ is a martingale, $\hat{Z}_0^R = \theta$.

Remark. This is a dynamical analogue of the following objects in the theory of Gibbs measures. Consider a nearest neighbour potential and the sequence of boxes in $\mathbf{Z}^d$, $E_R = [-R, R]^d$. Let $\rho_R(\eta_j, \cdot)$ be the distribution in E_j given the configuration η_j on ∂E_j, with ρ_R as Gibbs measure on E_R. Then $\rho_R(\eta_j, \cdot)$ is the analogue of $K_R^j(x, \cdot)$ above.

It can be shown by using the fact that for f convex, $\langle \mu_\theta^{R,j}, f \rangle$ decreases in R for every j, that

$$\mathcal{L}\left(\left(\hat{Z}_j^R \right)_{j \in \{0,1,\ldots,R\}} \right) \underset{R \to \infty}{\Longrightarrow} \mathcal{L}\left(\left(\hat{Z}_j^\infty \right)_{j \in \mathbf{N}} \right),$$

where $\left(\hat{Z}_j^\infty \right)_{j \in \mathbf{N}}$ is a martingale, $E(\hat{Z}_j^\infty) = \theta$. Moreover $\nu_\theta^{(R)} \underset{R \to \infty}{\Longrightarrow} \nu_\theta^{(\infty)}$, the limit of the *global equilibrium* as $N \to \infty$. The following dichotomy holds for every $\theta \in (0,1)$:

Case I: $\hat{Z}_j^\infty \in \{0,1\}, P(\hat{Z}_j^\infty = 1, \forall j \geq 0) = \theta, P(\hat{Z}_j^\infty = 0, \forall j \geq 0) = 1 - \theta$ (trivial entrance laws only)
$$\nu_\theta^{(\infty)} = \theta \delta_{\{x_\xi = 1, \forall \xi\}} + (1 - \theta) \delta_{\{x_\xi = 0, \forall \xi\}}.$$

Case II. $\hat{Z}_j^\infty \in (0,1) \forall j \geq 0$, $\hat{Z}_j^\infty \underset{j \to \infty}{\Longrightarrow} \theta$, and $\left\{ \mu_\theta^{\infty,j} : j \geq 0 \right\}_{\theta \in [0,1]}$ are extremal entrance laws,
$$\nu_\theta^{(\infty)} \text{ is nontrivial.}$$

Let us explain the meaning of this dichotomy in the context of our original system $\{x_{\xi,0}(t)\}_{\xi \in \Omega_N}$ in the limit $N \to \infty$ and $t \to \infty$. In case I if we choose $t \simeq N^R$ with R *arbitrarily large* then as $N \to \infty$ we will find the probability that a single component is either all close to 0 or all close to 1 converges to 1. Hence in case I the system of interacting diffusions forms growing clusters of components all 0 or all 1, while in case II the system is stable and preserves the initial density (spatial average in the thermodynamic limit) as $t \to \infty$. In case I we say that the system *clusters* and in case II that the system is *stable*.

What are the conditions on the $\{c_k\}$ to produce case I, respectively II? We expect of course that if the interaction is very weak then we are in case I whereas if the interaction is very strong that we are in case II. This follows in case II since strong interaction means that the kernels $K_j^R(x, dy)$ for fixed j and $R \to \infty$ converge rapidly to the kernel $\delta_x(dy)$. Consequently the process $\hat{Z}_k^R$ for $k \leq j$ can stay away from the traps 0,1 with probability converging to 1 rapidly as $R \to \infty$.

THEOREM 3.

If $\sum_k c_k^{-1} = \infty$, *then the system clusters.*

If $\sum_k c_k^{-1} < \infty$, *then the system is stable.*∎

Remark: In fact it follows from the results of Sawyer and Felsenstein [5] that the random walk on Ω_N with generator Q is recurrent (respectively, transient) iff $\sum c_k^{-1} = \infty$ ($< \infty$) provided that limsup of c_k^{-k} is less than N. The critical range is $c_k \sim k(\log k)^s$ since for $s > 1$ the sum exists and for $s \leq 1$ it diverges.

The proof of the theorem proceeds along the following lines (if $g(x) = x(1-x)$). One calculates the explicit form of $\Gamma_\theta^k(\cdot)$, that is, $F_k(\theta) = d_k \theta(1-\theta)$ with $d_0 = 1$, $d_{n+1} = \frac{c_n d_n}{c_n + d_n}$. From this fact it follows after some calculation that $E\big(\hat{Z}_R^R(1-\hat{Z}_R^R)\big) = \Big(\prod_{k=0}^{R}\big(1+\frac{d_k}{c_k}\big)\Big)^{-1}$ so that case I occurs if $\sum_k \frac{d_k}{c_k} = \infty$ and case II if $\sum_k \frac{d_k}{c_k} < \infty$. Using the recursion formula for d_n this condition is shown to be equivalent with $\sum c_k^{-1} = \infty, < \infty$.

(ii) In case II (clustering) the natural question to ask about the system $\{x_{\xi,0}(t)\}_\xi$ is how fast do clusters of values close to 0, respectively close to 1, grow. Can we capture this with the interaction chain? In case II, $\hat{Z}_j^R$ converges to a process which remains in one of the traps forever. Since $\hat{Z}_R^R = \theta \in (0,1)$, the procedure should be to rescale the index j as a function of R, that is consider $\hat{Z}_{j(R)}^R(j(R) \to \infty$ as $R \to \infty)$, in order to obtain a nontrivial limit process instead of the constant one. This translates into asking, if one uses time scale N^R, up to what block sizes $N^0, N, N^2, \ldots, N^{j(R)}$ is the configuration close to all 0 or close to all 1 with probability greater than zero but less than one?

Formally we proceed as follows. Let $\{f_\alpha(\cdot) : \alpha \in (0,1)\}$ be functions on $\mathbf{N}$ such that $f_0(R) = R$, $f_1(R) \geq 0$ and $f.(k)$ decreasing for every $k \in \mathbf{N}$. A simple example is $f_\alpha(k) = (1-\alpha)k$. The functions $\{f_\alpha(\cdot)\}_{\alpha\in[0,1]}$ have to be chosen such that $\big(\hat{Z}_{f_\alpha(R)}^R\big)_{\alpha\in[0,1]}$ converges to a process $(\hat{Z}_\alpha)_{\alpha\in[0,1]}$(recall $\hat{Z}_0 = \theta$ by construction) with the property that the $P(\hat{Z}_\alpha \in \{0,1\}) < 1$ if $\alpha < 1$. This means the functions $\{f_\alpha(\cdot)\}_{\alpha\in[0,1]}$ must be determined in terms of the sequence $(c_k)_{k\in\mathbf{N}}$. In some cases this is possible in analytically closed form as in the following.

THEOREM 4. *Assume that $c_k = c$ for all $k \in \mathbf{N}$. Set $f_\alpha(k) = (1-\alpha)k$. Then*

$$\mathcal{L}\left(\big(\hat{Z}_{f_\alpha(R)}^R\big)_{\alpha\in[0,1]}\right) \underset{R\to\infty}{\Longrightarrow} \mathcal{L}\left(\big(Y_{\log(1/(1-\alpha))}\big)_{\alpha\in[0,1]}\right)$$

where (Y_t) is the diffusion on $[0,1]$ with generator $\frac{1}{2}x(1-x)\big(\frac{\partial}{\partial x}\big)^2$ and $Y_0 = \theta$ (Fisher-Wright diffusion). ∎

This phenomenon is called *diffusive clustering* since clusters at time N^R are of size $N^{\alpha R}$ where α is a *random variable*. Cases in which $f_\alpha(k)/k \to 1$ correspond to large clusters and those in which $f_\alpha(k)/k \to 0$, $(\forall \alpha \in (0,1))$ correspond to slowly growing clusters. They occur for example if the c_k decay to 0 at polynomial speed, respectively if $c_k \to \infty$, but $\sum c_k^{-1} = \infty$.

References

[1] Cox, J.T. and Greven, A. (1991). Ergodic theorems for infinite systems of locally interacting diffusions, SFB Preprint no. 645, Heidelberg.

[2] Cox, J.T. and Griffeath, D. (1986). Diffusive clustering in the two dimensional voter model, Ann. Probab. 14, 347-370.

[3] Dawson, D.A. and Greven, A. (1990) Multiple time scale analysis of interacting diffusions, SFB Preprint 568, Heidelberg.

[4] Dawson, D.A. and Greven, A. (1991). Hierarchical models of interacting diffusions: multiple time scale phenomena, phase transition and pattern of cluster-formation, SFB Preprint 648, Heidelberg.

[5] Sawyer, S. and Felsenstein, J. (1983). Isolation by distance in a hierarchically clustered population, J. Appl. Prob. 20, 1-10.

[6] Shiga, T. (1980). An interacting system in population genetics, I and II, J. Math. Kyoto Univ. 20, 213-243, 723-733.

[7] Shiga, T. and Shimizu, A. (1980). Infinite dimensional stochastic differential equations and their applications, J. Math. Kyoto Univ. 20, 395-416.

Addresses:
D.A. Dawson
Department of Mathematics and Statistics
Carleton University,
Ottawa, Canada K1S 5B6.

A. Greven
Institut für Mathematische Stochastik
Lotzestr. 13
3400 Göttingen
FR Germany

Feynman's Operational Calculus As A Generalized Path Integral

B. DeFacio, G.W. Johnson, and M.L. Lapidus

Dedicated to Gopinath Kallianpur who is remarkably young in spite of his long experience and outstanding achievements.

Abstract. Feynman's heuristic prescription for forming functions of noncommuting operators is discussed along with methods for making his ideas rigorous. The emphasis is on one method and on the extent to which Feynman's operational calculus can be viewed as a generalized path integral.

1. Introduction.

Forming functions of operators is useful in many contexts. Perhaps the most spectacular functional calculus is provided by the Spectral Theorem for self—adjoint operators A. This theorem permits us to form the operator f(A) for any measurable function f on ℝ. Indeed, functions of several self—adjoint operators can be formed if the operators commute.

Serious difficulties arise in forming functions of two or more operators when they fail to commute. These difficulties have repercussions for physical theory. For example, they give rise to the problem of 'ambiguity of quantiziation' in quantum mechanics.

The exponentiation of operators is especially important since it is the key to modeling the evolution of various physical phenomena including many where probability theory is intimately involved. Semigroups of operators [8,9] is that theory whose specific concern is exponentiating operators. It has long been known that there is a close tie between Markov processes and certain semigroups of operators. There is an enormous literature on this subject; [3,6,14] are among the references available.

In his 1951 paper [4] "An operator calculus having applications in quantum electrodynamics" Feynman presented his heuristic ideas for a functional calculus for noncommuting operators. This subject is related in various ways to path integrals and can be thought of as generalizing certain aspects of such integrals. Indeed, this seems to be part of what Feynman had in mind. In fact, Feynman's papers on the path integral, quantum electrodynamics and the operator calculus, all of which appeared between 1948 and 1951, were cut from the same cloth. We will quote from the fascinating paper of Schweber [20], but the comments in parentheses are ours: "These papers on the space—time approach (or path integral approach) to nonrelativistic quantum mechanics, on quantum electrodynamics, and on his operator calculus must surely be placed near the top of any list of the most seminal and influential papers in theoretical physics during the twentieth century" [20, p.504]. Schweber also says [20, p.502], "In any case Feynman never felt order—by—order (the time—ordered integrals involved in the perturbation expansions obtained from applying Feynman's operational calculus) was anything but an approximation to the 'thing' and the 'thing' was the path integral". In the following paragraph [20, p.502] Schweber quotes Feynman "A way of saying what quantum electrodynamics was, was to say what the rule was for the diagram (the famous Feynman diagrams, a shorthand for keeping track of the terms of the perturbation series) — although I really thought behind it was my action form (and so, presumably, an associated path integral of some sort)."

We give next a brief discussion of Feynman's heuristic ideas [4] on forming functions of noncommuting operators:

(i)   Indicate the order of operation of noncommuting operators by attaching 'time' indices where operators with smaller time indices act before operators with larger time indices; that is, A(s)B(t) equals BA if s < t, AB if t < s and is undefined if s = t.   (ii)   With time indices—attached, form functions of the operators just as if they commuted.   (iii)   Finally restore the conventional ordering of the operators; that is, 'disentangle' the expression that results from applying (ii).

It is not always clear how to apply Feynman's 'rules', especially the disentangling, even on a heuristic level, and it is certainly not clear how to provide mathematical rigor.   In the next section we give a brief survey with references of some ways in which Feynman's ideas have been made mathematically precise.

In the last section we first carry out a simple heuristic calculation which should clarify the meaning of Feynman's 'rules'. Next we indicate how to make sense of the resulting perturbation series, thus illustrating one way [2] of making Feynman's ideas rigorous.   We also point out connections, not only in the results, but in the calculations themselves, with path integration.

2.   Some ways to make Feynman's ideas rigorous.

I.   Use known theories to achieve rigor in certain settings

One of the themes of this paper is that Feynman's operational calculus can be regarded as a generalized path integral.   This suggests that it is possible to use path integrals such as those introduced by Wiener and Feynman to give rigorous meaning to Feynman's operational calculus in the diffusion (probabilistic) and quantum mechanical settings, respectively.   This has been done explicitly in recent work [11—13,15,16].   The disentangling is carried out and made rigorous in the process of calculating appropriate

Wiener or Feynman integrals.

Both the Wiener and Feynman integrals are intimately associated with the Laplacian, the generator of Brownian motion. It seems clear that other path integrals associated with different generators and Markov processes could be used in a similar way in connection with Feynman's operational calculus.

The product integral is a time—ordered multiplicative integral of noncommuting operator—valued functions which can be used to make Feynman's ideas precise in certain contexts. See [17] and references therein.

II. Create a general theory within which the calculations envisioned by Feynman are rigorous at each stage.

Nelson [19] constructed a theory which makes rigorous not only the results of calculations like Feynman's but also the calculations themselves. A commutative framework is provided in which functions of operators can be formed and a formula for disentangling is given.

III. Useful and rigorous formulas via Feynman's ideas.

A common mathematical strategy is to use heuristic ideas to arrive at mathematical 'objects' which are useful in themselves and which can be made rigorous. This is the point of view adopted in the paper [2] which is now in preparation.

Much of Feynman's motivation in [4] seems to have been to find a method of calculation akin to path integration which would allow him to 'derive' formulas for evolving physical systems. He was especially interested in the perturbation series for quantum electrodynamics where there was no path integral available. Although Feynman's operational calculus extends well beyond exponential functions, exponentiation plays a special role in [4] and in [2] because of the interest in evolving systems. Part of the idea in [2] is to develop mathematical models for complicated physical systems which combine features of simpler systems and which are not readily treatable with standard models. The emphasis in [2] is on quantum models, but the method seems to be much more generally applicable.

The literature related to Feynman's operational calculus is quite diverse and involves varying levels of mathematical rigor. We finish this section by calling [1,18,7] to the reader's attention.

3. A simple illustration of Feynman's rules and of III above.

We begin here by stating our conditions precisely.

An understanding of the conditions depends on introductory ideas from the theory of semigroups of operators [8,9]. A good portion of what we will have to say is meaningful where the operators are all nxn matrices over $\mathbb{C}$, the complex numbers.

A. The conditions. Let H be a separable Hilbert space over $\mathbb{C}$, and let $\mathcal{L}(H)$ be the space of bounded linear operators on H. α will denote the infinitesimal generator of a (C_0)–contraction semigroup of operators in $\mathcal{L}(H)$. The family $\{\beta(s): 0 \leq s \leq t\}$ will be a commuting family of operators in $\mathcal{L}(H)$ and μ will be an associated finite Borel measure on $[0,t]$. We assume that β is Bochner integrable in the strong operator sense with respect to μ. We further assume (to simplify matters) that μ is a continuous measure. We also suppose that we have a second commuting family $\{\gamma(s): 0 \leq s \leq t\}$ and associated measure ν satisfying the same conditions imposed above on $\{\beta(s)\}$ and μ. We do <u>not</u> assume that the β's commute with the γ's nor that either of these sets commutes with the operators in the semigroup generated by α.

B.   <u>The heuristics and Feynman's rules.</u>   Rule (i) refers to "attaching time indices". How is this to be done? First of all, the family of operators, say $\{\beta(s)\}$, may come with time indices naturally attached. This happens in quantum mechanics when there is a time–dependent potential $V(s,\cdot)$; then $\beta(s)$ is the operator of multiplication by the function $V(s,\cdot)$. Alternately, $\{\beta(s)\}$ or $\{\gamma(s)\}$ could be a family of time–dependent convolution operators serving as nonlocal potentials. When no natural time indices come with the problem, it seems best most often to attach indices according to the uniform distribution on $[0,t]$. One writes

$$\beta = \frac{1}{t} \int_0^t \beta(s)ds, \text{ where } \beta(s) \equiv \beta.$$

This artifice turns out to be productive and, in fact, has been used exclusively (although not explicitly) in most of the work following Feynman and almost exclusively in Feynman's paper [4].

One sees in [11,12,15,16] that other possibilities can be of mathematical and physical interest. In connection with perturbation series in quantum mechanics, measures other than Lebesgue measure correspond to different arrangements of scattering times and different weights attached to those scatterings. The Kohn–Nirenberg correspondence which is used in the theory of pseudo–differential operators may be regarded as an alternative to Weyl quantization [5] corresponding to a different assignment of time indices.

We take the generator α independent of time and the first equality in (1) below comes from attaching time indices (governed by Lebesgue measure) to α. The individual exponentials, $\exp(-t\alpha)$, $\exp(\int_0^t \beta(s)\mu(ds))$ or $\exp(\int_0^t \gamma(s)\nu(ds))$, would give the evolution ofsystems where only α, β or γ, respectively, was involved. The idea is that $\exp(-t\alpha+\int_0^t\beta(s)\mu(ds)+\int_0^t\gamma(s)\nu(ds))$ is to give the evolution of the combined system where α, β, γ are mixed and weighted corresponding in this case to Leb.xμxν—measure. The second equality in (1) comes from following Rule (ii) of Feynman and proceeding as if we were in a commutative situation. Of course, this equality cannot be taken literally if we give the expressions their usual meaning.

$$\exp(-t\alpha+\int_0^t\beta(s)\mu(ds)+\int_0^t\gamma(s)\nu(ds))=\exp(-\int_0^t\alpha(s)ds+\int_0^t\beta(s)\mu(ds)+\int_0^t\gamma(s)\nu(ds))$$

(1)
$$=[\exp(-\int_0^t\alpha(s)ds)][\exp(\int_0^t\beta(s)\mu(ds))][\exp\int_0^t\gamma(s)\nu(ds))]$$

$$=[\exp(-\int_0^t\alpha(s)ds)][I+\int_0^t\beta(s)\mu(ds)+\frac{1}{2}(\int_0^t\beta(s)\mu(ds))^2+\cdots]$$

$$\cdot[I+\int_0^t\gamma(s)\nu(ds)+\frac{1}{2}(\int_0^t\gamma(s)\nu(ds))^2+\cdots]$$

$$= [\exp(-\int_0^t\alpha(s)ds)][I+\int_0^t\beta(s)\mu(ds)+\int_0^t\gamma(s)\nu(ds)+\frac{1}{2}(\int_0^t\beta(s)\mu(ds))^2$$

$$+\frac{1}{2}(\int_0^t\gamma(s)\nu(ds))^2+(\int_0^t\beta(s)\mu(ds))(\int_0^t\gamma(s)\nu(ds))+\cdots]$$

Rule (iii) tells us to 'disentangle'. This admonition gives little hint as to what should be done, and one gets the idea from [4] mainly from the examples. The key is to time—order the terms. Before returning to (1), we begin to disentangle by working with the last three terms in the second factor of the last expression in (1). Denoting the set $\{(s_1,s_2):\ 0<s_1<s_2<t\}$ by $s_1< s_2$,

(2)
$$(\int_0^t\beta(s)\mu(ds))^2 = \int_0^t\int_0^t\beta(s_1)\beta(s_2)(\mu x\mu)(ds_1 ds_2)$$

$$= \iint_{s_1<s_2} \beta(s_2)\beta(s_1)(\mu x\mu)(ds_1 ds_2)+ \iint_{s_2<s_1} \beta(s_1)\beta(s_2)(\mu x\mu)(ds_1 ds_2)$$

$$= 2 \iint_{s_1<s_2} \beta(s_2)\beta(s_1)(\mu x\mu)(ds_1 ds_2).$$

Similarly,

(3) $\qquad (\int_0^t \gamma(s)\nu(ds))^2 = 2 \iint_{s_1<s_2} \gamma(s_2)\gamma(s_1)(\nu\times\nu)(ds_1 ds_2).$

Also, using Rule (ii) again,

(4) $\qquad (\int_0^t \beta(s)\mu(ds))(\int_0^t \gamma(s)\nu(ds)) =$

$\iint_{s_1<s_2} \gamma(s_2)\beta(s_1)(\mu\times\nu)(ds_1 ds_2) + \iint_{s_2<s_1} \beta(s_1)\gamma(s_2)(\mu\times\nu)(ds_1 ds_2).$

Now, using (1) through (4) we obtain

(5) $\qquad \exp(-t\alpha + \int_0^t \beta(s)\mu(ds) + \int_0^t \gamma(s)\nu(ds))$

$= \exp(-\int_0^t \alpha(s)ds) + \int_0^t \exp(-\int_{s_1}^t \alpha(s)ds)\beta(s_1)\exp(-\int_0^{s_1} \alpha(s)ds)\mu(ds_1)$

$+ \int_0^t \exp(-\int_{s_1}^t \alpha(s)ds)\gamma(s_1)\exp(-\int_0^{s_1} \alpha(s)ds)\nu(ds_1)$

$+ \iint_{s_1<s_2} \exp(-\int_{s_2}^t \alpha(s)ds)\beta(s_2)\exp(-\int_{s_1}^{s_2} \alpha(s)ds)\beta(s_1)$

$\exp(-\int_0^{s_1} \alpha(s)ds)(\mu\times\mu)(ds_1 ds_2) + \iint_{s_1<s_2} \exp(-\int_{s_2}^t \alpha(s)ds)\gamma(s_2)$

$\exp(-\int_{s_1}^{s_2} \alpha(s)ds)\gamma(s_1)\exp(-\int_0^{s_1} \alpha(s)ds)(\nu\times\nu)(ds_1 ds_2)$

$+ \iint_{s_1<s_2} \exp(-\int_{s_2}^t \alpha(s)ds)\gamma(s_2)\exp(-\int_{s_1}^{s_2} \alpha(s)ds)\beta(s_1)$

$\exp(-\int_0^{s_1} \alpha(s)ds)(\mu\times\nu)(ds_1 ds_2) + \iint_{s_1<s_2} \exp(-\int_{s_2}^t \alpha(s)ds)\beta(s_2)$

$$\exp(-\int_{s_1}^{s_2} \alpha(s)ds)\,\gamma(s_1)\exp(-\int_0^{s_1}\alpha(s)ds)(\mu x\nu)(ds_1 ds_2)+\cdots.$$

Once all the terms involving products are time–ordered, the disentangling is complete and the idea is that the expression should again be taken literally. An additional 'rule' which one learns from Feynman's examples is that integrals in time–ordered terms may be calculated. Using this, we finally obtain

(6)
$$\exp(-t\alpha+\int_0^t\beta(s)\mu(ds)+\int_0^t\gamma(s)\nu(ds))$$

$$=\exp(-t\alpha)+\int_0^t\exp(-(t-s_1)\alpha)\beta(s_1)\exp(-s_1\alpha)\mu(ds_1)$$

$$+\ \int_0^t\exp(-(t-s_1)\alpha)\,\gamma(s_1)\exp(-s_1\alpha)\nu(ds_1)$$

$$+\ \iint_{s_1<s_2}\exp(-(t-s_2)\alpha)\beta(s_2)\exp(-(s_2-s_1)\alpha)\beta(s_1)\exp(-s_1\alpha)(\mu x\mu)(ds_1 ds_2)$$

$$+\ \iint_{s_1<s_2}\exp(-(t-s_2)\alpha)\,\gamma(s_2)\exp(-(s_2-s_1)\alpha)\,\gamma(s_1)\exp(-s_1\alpha)(\nu x\nu)(ds_1 ds_2)$$

$$+\ \iint_{s_1<s_2}\exp(-(t-s_2)\alpha)\,\gamma(s_2)\exp(-(s_2-s_1)\alpha)\beta(s_1)\exp(-s_1\alpha)(\mu x\nu)(ds_1 ds_2)$$

$$+\ \iint_{s_1<s}\exp(-(t-s_2)\alpha)\beta(s_2)\exp(-(s_2-s_1)\alpha)\,\gamma(s_1)\exp(-s_1\alpha)(\nu x\mu)(ds_1 ds_2)+$$

$$\cdots.$$

C. The Theorem.

THEOREM. We assume that the conditions described in subsection A are satisfied on $[0,T]$. Then for every $t\in[0,T]$: (i) The integrals in the terms of the series (6) exist in the strong operator sense. (ii) The series converges in operator norm. (iii) Regarding the series (6) as defining the operator

$$\exp(-t\alpha+\int_0^t\beta(s)\mu(ds)+\int_0^t\gamma(s)\nu(ds)),$$ we have the inequality

$$(7)\|\exp(-t\alpha+\int_0^t\beta(s)\mu(ds)+\int_0^t\gamma(s)\nu(ds))\|\leq\exp(\int_0^t\|\beta(s)\|\mu(ds)+\int_0^t\|\gamma(s)\|\nu(ds)).$$

(iv) If we let

$$L_t = L_t(\alpha;\beta,\mu;\gamma,\nu) := \exp(-t\alpha + \int_0^t \beta(s)\mu(ds) + \int_0^t \gamma(s)\nu(ds)), \text{ then } L_t \text{ satisfies}$$

the integral equation

$$(8) \quad L_t = \exp(-t\alpha) + \int_0^t \exp(-(t-s)\alpha)\beta(s)L_s\mu(ds) + \int_0^t \exp(-(t-s)\alpha)\gamma(s)L_s\nu(ds).$$

Discussion of the proof. The key to (i) is the fact [10] established in connection with [2] that the product of strong operator measurable functions is strong operator measurable.

To establish (ii) and (iii), one moves norms inside the integrals involved in (6) and then reverses the steps of the heuristic disentangling process. These steps can now be justified since the functions that appear are real—valued and so commute. The combinatorics is somewhat more complicated for higher order terms especially when there are more than two families of operators perturbing the semigroup.

The assumption that μ and ν are continuous is used in connection with (2)—(4). For example, the integral over $[0,t]\times[0,t]$ can be written as the sum over the sets where $s_1 < s_2$ and $s_2 < s_1$ because the $\mu \times \nu$ measure of the diagonal is 0. In [11] the assumption of continuity is not made, and the combinatorics is far more complicated. The paper [2] concentrates on continuous measures but includes discontinuous examples.

One of the central ideas in establishing (iv) is that the second and third expressions on the right—hand side of (8) give all the terms in the series that end with β, respectively, γ perturbations. This is the entire series except for $\exp(-t\alpha)$ which is the first term on the right—hand side. A final point: The argument for (8)(and above) is made harder not easier by the commutativity within the families $\{\beta(s)\}$ and $\{\gamma(s)\}$. The case of complete noncommutativity is easier since then precisely <u>all</u> permutations of the appropriate variables have to be considered.

D. Special cases and comparisons   If we take $H = L^2(\mathbb{R}^N)$, $\alpha = iH_0$(where $H_0 = -\frac{1}{2}\Delta$ is the free Hamiltonian), $\beta(s) \equiv iV$ where V represents the operator of multiplication by a time—independent potential V, and $\gamma(s) \equiv 0$, then (6) is the usual perturbation series (or Dyson series) from nonrelativistic quantum mechanics. The series equals the semigroup $\exp\{-t[i(H_0+V)]\}$ and $\exp\{-it(H_0+V)\}\psi_0$ satisfies the Schrodinger equation along with appropriate initial data. The semigroup also satisfies the integral equation model of the evolution:

$$L_t\psi_o = \exp(-itH_o)\psi_o + \int_0^t \exp(-i(t-s)H_o)(iV)L_s\psi_o ds.$$

With the situation as above except that $\alpha=H_o$ and $\beta(s)\equiv V$, the series equals the heat semigroup $\exp\{-t(H_o+V)\}$ and $\exp\{-t(H_o+V)\}\psi_o$ satisfies the heat (or diffusion) equation as well as the integral equation $L_t\psi_o=\exp(-tH_o)\psi_o$

$+\int_0^t \exp(-(t-s)H_o)VL_s\psi_o ds.$ The Feynman–Kac formula asserts that

$$(\exp(-t(H_o+V))\psi_o)(\xi)= \int_{C_o[0,t]} \exp\{-\int_0^t V(x(s)+\xi)ds\}\psi_o(x(t)+\xi)m(dx)$$

where m denotes Wiener measure. In this case, the series (6) can also be arrived at by expanding the exponential in the integrand above and calculating the resulting Wiener integrals. <u>Not only do</u> <u>we obtain the same series, but the calculations parallel the earlier</u> <u>heuristic calculations thus reinforcing the idea that Feynman's</u> <u>operational calculus provides a kind of generalized functional integral.</u>

By expanding the exponentials in

$$\int_{C_o[0,t]} \exp\{-\int_0^t V_1(x(s)+\xi)ds\}\mu - \int_0^t V_2(x(s)+\xi)\nu(ds)\}\psi_o(x(t)+\xi)m(dx)$$

and calculating the resulting series of Wiener integrals, we obtain the series (6) in the case $\alpha=H_o$, $\beta(s)\equiv V_1$ and $\gamma(s)\equiv V_2$. <u>The</u> <u>calculations again parallel the earlier heuristic calculations.</u>

Further examples and results related to this paper can be found in [11,12]. The reader should also consult the papers [15,16].

References

1.   H. Araki, Expansionals in Banach algebras, Ann. Sci. Ecole Norm. Sup. (4) 6(1973), 67–84.

2.   B. DeFacio, G.W. Johnson and M.L. Lapidus, Feynman's operational calculus and quantum models, in preparation.

3.   Feller, The parabolic differential equations and associated semigroup of transformations, Ann. of Math. 55(1952), 468–519.

4.   R.P. Feynman, An operator calculus having applications in quantum electrodynamics, Phys. Rev. 84(1951), 108–128.

5.   G.B. Folland, Harmonic Analysis in Phase Space, Ann. of Math. Studies, Princeton U. Press, Princeton, 1980.

6.   M. Fukushima, Dirichlet Forms and Markov Processes, North Holland Pub., Amsterdam, 1980.

7.   T.L. Gill and W.W. Zachary, Time—ordered operators and Feynman—Dyson algebras, J. Math. Phys. 28(1987), 1459—1470.

8.   J.A. Goldstein, Semigroups of Linear Operators and Applications, Oxford U. Press, New York, 1985.

9.   E. Hille and R.S. Phillips, Functional Analysis and Semi—groups, Amer. Math. Soc. Colloq. Publ. Vol XXXI, rev. ed., Amer. Math. Soc., Providence, 1957.

10.   G.W. Johnson, The product of strong operator measurable functions is strong operator measurable, Proc. Amer. Math. Soc., accepted.

11.   G.W. Johnson and M.L. Lapidus, Generalized Dyson series, generalized Feynman diagrams, the Feynman integral and Feynman's operational calculus, Mem. Amer. Math. Soc. 62(1986), 1—78.

12.   G.W. Johnson and M.L. Lapidus, Noncommutative operations on Wiener functionals and Feynman's operational calculus, J. Funct. Anal. 81(1988), 74—99.

13.   G.W. Johnson and M.L. Lapidus, The Feynman Integral and Feynman's Operational Calculus, Oxford U. Press, in prep.

14.   J. Lamperti, Stochastic Processes, Springer, New York, 1977.

15.   M.L. Lapidus, The Feynman—Kac formula with a Lebesgue—Stieltjes measure and Feynman's operational calculus, Stud. Appl. Math. 76(1987), 93—132.

16.   M.L. Lapidus, The Feynman—Kac formula with a Lebesgue—Stieltjes measure: An integral equation in the general case, Integ. Equs. and Op. Theory 12(1989), 163—210.

17.   M.L. Lapidus, Strong product integration of measures and the Feynman—Kac formula with a Lebesgue—Stieltjes measure, Supp. Rend. Circ. Mat. Di Palermo 17(1987), 271—312.

18.   V.P. Maslov, Operational Methods, (English transl., revised from the 1973 Russian ed.), Mir, Moscow, 1976.

19.   E. Nelson, Operants: A functional calculus for noncommuting operators, in Proc., Conf. in Honor of Marshall Stone (F.E. Browder, ed.), 172—187, Springer, New York, 1970.

20.   S.S. Schweber, Feynman and the visualization of space—time processes, Rev. Mod. Phys., 58(1986), 449—508.

B. DeFacio
Physics, U. of Missouri
Columbia, MO   65211, USA

G.W. Johnson
Math, U. of Nebraska
Lincoln, NE   68588, USA

M.L. Lapidus
Math.,U. of California, Riverside
Riverside, CA   92521, USA

Forward and Backward Equations for an Adjoint Process

ROBERT J. ELLIOTT and HAILIANG YANG

Abstract. A Markov chain is observed only through a noisy continuous observation process. A related optimal control problem is formulated in separated form by considering the related Zakai equation. An adjoint process is defined and shown to satisfy a forward stochastic partial differential equation, and also a system of backward parabolic equations.

1. Introduction.

Without loss of generality the state space of a finite state Markov chain can be taken to be the set of unit basis vectors in R^N. We suppose such a chain X_t, $0 \le t \le T$, is observed through the noisy process y, where

$$(1.1) \qquad y_t = \int_0^t h(X_s)ds + w_t.$$

Here w is a Brownian motion independent of X.

For simplicity a terminal cost of the form $\langle \ell, X_T \rangle$ is considered and, following Davis, [5], the control problem is formulated in separated form by considering an unnormalized conditional distribution of X_t.

An adjoint process is introduced and shown to satisfy forward and backward equations. Early works of Bismut [3], [4] have discussed the adjoint process using different methods. Some of our techniques are related to those of Bensoussan [1]. A similar problem for a controlled Markov chain for which only the jump times, but not the jump locations, can be observed is discussed in [7].

2. The System.

The formulation in this section is well known. Let $\{X_t\}$, $t \in [0, T]$ be a Markov chain, defined on a probability space (Ω, F, P), whose state space is the set

$$S = \{e_1, \ldots, e_N\}$$

Partially supported by NSERC Grant A-7964 and the U.S. Army Research Office under contract DAAL03-87-0102.

where $e_i = (0, \ldots, 1, 0, \ldots, 0)'$, $i = 1, 2, \ldots, N$, is a unit vector of R^N.

Write $p_t^i = P(X_t = e_i)$, $0 \leq i \leq W$. We shall suppose that for some family of matrices which depend on the control parameter $A_t(u)$, $p_t = (p_t^1, \ldots, p_t^N)'$ satisfies the forward Kolmogorov equation

$$(2.1) \qquad\qquad \frac{dp_t}{dt} = A_t(u)p_t.$$

$A = (a_{ij}(t, u))$, $t \geq 0$, is the family of Q matrices of the process. We shall suppose the $a_{ij}(t, u)$ are differentiable in u.

Suppose y_t is a Brownian motion process on (Ω, F, P) independent of X_t and write Y_t for the right continuous, complete filtration generated by y. The set $\underline{U}$ of admissible controls will be the set of Y-predictable functions with values in a compact, convex set $U \subset R^k$. Suppose h is a real valued function on S, (so h is just given by a vector $h = (h(e_1), \ldots, h(e_N))$.

For $u \in \underline{U}$ write $\Lambda_{s,t}^u = \exp\left\{ \int_s^t h(X_r^u)dy_r - \frac{1}{2} \int_s^t |h(X_r^u)|^2 dr \right\}$ and define a new probability measure P^u by

$$(2.2) \qquad\qquad \frac{dP^u}{dP} = \Lambda_{0,T}^u.$$

Then according to Girsanov's Theorem, P^u is a probability measure, and under P^u the process W_t is a Brownian motion, where W_t is defined by

$$(2.3) \qquad\qquad y_t = \int_0^t h(X_s^u)ds + W_t.$$

Also $\{X_t\}$ and $\{W_t\}$ are independent, and $\{X_t\}$ has the same distribution as under measure P. Note that under P^u the process y represents a noisy observation of $\int_0^t h(X_s^u)ds$ as in (1.1).

From Davis [5] or Elliott [6] we know that if $\hat{p}_t^i$ is the Y_t optional projection of $I_{\{X_t=e_i\}}$ under P^u, then $\hat{p}_t^i = E^u[I_{\{X_t=e_i\}} \mid Y_t] = P^u[X_t = e_i \mid Y_t]$ a.s. and $\hat{p}_t = (\hat{p}_t^1, \ldots, \hat{p}_t^w)'$ satisfies

$$(2.4) \qquad\qquad d\hat{p}_t = A(u_t)\hat{p}_t dt + (H - \tilde{h}'\hat{p}_t I)\hat{p}_t dv_t,$$

where $\tilde{h}' = (h(e_1), \ldots, h(e_N))$, H is the $N \times N$ diagonal matrix with elements $H_{ii} = h(e_i)$ and I is the $N \times N$ identity matrix. v_t is the innovations process, given by

$$(2.5) \qquad\qquad dv_t = dy_t - \tilde{h}'\hat{p}_t dt.$$

Now $\Lambda_{0,T} = \Lambda_{0,t} \cdot \Lambda_{t,T}$ so

$$\hat{p}_t^i = E^u[I_{\{X_t = e_i\}} \mid Y_t]$$

$$= \frac{E[\Lambda_{0,t}^u I_{\{X_t = e_i\}} \mid Y_t]}{E[\Lambda_{0,t}^u \mid Y_t]}.$$

Let q_t^i be the Y_t optimal projection of $\Lambda_{0,t}^u I_{\{X_t = e_i\}}$. Then

$$q_t^i = E[\Lambda_{0,t}^u I_{\{X_t = e_i\}} \mid Y_t]$$

a.s. and $q_t = (q_t^1, \ldots, q_t^N)'$, the unnormalized density of X_t conditional on Y_t, satisfies

(2.6) $$q_t = \hat{p}_t E[\Lambda_{0,t}^u \mid Y_t] = \hat{p}_t \overline{\Lambda}_{0,t}^u.$$

Here $\overline{\Lambda}^u$ is the Y-optional projection of Λ^u under measure P, so that $\overline{\Lambda}_t^u = E[\Lambda_{0,t}^u \mid Y_t]$ a.s. Now $\overline{\Lambda}_{0,t}^u$, satisfies

(2.7) $$d\overline{\Lambda}_{0,t}^u = \overline{\Lambda}_{0,t}^u \hat{h}(X_t) dy_t.$$

Here

$$\hat{h}(X_t) = E[\Lambda_{0,t}^u h(X_t) \mid Y_t] / E[\Lambda_{0,t}^u \mid Y_t]$$

$$= E^u[h(X_t) \mid Y_t].$$

Therefore, by Itô's rule from (2.6), (2.4) and (2.7)

$$dq_t = \hat{p}_t \overline{\Lambda}_{0,t} \hat{h}(X_t) dy_t + A\hat{p}_t \overline{\Lambda}_{0,t} dt$$

$$+ (H - \tilde{h}' \hat{p}_t I) \hat{p}_t \overline{\Lambda}_{0,t} dv_t + (H - \tilde{h}' \hat{p}_t I) \hat{p}_t \overline{\Lambda}_{0,t} \hat{h}(X_t) dt.$$

Since

$$\tilde{h}' \hat{p}_t = \sum_{i=1}^N h(e_i) E^u[I_{\{X_t = e_i\}} \mid Y_t]$$

$$= E^u\Big[\sum_{i=1}^N h(e_i) I_{\{X_t = e_i\}} \mid Y_t\Big] = E^u[h(X_t) \mid Y_t].$$

we see

$$dq_t = q_t \hat{h}(X_t) dy_t + Aq_t dt + Hq_t dy_t$$

$$- \hat{h}(X_t) q_t dy_t - (H - \hat{h}(X_t) I) q_t \hat{h}(X_t) dt$$

$$+ (H - \hat{h}(X_t) I) q_t \hat{h}(X_t) dt.$$

That is, q_t satisfies the Zakai equation

(2.8) $$dq_t = A_t(u)q_t dt + Hq_t dy_t.$$

Cost: The cost function will be

$$J(u) = E^u[\langle \ell, X^u_T \rangle] = E[\Lambda^u_{0,T} \langle \ell, X^u_T \rangle]$$

$$= E[\langle \ell, \Lambda^u_T X^u_T \rangle] = E[\langle \ell, E[\Lambda^u_T X^u_T \mid Y_t] \rangle]$$

$$= E[\langle \ell, q^u_T \rangle].$$

The control problem has, therefore, been formulated in separated form: find $u \in \underline{U}$ which minimizes

(2.9) $$J(u) = E[\langle \ell, q^u_T \rangle]$$

where q is given by (2.8).

3. Differentiation.

For $u \in \underline{U}$ write $\Phi^u(t,s)$ for the fundamental matrix solution of

(3.1) $$d\Phi^u(t,s) = A_t(u)\Phi^u(t,s)dt + H\Phi^u(t,s)dy_t$$

with initial condition $\Phi^u(s,s) = I$, the $N \times N$ identity matrix.

LEMMA 3.1. *For $u \in \underline{U}$, consider the matrix Ψ^u defined by the equation*

$$\Psi^u(t,s) = I - \int_s^t \Psi^u(r,s)A_r(u)dr$$

(3.2)
$$- \int_s^t \Psi^u H dy_r + \int_s^t \Psi^u H^2 dr.$$

Then $\Psi^u \Phi^u = I$ for $t \geq s$.

Proof: Using the Itô rule we see $d(\Psi\Phi) = 0$, $\Psi(s,s)\Phi(s,s) = I$.

We shall suppose there is an optimal control $u^* \in \underline{U}$. Write q^* for q^{u^*}, Φ^* for Φ^{u^*} etc. Consider any other control $v \in \underline{U}$. Then for $\theta \in [0,1]$, $u_\theta(t) = u^*(t) + \theta(v(t) - u^*(t)) \in \underline{U}$.

Now

(3.3) $$J(u_\theta) \geq J(u^*).$$

Therefore, if the Gâteaux derivative $J'(u^*)$ of J, as a functional on the Hilbert space $H = L^2[\Omega \times [0,T], R^k]$, is well defined, differentiating (3.3) in θ and letting $\theta = 0$, we have

(3.4) $$\langle J'(u^*), \ v(t) - u^*(t) \rangle \geq 0$$

for all $v \in \underline{U}$.

LEMMA 3.2.   *Suppose $v \in \underline{U}$ is such that $u_\theta^* = u^* + \theta v \in \underline{U}$ for $\theta \in [0, \alpha]$.*
Write $q_t(\theta)$ for the solution $q_t(u_\theta^*)$ of (2.8). Then $z_t = \left.\dfrac{\partial q_t(\theta)}{\partial \theta}\right|_{\theta=0}$ exists
and is the unique solution of the equation

$$(3.5) \qquad z_t = \int_0^t \left(\frac{\partial A}{\partial u}(u^*)\right) v_r q_r^* dr + \int_0^t A(u_r^*) z_r \, dr + \int_0^t H z_r \, dy_r.$$

Proof: Differentiating under the integrals gives the result. This is
justified by the result of [4].

LEMMA 3.3.   *Write*

$$(3.6) \qquad \eta_{0,t} = \int_0^t \Psi^*(r, 0)\left(\frac{\partial A}{\partial u}(u^*)\right) v_r q_r^* dr.$$

Then $z_t = \Phi^*(t, 0)\eta_{0,t}$.

Proof: By Itô's rule we see $\Phi^*(t, 0)\eta_{0,t}$ satisfies the equation (3.5).

COROLLARY 3.4.   *Because $J(u_\theta^*) = E[\langle \ell, q_T^{u_\theta^*}\rangle]$ we see*

$$(3.7) \qquad \left.\frac{\partial J}{\partial \theta}(u_\theta^*)\right|_{\theta=0} = E[\langle \ell, \Phi^*(T, 0)\eta_{0,T}\rangle].$$

Write $M_t = E[\Phi^*(T, 0)'\ell \mid Y_t]$. Then M_t is a square integrable martingale on the Y-filtration; hence, (see [6]), M_t has representation

$$(3.8) \qquad M_t = E[\Phi^*(T, 0)'\ell] + \int_0^t \gamma_r \, dy_r$$

where γ is a $\{Y_t\}$ predictable process, such that

$$\int_0^T E|\gamma_r^2| dr < \infty.$$

DEFINITION 3.5.   *The adjoint process is*

$$p_t := \Psi^*(t, 0)' M_t$$

where the prime $'$ denotes the transpose of the matrix.

THEOREM 3.6.

$$(3.9) \qquad \left.\frac{\partial J(u_\theta^*)}{\partial \theta}\right|_{\theta=0} = \int_0^T E\left[\left\langle p_r, \frac{\partial A}{\partial u}(u^*) v_r q_r^*\right\rangle\right] dr.$$

Proof. Using (3.6) and (3.8)

$$\langle M_T, \eta_{0,T} \rangle = \int_0^T \left\langle M_r, \Psi^*(r,0) \frac{\partial A}{\partial u}(u^*) v_r q_r^* \right\rangle dr + \int_0^T \langle \gamma_r, \eta_{0,r} \rangle dy_r.$$

From (3.7)

$$\left. \frac{\partial J(u_\theta^*)}{\partial \theta} \right|_{\theta=0} = E[\langle M_T, \eta_{0,T} \rangle],$$

so the result follows.   □

Under integrability conditions J' is in H, and so a Gâteaux derivative.

Now consider perturbations of u^* of the form $u_\theta(t) = u^*(t) + \theta(v(t) - u^*(t))$ for $\theta \in [0,1]$, and any $v \in \underline{U}$. Then

$$\left. \frac{\partial J(u_\theta)}{\partial \theta} \right|_{\theta=0} = \langle J'(u^*), \ v - u^* \rangle \geq 0.$$

for all $v \in \underline{U}$. So we have the following

THEOREM 3.7. Suppose $u^* \in \underline{U}$ is an optimal control. Then a.s. in w and a.e. in t

$$(3.10) \qquad \left\langle p_t, \frac{\partial A}{\partial u}(u^*)(v_t - u_t^*)q_t^* \right\rangle \geq 0.$$

4. Equations for the Adjoint Process.

Suppose the optimal control u^* is a Markov, feedback control in the state variable q.

We have the following expression for the integrand in (3.8).

LEMMA 4.1.

$$(4.1) \qquad \gamma_r = \Phi^*(r,0)' \frac{\partial p_r}{\partial q} H q_r + \Phi^*(r,0)' H p_r.$$

Proof: $\Phi^*(t,0)' p_t = M_t = E[\Phi^*(T,0)'\ell] + \int_0^t \gamma_r \, dy_r$. If u^* is Markov, q^* is also Markov. Write $q = q_t^*$, $\Phi = \Phi^*(t,0)$, then by the Markov property

$$E[\Phi^*(T,0)'\ell \mid Y_t] = E[\Phi'\Phi^*(T,t)'\ell \mid q, \Phi]$$

$$= \Phi' E[\Phi^*(T,t)'\ell \mid q].$$

So $p_t = E[\Phi^*(T,t)'\ell \mid q]$ is a function of q only. Therefore,

$$p_t = p_0 + \int_0^t \frac{\partial p}{\partial q}(Aq_r\,dr + Hq_r\,dy_r)$$

$$+ \int_0^t \frac{\partial p_r}{\partial r}\,dr + \frac{1}{2}\sum_{i,j=1}^N \int_0^t \frac{\partial^2 p_r}{\partial q^i \partial q^j}\,h(e_i)h(e_j)q_r^i q_r^j\,dr$$

$$= p_0 + \int_0^t \left[\frac{\partial p_r}{\partial q}\,Aq_r + \frac{1}{2}\sum_{i,j=1}^N \frac{\partial^2 p_r}{\partial q^i \partial q^j}h(e_i)h(e_j)q_r^i q_r^j + \frac{\partial p_r}{\partial r}\right]dr$$

$$+ \int_0^t \frac{\partial p_r}{\partial q}\,Hq_r\,dy_r.$$

Then

$$M_t = \Phi^*(t,0)'p_t$$

$$= p_0 + \int_0^t \Phi^*(r,0)' \left[\frac{\partial p_r}{\partial q}\,Aq_r + \frac{1}{2}\sum_{ij=1}^N \frac{\partial^2 p_r}{\partial q_r^i \partial q_r^j}h(e_i)h(e_j)q_r^i q_r^j + \frac{\partial p_r}{\partial r}\right]dr$$

$$+ \int_0^t \Phi^*(r,0)'\frac{\partial p_r}{\partial q}\,Hq_r\,dy_r + \int_0^t \Phi^*(r,0)'A(u)'p_r\,dr$$

(4.2)

$$+ \int_0^t \Phi^*(r,0)'Hp_r\,dy_r + \int_0^t \Phi^{*'}H\frac{\partial p_r}{\partial q}\,Hq_r\,dr.$$

Since M_t is a Martingale the sum of the dr integrals in (4.2) must be 0, and, therefore,

$$\gamma_r = \Phi^*(r,0)'\frac{\partial p_r}{\partial q}\,Hq_r + \Phi^*(r,0)'Hp_r.$$

$\square$

THEOREM 4.2.

(4.3) $\quad p_t = E[\Phi^*(T,0)'\ell] + \int_0^t \frac{\partial p_r}{\partial q}\,Hq_r\,dy_r - \int_0^t \left(A'p_r + H\,\frac{\partial p_r}{\partial q}\,Hq_r\right)dr.$

Proof:

$$p_t = \Psi^*(t,0)' M_t = E[\Phi^*(T,0)'\ell]$$

$$+ \int_0^t \Psi^{*\prime} \left(\Phi^{*\prime} \frac{\partial p_r}{\partial q} H q_r + \Phi^{*\prime} H p_r \right) dy_r$$

$$- \int_0^t A' \Psi^{*\prime} M_r dr - \int_0^t H \Psi^{*\prime} M_r dy_r$$

$$+ \int_0^t H^2 \Psi^{*\prime} M_r dr - \int_0^t H \Psi^{*\prime} \left(\Phi^{*\prime} \frac{\partial p_r}{\partial q} H q_r + \Phi^{*\prime} H p_r \right) dr$$

$$= E[\Phi^*(T,0)'\ell] + \int_0^t \left(\frac{\partial p_r}{\partial q} H q_r + H p_r \right) dy_r$$

$$- \int_0^t A' p_r dr - \int_0^t H p_r dy_r + \int_0^t H^2 p_r dr$$

$$- \int_0^t \left(H \frac{\partial p_r}{\partial q} H q_r + H^2 p_r \right) dr.$$

So

$$p_t = E[\Phi^*(T,0)'\ell] + \int_0^t \frac{\partial p_r}{\partial q} H q_r dy_r$$

$$- \int_0^t \left(A' p_r + H \frac{\partial p_r}{\partial q} H q_r \right) dr.$$

□

From (4.2), equating the dr integrals to zero we also obtain the following result.

THEOREM 4.3. p_t satisfies the backward parabolic system
(4.4)

$$\frac{\partial p_t}{\partial t} + \frac{\partial p_t}{\partial q} A q_t + H \frac{\partial p_t}{\partial q} H q_t + \frac{1}{2} \sum_{ij=1}^N \frac{\partial^2 p_t}{\partial q^i \partial q^j} h(e_i) h(e_j) q_t^i q_t^j + A(u^*)' p_t = 0.$$

with terminal condition

$$p_T = \ell.$$

REMARKS 4.4. In [3] Bismut considers a forward equation, with a terminal condition, for the adjoint process.

References.

[1] A. Bensoussan, Lectures on stochastic control. In *Lecture Notes in Mathematics*, Vol. 972. Springer-Verlag, Berlin, Heidelberg, New York, 1982.

[2] J.M. Bismut, Linear quadratic optimal stochastic control with random coefficients. *S.I.A.M. J. Control and Optimization*, Vol. 14, No. 3 (1976), 419–444.

[3] J.M. Bismut, An introductory approach to duality in optimal stochastic control. *S.I.A.M. Review* 20 (1978), 62–78.

[4] J.N. Blagovescenskii and M.I. Freidlin, Some properties of diffusion process depending on a parameter. *Dokl. Akad. Nauk.* 138 (1961); *Soviet Math.*, 2 (1961), 633–636.

[5] M.H.A. Davis, Nonlinear semigroups in the control of partially observable stochastic systems. In *Lecture Notes in Mathematics*, Vol. 695, Springer-Verlag, Berlin, Heidelberg, New York, 1977, 37–49.

[6] R.J. Elliott, *Stochastic Calculus and Applications.* Springer-Verlag, Berlin, Heidelberg, New York, 1982.

[7] R.J. Elliott, A partially observed control problem for Markov chains. To appear.

The Transition Function of a Measure-Valued Branching Diffusion with Immigration

S. N. ETHIER AND R. C. GRIFFITHS

Abstract. Let S be a compact metric space, let $\theta \geq 0$, let ν_0 be a Borel probability measure on S, and let λ be real. An explicit formula is found for the transition function of the measure-valued branching diffusion with type space S, immigration intensity $\theta/2$, immigrant-type distribution ν_0, and criticality parameter λ. If $\lambda > 0$, the formula shows that the process is strongly ergodic.

1. Introduction.

Let S be a compact metric space, let $\theta \geq 0$, let ν_0 belong to $\mathcal{P}(S)$, the set of Borel probability measures on S, and let λ be real. Consider the measure-valued branching diffusion with immigration (MBDI) in $\mathcal{M}(S)$, the set of finite positive Borel measures on S with the topology of weak convergence, characterized in terms of the generator

$$(1.1) \quad \mathcal{L} = \tfrac{1}{2} \int_S \mu(dx) \frac{\delta^2}{\delta\mu(x)^2} + \tfrac{1}{2}\theta \int_S \nu_0(dx) \frac{\delta}{\delta\mu(x)} - \tfrac{1}{2}\lambda \int_S \mu(dx) \frac{\delta}{\delta\mu(x)},$$

where $\delta\varphi(\mu)/\delta\mu(x) = \lim_{\varepsilon \to 0+} \varepsilon^{-1}\{\varphi(\mu + \varepsilon\delta_x) - \varphi(\mu)\}$. Here $\mathcal{D}(\mathcal{L}) = \{\varphi : \varphi(\mu) \equiv F(\langle f_1, \mu\rangle, \ldots, \langle f_k, \mu\rangle), F \in C_c^2(\mathbf{R}^k), f_1, \ldots, f_k \in C(S), k \geq 1\}$, and $\langle f, \mu\rangle = \int_S f \, d\mu$. We interpret S as the type space, θ as twice the immigration intensity, ν_0 as the distribution of the type of a new immigrant, and λ as a criticality parameter.

The study of measure-valued branching diffusions was initiated by Watanabe (1968) and Dawson (1975). Such diffusions are now known as Dawson–Watanabe superprocesses.

If S consists of a single point, then $\mathcal{M}(S)$ can be identified with $[0, \infty)$, and $\mathcal{L}$ reduces to

$$(1.2) \qquad L = \tfrac{1}{2} z \frac{d^2}{dz^2} + \tfrac{1}{2}(\theta - \lambda z)\frac{d}{dz},$$

Research supported in part by NSF grant DMS-9102925.

where $\mathcal{D}(L) = C_c^2([0, \infty))$. This is the generator of a one-dimensional continuous-state branching diffusion with immigration. Its transition density was found by Feller (1951).

It is our aim here to find an explicit formula for the transition function of the MBDI corresponding to (1.1). Shiga (1990) has shown that this process starting at $\mu \in \mathcal{M}(S)$ can be constructed from a pair of independent Poisson random measures associated with an excursion law of the one-dimensional continuous-state branching diffusion corresponding to (1.2) with $\theta = 0$. Our derivation does not rely on Shiga's construction.

To state the main result, we need to introduce some notation. Given $\alpha > 0$ and λ real, define $q_n^{\alpha, \lambda} : (0, \infty) \mapsto (0, 1)$ for $n = 0, 1, 2, \ldots$ by

$$(1.3) \qquad q_n^{\alpha, \lambda}(t) = \mathbf{P}\{N_{\alpha/C_\lambda(t)} = n\},$$

where $\{N_t, \ t \geq 0\}$ is a standard rate 1 Poisson process and $C_\lambda : (0, \infty) \mapsto (0, \infty)$ is defined by

$$(1.4) \qquad C_\lambda(t) = \begin{cases} \lambda^{-1}(e^{\lambda t/2} - 1) & \text{if } \lambda \neq 0, \\ t/2 & \text{if } \lambda = 0. \end{cases}$$

Given $\theta > 0$, $\nu_0 \in \mathcal{P}(S)$, and $\beta > 0$, let $\gamma_1 > \gamma_2 > \cdots$ be the points of an inhomogeneous Poisson point process on $(0, \infty)$ with intensity function $\theta z^{-1} e^{-z}$ $(z > 0)$, let $\xi_1, \xi_2, \ldots$ be i.i.d. ν_0 and independent of $\gamma_1, \gamma_2, \ldots$, and define $\Gamma_{\theta, \nu_0}^\beta \in \mathcal{P}(\mathcal{M}(S))$ by

$$(1.5) \qquad \Gamma_{\theta, \nu_0}^\beta(\cdot) = \mathbf{P}\left\{\beta \sum_{i=1}^\infty \gamma_i \delta_{\xi_i} \in \cdot\right\};$$

$\Gamma_{\theta, \nu_0}^\beta$ is the distribution of a gamma random measure (see Section 2). Given $\nu_0 \in \mathcal{P}(S)$ and $\beta > 0$, define $\Gamma_{0, \nu_0}^\beta \in \mathcal{P}(\mathcal{M}(S))$ by

$$(1.6) \qquad \Gamma_{0, \nu_0}^\beta(\cdot) = \delta_0(\cdot),$$

where the 0 on the right denotes the zero measure. Finally, for each $n \geq 1$, define $\eta_n : S^n \mapsto \mathcal{P}(S)$ by letting $\eta_n(x_1, \ldots, x_n)$ be the empirical measure determined by $(x_1, \ldots, x_n) \in S^n$:

$$(1.7) \qquad \eta_n(x_1, \ldots, x_n) = n^{-1}(\delta_{x_1} + \cdots + \delta_{x_n}).$$

THEOREM 1.1. *Let S be a compact metric space, and let $\theta \geq 0$, $\nu_0 \in \mathcal{P}(S)$, and $\lambda \in \mathbf{R}$. Then the transition function $P(t, \mu, d\nu)$ of the MBDI in $\mathcal{M}(S)$*

with parameters S, θ, ν_0, and λ is given for each $t > 0$ and $\mu \in \mathcal{M}(S) - \{0\}$
by

$$(1.8) \quad P(t, \mu, \cdot) = q_0^{\mu(S), \lambda}(t)\, \Gamma_{\theta, \nu_0}^{C - \lambda(t)}(\cdot)$$

$$+ \sum_{n=1}^{\infty} q_n^{\mu(S), \lambda}(t) \int_{S^n} (\mu/\mu(S))^n (d x_1 \times \cdots \times d x_n)$$

$$\Gamma_{n+\theta, (n+\theta)^{-1}\{n \eta_n(x_1, \ldots, x_n) + \theta \nu_0\}}^{C - \lambda(t)}(\cdot).$$

Also, $P(t, 0, \cdot) = \Gamma_{\theta, \nu_0}^{C - \lambda(t)}(\cdot)$ for each $t > 0$.

In particular, for each $t > 0$ and $\mu \in \mathcal{M}(S)$, $P(t, \mu, \cdot)$ is a mixture of
probability distributions of the form (1.5) [or (1.6)].

Shiga, Shimizu, and Tanaka (1987) showed that, if $\lambda > 0$, the MBDI
of Theorem 1.1 is weakly ergodic. The theorem allows us to prove that it
is strongly ergodic. In fact, we have the following estimate of the rate of
convergence to equilibrium.

COROLLARY 1.2. *Assume, in addition to the assumptions of Theorem 1.1,
that $\lambda > 0$. Then, for each $t > 0$ and $\mu \in \mathcal{M}(S)$,*

$$(1.9) \quad \|P(t, \mu, \cdot) - \Gamma_{\theta, \nu_0}^{\lambda^{-1}}(\cdot)\|_{\mathrm{var}} \le 1 - e^{-\mu(S)/C_\lambda(t)} + 1 - (1 - e^{-\lambda t/2})^\theta,$$

where $\| \cdot \|_{\mathrm{var}}$ denotes the total variation norm.

The proofs of the theorem and the corollary are deferred to Section
3. Section 2 contains two preliminary lemmas, and Section 4 discusses the
relationship between the MBDI of Theorem 1.1 and a Fleming–Viot (1979)
probability-measure-valued diffusion process.

2. Laplace functionals.

The compact metric space S is fixed throughout. We denote by $C_+(S)$
the set of all nonnegative continuous functions on S.

LEMMA 2.1. *Let $\theta \ge 0$, $\nu_0 \in \mathcal{P}(S)$, and $\beta > 0$. Then, for each $f \in C_+(S)$,*

$$(2.1) \quad \int_{\mathcal{M}(S)} e^{-\langle f, \nu \rangle} \, \Gamma_{\theta, \nu_0}^{\beta}(d\nu) = \exp\{-\theta \langle \log(1 + \beta f), \nu_0 \rangle\}.$$

Proof. The case $\theta = 0$ is trivial, so we assume that $\theta > 0$. Let $\gamma_1 > \gamma_2 > \cdots$
and $\xi_1, \xi_2, \ldots$ be as in (1.5), and let $f \in C_+(S)$. Then

$$(2.2) \quad \int_{\mathcal{M}(S)} e^{-\langle f, \nu \rangle} \Gamma^{\beta}_{\theta, \nu_0}(d\nu) = \mathbf{E}\left[\exp\left\{-\left\langle f, \beta \sum_{i=1}^{\infty} \gamma_i \delta_{\xi_i} \right\rangle\right\}\right]$$

$$= \mathbf{E}\left[\prod_{i=1}^{\infty} \langle e^{-\beta f \gamma_i}, \nu_0 \rangle \right]$$

$$= \exp\left\{-\int_0^{\infty} \left(1 - \langle e^{-\beta f z}, \nu_0 \rangle\right) \theta z^{-1} e^{-z} \, dz \right\}$$

$$= \exp\{-\theta \langle \log(1 + \beta f), \nu_0 \rangle\},$$

where the third equality uses a basic property of Poisson point processes, and the fourth equality uses the identity

$$(2.3) \qquad \int_0^{\infty} (1 - e^{-sz}) z^{-1} e^{-z} \, dz = \log(1 + s), \qquad s \geq 0,$$

which follows from the facts that both sides have the same derivative with respect to s and are equal at $s = 0$.

An alternative description of $\Gamma^{\beta}_{\theta, \nu_0}$ can be inferred from Exercise 6.1.2 of Daley and Vere-Jones (1988): it is the distribution of a gamma random measure ν with the property that if $n \geq 2$ and $\Lambda_1, \ldots, \Lambda_n \in \mathcal{B}(S)$ are disjoint, then $\nu(\Lambda_1), \ldots, \nu(\Lambda_n)$ are independent with $\nu(\Lambda_i)$ having a gamma$(\theta \nu_0(\Lambda_i), \beta^{-1})$ distribution for $i = 1, \ldots, n$.

The following result is well known; see e.g. Shiga (1990).

LEMMA 2.2. *Under the assumptions of Theorem 1.1,*

$$(2.4) \quad \int_{\mathcal{M}(S)} e^{-\langle f, \nu \rangle} P(t, \mu, d\nu)$$

$$= \exp\left\{-\left\langle \frac{e^{-\lambda t/2} f}{1 + C_{-\lambda}(t) f}, \mu \right\rangle - \theta \langle \log(1 + C_{-\lambda}(t) f), \nu_0 \rangle \right\}$$

for all $f \in C_+(S)$, $t > 0$, and $\mu \in \mathcal{M}(S)$.

Proof. For completeness, we provide a proof; our argument follows Shiga (1991).

Give $\Omega = C_{\mathcal{M}(S)}[0, \infty)$ the topology of uniform convergence on compact sets, and let $\mathcal{F}$ be the Borel σ-field. Denote by $\{\mu_t, \ t \geq 0\}$ the canonical coordinate process (i.e., $\mu_t(\omega) = \omega(t)$ for all $t \geq 0$), and let $\{\mathcal{F}_t\}$ be the associated filtration. For each $\mu \in \mathcal{M}(S)$, let $P_\mu \in \mathcal{P}(\Omega)$ denote the unique solution of the martingale problem for $\mathcal{L}$ starting at μ.

Fix $f \in C_+(S)$ and $\mu \in \mathcal{M}(S)$. Define $u \in C_{C_+(S)}[0, \infty)$ to be the unique solution of the differential equation

$$(2.5) \qquad \frac{d}{dt}u(t) = -\tfrac{1}{2}(u(t)^2 + \lambda u(t)), \qquad u(0) = f.$$

Fix $T > 0$. Then

$$(2.6) \quad e^{-\langle u(T-t), \mu_t \rangle} + \tfrac{1}{2}\theta \int_0^t \langle u(T-s), \nu_0 \rangle\, e^{-\langle u(T-s), \mu_s \rangle}\, ds, \quad 0 \le t \le T,$$

is an $\{\mathcal{F}_t\}$-martingale on $(\Omega, \mathcal{F}, P_\mu)$ by Lemma 4.3.4 of Ethier and Kurtz (1986). By the second part of Lemma 4.3.2 of the same reference (with $E = [0, \infty) \times \mathcal{M}(S)$ and $X(t) = (t, \mu_t)$ for all $t \ge 0$), the same is true of

$$(2.7) \quad \exp\left\{-\langle u(T-t), \mu_t \rangle + \tfrac{1}{2}\theta \int_0^t \langle u(T-s), \nu_0 \rangle\, ds\right\}, \quad 0 \le t \le T,$$

and it follows that

$$(2.8) \quad \mathbf{E}^{P_\mu}\left[e^{-\langle f, \mu_T \rangle}\right] = \exp\left\{-\langle u(T), \mu \rangle - \tfrac{1}{2}\theta \int_0^T \langle u(T-s), \nu_0 \rangle\, ds\right\}.$$

Finally, the solution of (2.5) is $u(t) = e^{-\lambda t/2} f/(1 + C_{-\lambda}(t)f)$, which when substituted into (2.8) implies (2.4).

Under the assumptions of Corollary 1.2, a straightforward calculation using both lemmas gives

$$(2.9) \quad \int_{\mathcal{M}(S)} e^{-\langle f, \mu \rangle}\left\{\int_{\mathcal{M}(S)} e^{-\langle g, \nu \rangle}\, P(t, \mu, d\nu)\right\} \Gamma_{\theta, \nu_0}^{\lambda^{-1}}(d\mu)$$
$$= \exp\{-\theta\langle \log[1 + \lambda^{-1}(f+g) + \lambda^{-1}C_{-\lambda}(t)fg], \nu_0 \rangle\}$$

for all $f, g \in C_+(S)$ and $t > 0$. As pointed out by Shiga, Shimizu, and Tanaka (1987) and more explicitly by Shiga (1991), the symmetry of (2.9) in f and g implies that, if $\lambda > 0$, the MBDI of Theorem 1.1 is reversible with respect to its unique stationary distribution.

3. Proofs.

Proof of Theorem 1.1. Denote the right side of (1.8) by $Q(t, \mu, \cdot)$. Then for each $f \in C_+(S)$, $t > 0$, and $\mu \in \mathcal{M}(S) - \{0\}$, we have

$$(3.1) \quad \int_{\mathcal{M}(S)} e^{-\langle f, \nu \rangle} Q(t, \mu, d\nu)$$

$$= \left\{ q_0^{\mu(S), \lambda}(t) + \sum_{n=1}^{\infty} q_n^{\mu(S), \lambda}(t) \int_{S^n} (\mu/\mu(S))^n (dx_1 \times \cdots \times dx_n) \right.$$

$$\left. \exp\{-n\langle \log(1 + C_{-\lambda}(t)f), \eta_n(x_1, \ldots, x_n)\rangle\} \right\}$$

$$\exp\{-\theta\langle \log(1 + C_{-\lambda}(t)f), \nu_0\rangle\}$$

$$= \sum_{n=0}^{\infty} \frac{(\mu(S)/C_\lambda(t))^n}{n!} e^{-\mu(S)/C_\lambda(t)} \langle (1 + C_{-\lambda}(t)f)^{-1}, \mu/\mu(S)\rangle^n$$

$$\exp\{-\theta\langle \log(1 + C_{-\lambda}(t)f), \nu_0\rangle\}$$

$$= \exp\left\{ -\left\langle \frac{e^{-\lambda t/2} f}{1 + C_{-\lambda}(t)f}, \mu \right\rangle - \theta\langle \log(1 + C_{-\lambda}(t)f), \nu_0\rangle \right\},$$

where the first equality uses Lemma 2.1. By Lemma 2.2, $P(t, \mu, \cdot) = Q(t, \mu, \cdot)$ for all $t > 0$ and $\mu \in \mathcal{M}(S) - \{0\}$. The case in which $\mu = 0$ is immediate from Lemmas 2.1 and 2.2.

Proof of Corollary 1.2. Suppose that $\theta > 0$. Let the $\mathcal{M}(S) - \{0\}$-valued random variable ν have distribution $\Gamma_{\theta, \nu_0}^{\lambda^{-1}}$. Then the $\mathcal{P}(S)$-valued random variable $\nu/\nu(S)$, whose distribution we denote by Π_{θ, ν_0}, is independent of the positive random variable $\lambda\nu(S)$ (see Kingman (1975)), which has the gamma$(\theta, 1)$ distribution, that is, Lebesgue density $h(z) = \Gamma(\theta)^{-1} z^{\theta-1} e^{-z}$ on $(0, \infty)$. Thus, for $0 < \varepsilon < 1$,

$$(3.2) \quad \|\Gamma_{\theta, \nu_0}^{\lambda^{-1}(1-\varepsilon)}(\cdot) - \Gamma_{\theta, \nu_0}^{\lambda^{-1}}(\cdot)\|_{\mathrm{var}}$$

$$= \sup_{B \in \mathcal{B}(\mathcal{M}(S))} |\mathbf{P}\{(1 - \varepsilon)\nu \in B\} - \mathbf{P}\{\nu \in B\}|$$

$$= \sup_{B \in \mathcal{B}(\mathcal{M}(S))} \left| \int_{\mathcal{P}(S)} [\mathbf{P}\{(1 - \varepsilon)\nu(S)\pi \in B\} - \mathbf{P}\{\nu(S)\pi \in B\}] \Pi_{\theta, \nu_0}(d\pi) \right|$$

$$\leq \sup_{C \in \mathcal{B}((0, \infty))} |\mathbf{P}\{(1 - \varepsilon)\lambda\nu(S) \in C\} - \mathbf{P}\{\lambda\nu(S) \in C\}|$$

$$= \frac{1}{2} \int_0^\infty \left| h\left(\frac{z}{1-\varepsilon}\right) \frac{1}{1-\varepsilon} - h(z) \right| dz$$

$$= \frac{1}{2}\Gamma(\theta)^{-1} \int_0^\infty z^{\theta-1} e^{-z} |(1 - \varepsilon)^{-\theta} e^{-\varepsilon z/(1-\varepsilon)} - 1| \, dz$$

$$\leq \frac{1}{2}\Gamma(\theta)^{-1} \int_0^\infty z^{\theta-1} e^{-z} \{[(1 - \varepsilon)^{-\theta} - 1]e^{-\varepsilon z/(1-\varepsilon)} + 1 - e^{-\varepsilon z/(1-\varepsilon)}\} \, dz$$

$$= \frac{1}{2}\left\{ [(1 - \varepsilon)^{-\theta} - 1]\left(1 + \frac{\varepsilon}{1-\varepsilon}\right)^{-\theta} + 1 - \left(1 + \frac{\varepsilon}{1-\varepsilon}\right)^{-\theta} \right\}$$

$$= 1 - (1 - \varepsilon)^{\theta}.$$

Now by Theorem 1.1, regardless of whether $\theta > 0$,

$$(3.3) \qquad \|P(t,\mu,\cdot) - \Gamma_{\theta,\nu_0}^{\lambda^{-1}}(\cdot)\|_{\mathrm{var}}$$
$$\leq 1 - e^{-\mu(S)/C_\lambda(t)} + \|\Gamma_{\theta,\nu_0}^{\lambda^{-1}(1-e^{-\lambda t/2})}(\cdot) - \Gamma_{\theta,\nu_0}^{\lambda^{-1}}(\cdot)\|_{\mathrm{var}}$$

for all $t > 0$ and $\mu \in \mathcal{M}(S)$, so the result follows from this and (3.2).

4. Relationship to a Fleming–Viot diffusion.

Let S be a compact metric space, and let $\theta \geq 0$ and $\nu_0 \in \mathcal{P}(S)$. Define the bounded linear operator A on $C(S)$ by

$$(4.1) \qquad (Af)(x) = \tfrac{1}{2}\theta \int_S (f(\xi) - f(x))\,\nu_0(d\xi),$$

and consider the Fleming–Viot (1979) diffusion (FVD) in $\mathcal{P}(S)$ characterized in terms of the generator

$$(4.2) \qquad (\mathcal{L}^\circ \varphi)(\mu) = \tfrac{1}{2}\int_S \int_S \mu(dx)(\delta_x(dy) - \mu(dy))\frac{\delta^2\varphi(\mu)}{\delta\mu(x)\,\delta\mu(y)}$$
$$+ \int_S \mu(dx)A\Big(\frac{\delta\varphi(\mu)}{\delta\mu(\cdot)}\Big)(x).$$

Here $\mathcal{D}(\mathcal{L}^\circ) = \{\varphi : \varphi(\mu) \equiv F(\langle f_1,\mu\rangle,\dots,\langle f_k,\mu\rangle),\ F \in C^2(\mathbf{R}^k),\ f_1,\dots,f_k \in C(S),\ k \geq 1\}$. We interpret S as the type space, θ as twice the mutation intensity, and ν_0 as the distribution of the type of a new mutant.

Recently, an explicit formula was found for the transition function of the above FVD. To state that result, we need to introduce some additional notation. Given $\theta \geq 0$, define $d_n^\theta : (0,\infty) \mapsto [0,1)$ for $n = 0,1,2,\dots$ by

$$(4.3) \qquad d_n^\theta(t) = \mathbf{P}\{\dot{D}_t^\theta = n\},$$

where $\{D_t^\theta,\ t \geq 0\}$ is a pure death process in $\mathbf{Z}_+ \cup \{\infty\}$ starting at ∞ with death rate $n(n-1+\theta)/2$ from state n for each $n \geq 0$; ∞ is an entrance boundary. (An explicit formula is available for $d_n^\theta(t)$; see e.g. Tavaré (1984).) Given $\theta > 0$ and $\nu_0 \in \mathcal{P}(S)$, let $\gamma_1 > \gamma_2 > \cdots$ and $\xi_1, \xi_2, \dots$ be as in (1.5), and define $\Pi_{\theta,\nu_0} \in \mathcal{P}(\mathcal{P}(S))$ by

$$(4.4) \qquad \Pi_{\theta,\nu_0}(\cdot) = \mathbf{P}\Big\{\Big(\sum_{i=1}^\infty \gamma_i\Big)^{-1}\sum_{i=1}^\infty \gamma_i\delta_{\xi_i} \in \cdot\Big\}.$$

Given $\nu_0 \in \mathcal{P}(S)$, define $\Pi_{0,\nu_0} \in \mathcal{P}(\mathcal{P}(S))$ arbitrarily.

PROPOSITION 4.1. (Ethier and Griffiths (1993)) *Let S be a compact metric space, and let $\theta \geq 0$ and $\nu_0 \in \mathcal{P}(S)$. Then the transition function $P^\circ(t, \mu, d\nu)$ of the FVD in $\mathcal{P}(S)$ with parameters S, θ, and ν_0 is given for each $t > 0$ and $\mu \in \mathcal{P}(S)$ by*

$$(4.5) \quad P^\circ(t, \mu, \cdot) = d_0^\theta(t)\,\Pi_{\theta,\nu_0}(\cdot)$$
$$+ \sum_{n=1}^{\infty} d_n^\theta(t) \int_{S^n} \mu^n(dx_1 \times \cdots \times dx_n)$$
$$\Pi_{n+\theta,(n+\theta)^{-1}\{n\eta_n(x_1,\ldots,x_n)+\theta\nu_0\}}(\cdot).$$

Notice the similarity between (1.8) and (4.5). Moreover, the time-changed Poisson process appearing in (1.3) is a time-inhomogeneous pure death process with death rate $n/2C_{-\lambda}(t)$ from state $n \geq 0$ at time $t > 0$.

There is a close relationship between the MBDI of Theorem 1.1 and the FVD of Proposition 4.1. That relationship can be described in part as follows.

Recall from the proof of Lemma 2.2 that $\Omega = C_{\mathcal{M}(S)}[0, \infty)$ has the topology of uniform convergence on compact sets, that $\mathcal{F}$ is the Borel σ-field, and that $\{\mu_t,\ t \geq 0\}$ is the canonical coordinate process. Define $\zeta : \Omega \mapsto [0, \infty]$ by

$$(4.6) \qquad\qquad \zeta = \inf\{t \geq 0 : \overrightarrow{\mu_t}(S) = 0\},$$

where $\inf \emptyset = \infty$. Let

$$(4.7) \qquad\qquad \Omega_0 = \left\{ \int_0^\zeta \frac{1}{\mu_s(S)}\, ds = \infty \right\} \subset \Omega,$$

and define $\tau(t) : \Omega_0 \mapsto [0, \infty)$ for all $t \geq 0$ by

$$(4.8) \qquad\qquad t = \int_0^{\tau(t)} \frac{1}{\mu_s(S)}\, ds.$$

PROPOSITION 4.2. (Shiga (1990)) *Under the assumptions of Theorem 1.1, let $\mu \in \mathcal{M}(S) - \{0\}$ and denote by P_μ the unique solution of the martingale problem for $\mathcal{L}$ starting at μ. Then $P_\mu(\Omega_0) = 1$ and the P_μ-distribution of $\{\mu_{\tau(t)}/\mu_{\tau(t)}(S),\ t \geq 0\}$ on $C_{\mathcal{P}(S)}[0, \infty)$ solves the martingale problem for $\mathcal{L}^\circ$ starting at $\mu/\mu(S)$.*

In short, normalization and a random time change transform the MBDI of Theorem 1.1 into the FVD of Proposition 4.1. It would be interesting to derive (4.5) directly from (1.8) using Proposition 4.2, but this remains an open problem.

References.

DALEY, D. J. and VERE-JONES, D. (1988). *An Introduction to the Theory of Point Processes.* Springer–Verlag, New York.

DAWSON, D. A. (1975). Stochastic evolution equations and related measure processes. *J. Multivariate Anal.* **5** 1–52.

ETHIER, S. N. and GRIFFITHS, R. C. (1993). The transition function of a Fleming–Viot process. *Ann. Probab.*, to appear.

ETHIER, S. N. and KURTZ, T. G. (1986). *Markov Processes: Characterization and Convergence.* Wiley, New York.

FELLER, W. (1951). Two singular diffusion problems. *Ann. Math.* **54** 173–188.

FLEMING, W. H. and VIOT, M. (1979). Some measure-valued Markov processes in population genetics theory. *Indiana Univ. Math. J.* **28** 817–843.

KINGMAN, J. F. C. (1975). Random discrete distributions. *J. Roy. Statist. Soc.* B **37** 1–22.

SHIGA, T. (1990). A stochastic equation based on a Poisson system for a class of measure-valued diffusion processes. *J. Math. Kyoto Univ.* **30** 245–279.

SHIGA, T. (1991). Infinite-dimensional diffusion processes arising in population genetics. Lecture notes based on lectures given at Taiwan University.

SHIGA, T., SHIMIZU, A., and TANAKA, H. (1987). Some measure-valued diffusion processes associated with genetical diffusion models. Unpublished manuscript.

TAVARÉ, S. (1984). Line-of-descent and genealogical processes, and their applications in population genetics models. *Theoret. Popn. Biol.* **26** 119–164.

WATANABE, S. (1968). A limit theorem of branching processes and continuous state branching processes. *J. Math. Kyoto Univ.* **8** 141–167.

DEPARTMENT OF MATHEMATICS
UNIVERSITY OF UTAH
SALT LAKE CITY, UTAH 84112, USA

DEPARTMENT OF MATHEMATICS
MONASH UNIVERSITY
CLAYTON, VICTORIA 3168, AUSTRALIA

Scattering theory for unitary cocycles
by
F. Fagnola and Kalyan B. Sinha

ABSTRACT
We compare the asymptotic behaviours of two unitary quantum stochastic cocycles. We give a sufficient condition in order the scattering operator to exists and some applications.

1. Introduction

We shall follow the notations of the book [6]. The symmetric (Bosonic) Fock space $\mathcal{H}$ over $L^2(\mathbb{R}_+)$ (or $L^2(\mathbb{R})$ in some cases) is isometrically isomorphic to $L^2(\mathbb{P})$, where $\mathbb{P}$ is the Wiener measure on the space of Brownian paths $\omega(t)$ (standard Brownian motion), via the Wiener-Segal isometry

$$e(f) \mapsto \exp\left(\int_0^\infty f(s)d\omega(s) - \frac{1}{2}\int_0^\infty (f(s))^2 ds\right).$$

Here $e(f)$ denotes the exponential or coherent vector corresponding to $f \in L^2(\mathbb{R}_+)$. The set of exponential vectors is total in $\mathcal{H}$. Then one has the three basic quantum martingales viz. $\Lambda(t)$ the gauge, $A(t)$ the annihilation and $A^\dagger(t)$ the creation processes.

We also have another separable Hilbert space h called the initial Hilbert space, and a triple (S, L, H) of bounded operators in h with S unitary, H selfadjoint. Then it is known [3],[6] that the quantum stochastic differential equation (q.s.d.e.):

$$\begin{cases} dU_t = U_t\left((S-I)d\Lambda(t) + L^*SdA(t) - LdA^\dagger(t) + (iH - \frac{1}{2}L^*L)dt\right), \\ U_0 = I \end{cases}$$

$$(1.1)$$

has unique adapted unitary solution in $h \otimes \mathcal{H}$. The triple (S, L, H) is called the generator of the evolution U.

For $f \in L^2(\mathbb{R}_+)$, we define the right shift as follows:

$$\theta_t f(s) = \begin{cases} 0 & \text{if } s < t \\ f(s-t) & \text{if } s \geq t, \end{cases} \qquad (1.2)$$

$S_t e(f) = e(\theta_t f)$ and extend linearly.

It is clear that $\{\theta_t\}_{t \geq 0}$ and $\{S_t\}_{t \geq 0}$ are two strongly continuous semigroup of isometries in $L^2(I\!\!R_+)$ and $\mathcal{H}$ respectively. Then it is also known [3],[6] that the solution U of (1.1) is a stochastic cocycle i.e.

$$U_{t+s} = U_t S_t U_s S_t^* \qquad (1.3)$$

for all $t, s \geq 0$. Notice that, if we have the Fock space over $L^2(I\!\!R)$ instead of $L^2(I\!\!R_+)$, then we may replace $\{\theta_t\}_{t \geq 0}$ by the translation group and have corresponding unitary group $\{S_t\}_{t \in I\!\!R}$. In such a case also, U will satisfy (1.3) with the new $\{S_t\}_{t \in I\!\!R}$.

If in the generator triple (S, L, H), S remains unitary while L and H are closed and self-adjoint respectively, though unbounded, then the problems of uniqueness and unitarity of solutions of (1.1) become considerably complex. We only use the results contained in [1], [2].

For a pair of deterministic unitary evolutions, scattering theory aims to relate the two associated self-adjoint generators by comparing the time-asymptotic behaviours of the evolutions (see e.g. [4]). In analogy, for a pair of q.s.d.e.'s as in (1.1), we compare the asymptotic behaviours of the associated unitary evolutions. This is done in section 2 where we give a general condition which is the quantum analogue of a well known classical condition (see [5] Theorem 3.7 p.535). In section 3 we consider some applications.

2. Scattering Theory

We already have one q.s.d.e. (1.1) whose solution U is an evolution. Let V be another such evolution satisfying q.s.d.e.

$$\begin{cases} dV_t = V_t \left((W - I)d\Lambda(t) + M^* W dA(t) - M dA^\dagger + \left(iK - \frac{1}{2} M^* M \right) dt \right) \\ V_0 = I, \end{cases}$$

$$(2.1)$$

so that the triple (W, M, K) is the generator of V. As in [4], we are led to the following natural hypotheses of scattering theory.

A1 : the q.s.d.e.'s (1.1) and (2.1) have unique unitary solutions,

A2 : the limit s-$\lim_{t \to \infty} V_t U_t^* \equiv \Omega_+$ exists in $h \otimes \mathcal{H}$.

Then we have the following

Theorem 2.1. *The following facts hold:*

(i) *Assume A1. Then $\{V_t S_t\}_{t \geq 0}$ and $\{U_t S_t\}_{t \geq 0}$ are a pair of strongly continuous semigroups of isometries in $h \otimes \mathcal{H}$.*

(ii) *Assume A1 and A2. Then Ω_+ is an isometry and has the intertwining property:*

$$V_t S_t \Omega_+ = \Omega_+ U_t S_t.$$

(iii) *Assume A1 and A2, and let B and A be the generators of the isometric semigroups $V_t S_t$ and $U_t S_t$ respectively. Then Ω_+ maps $D(A)$ into $D(B)$ and $B\Omega_+ = \Omega_+ A$.*

Remark : In the case of $L^2(\mathbb{R})$ in lieu of $L^2(\mathbb{R}_+)$, $\{V_t S_t\}_{t\geq 0}$ and $\{U_t S_t\}_{t\geq 0}$ are strongly continuous group of unitaries. Now we can define two isometries $\Omega_\pm$ as in A2 corresponding to the two strong limits as $t \to \pm\infty$. Then in Theorem 2.1 (ii) and (iii) we shall have two intertwining properties with $\Omega_\pm$ in place of Ω_+.

Proof (of Theorem 2.1): (i) The strong continuity of the solutions of (1.1) and (2.1) are obvious from the q.s.d.e.'s and A1. Recalling that $\{S_t\}_{t\geq 0}$ is a strongly continuous semigroup of isometries, we have by (1.3) that

$$U_t S_t U_s S_s = U_t S_t U_s S_t^* S_t S_s = U_{t+s} S_{t+s},$$

proving the semigroup property.

The intertwining property (ii) follows from A2. Infact, we have by (1.3)

$$
\begin{aligned}
V_t S_t \Omega_+ &= \text{s-} \lim_{s\to\infty} V_t S_t V_s U_s^* \\
&= \text{s-} \lim_{s\to\infty} (V_t S_t V_s S_t^*)(S_t U_s^* S_t^* U_t^*) U_t S_t \\
&= \text{s-} \lim_{s\to\infty} (V_{t+s} U_{t+s}^*) U_t S_t = \Omega_+ U_t S_t.
\end{aligned}
$$

The last part (iii) is an easy consequence of differentiating both sides of (ii) at $t = 0$ and applying Stone's theorem. $\square$

The hypothesis A2 is satisfied under the following sufficient condition which is the quantum analogue of a well known classical condition (see [5] Theorem 3.7 p.535).

Theorem 2.2. *Suppose that there exists a dense subset $\mathcal{D}$ of h such that, for all $u \in h$ and all $t \geq 0$ the vector $U_t^* ue(0)$ belongs to the domain of the operators $L, L^*, L^* L, H, M, K, M^* W, M^* W S^* L$. Then the following conditions*

$$\int_0^\infty \|(S^* L - W^* M) U_s^* ue(0)\|^2 \, ds < +\infty \tag{2.2}$$

$$\int_0^\infty \left\| \left(i(H - K) - \frac{1}{2} L^* L - \frac{1}{2} M^* M + M^* W S^* L \right) U_s^* ue(0) \right\| \, ds < +\infty. \tag{2.3}$$

for all $u \in \mathcal{D}$, imply that the limit s-$\lim_{t\to\infty} V_t U_t^*$ exists in $h \otimes \mathcal{H}$.

Proof. It suffices to show that $\{V_t U_t^* xe(f)\}$ is strongly Cauchy for all x in a suitable dense subset of h and all f in $L^2(\mathbb{R}_+)$ with compact essential support. For all such f, let t_f be a positive real number such that $\text{supp}(f) \subseteq$

$[0, t_f]$. By the cocycle property (1.3), for all $t \geq 0$ and all $x \in h$, we have then

$$V_{t+t_f} U_{t+t_f}^* xe(f) = V_{t_f} S_{t_f} V_t U_t^* S_{t_f}^* U_{t_f}^* xe(f).$$

Note that $S_{t_f}^* U_{t_f}^* xe(f) \in h \otimes e(0)$. Thus it suffices to show that $\{V_t U_t^* ue(0)\}$ is strongly Cauchy for all $u \in \mathcal{D}$. We fix $u \in \mathcal{D}$ and apply Ito's formula to get

$$\|(V_t U_t^* - V_s U_s^*) ue(0)\|^2 \leq 2 \int_s^t \|(S^* L - W^* M) U_r^* ue(0)\|^2 \, dr$$

$$+ 2 \left(\int_s^t \|(i(H - K) - \frac{1}{2} L^* L - \frac{1}{2} M^* M + M^* W S^* L) U_r^* ue(0)\| dr \right)^2$$

Hence the conclusion follows from (2.2) and (2.3). □

3. Applications

In this section we consider a few situations where Ω_+ exists.

(i) *Perturbation of diffusion*

We take $h \equiv L^2(\mathbb{R})$, p the self-adjoint extension of $-id/dx$ in h, $U_t = \exp(-ip\,\omega(t))$ and assume that V_t is the solution of the q.s.d.e.

$$\begin{cases} dV_t = V_t \left(-ip d\omega(t) + (iv - \frac{1}{2} p^2) dt \right) \\ V_0 = I \end{cases} \tag{3.1}$$

where $\omega(t) = i(A^\dagger(t) + A(t))$. It is known [4] that if v is the operator of multiplication by the real function v and if $v \in L^2(\mathbb{R})$, then v is p-bounded and has p^2-bound 0 so that $(-\frac{1}{2} p^2 + iv)$ is the generator of a contraction semigroup, viz. $\mathbb{E} V_t$, where $\mathbb{E}$ denotes the vacuum expectation. The existence and unitarity of the solution of (3.1) follows from the recent work in [1] when e.g. $v \in C^2(\mathbb{R})$ with v' and $v'' \in L^2(\mathbb{R})$. In such a case we have the following result for scattering w.r.t. the pair V and U.

Proposition 3.1. *Let U and V be as above. Assume furthermore that there exists positive constants c, δ, such that $|v(x)| \leq c(1 + x^2)^{-\frac{1}{2} - \delta}$ for all $x \in \mathbb{R}$. Then Ω_+ exist.*

Proof. It suffices to apply the Theorem 2.2. We set

$$\mathcal{D} = \{u \in L^2(\mathbb{R})| \text{ its Fourier transform } \hat{u} \in C_0^\infty(\mathbb{R}\backslash\{0\}) \}.$$

Note that $\mathcal{D}$ is dense in $L^2(\mathbb{R})$ and that if $u \in \mathcal{D}$, then there is a $C_0^\infty(0, \infty)$-function ϕ such that $u = \phi(-\Delta)u$, where Δ is the 1-dimensional Laplacian. The condition (2.2) is obviously satisfied since $L = M$ and $W = S$ and the left-hand side vanishes. Let us check the condition (2.3). For all $u \in \mathcal{D}$ and all $r \geq 0$ we have

$$
\begin{aligned}
\|vU_r^* ue(0)\|^2 &= \int dx\, d\mathbb{P}(\omega)|v(x + \omega(r))u(x)|^2 \\
&= \int |u(x)|^2 (e^{-r\Delta/2}v^2)(x)dx \\
&= (u, (e^{-r\Delta/2}(v^2))u) = (u, \phi(-\Delta)e^{-r\Delta/2}(v^2)u) \\
&= \int (1+x^2)^n |u(x)|^2 [\phi(-\Delta)e^{-r\Delta/2}(v^2)](x)(1+x^2)^{-n}dx \\
&\leq \|(1+x^2)^{-n}\phi(-\Delta)e^{-r\Delta/2}v^2\|\ \|(1+x^2)^{n/2}u\|^2 \\
&\leq C(u)\|(1+x^2)^{-n}\phi(-\Delta)e^{-r\Delta/2}(1+x^2)^{-1-2\delta}\| \\
&\leq C'(u)(1+r)^{-2-4\delta}
\end{aligned}
$$

by using an estimate given in [4] p.534, if we choose $n > 1 + 2\delta$ with some constants $C(u)$ and $C'(u)$ are constants depending on u. This shows also that the condition (2.3) is satisfied. $\square$

(ii) *Pure second quantization*

Let h be any separable Hilbert space and let S and W be two unitary operators in h. Assume furthermore, that U and V satisfy :

$$
\begin{cases} dU_t = U_t(S - I)d\Lambda(t) \\ U_0 = I \end{cases}, \qquad
\begin{cases} dV_t = V_t(W - I)d\Lambda(t) \\ V_0 = I \end{cases}.
$$

Clearly the conditions (2.2) and (2.3) are verified since $L = M = H = K = 0$. Thus Ω_+ exists by Theorem 2.2. $\square$

(iii) *Pure birth processes*

Let $h = l^2(\mathbb{Z})$ with canonical orthonormal basis $(e_k)_{k \in \mathbb{Z}}$. Consider a sequence $(\lambda(k))_{k \in \mathbb{Z}}$ of complex numbers satisfying the condition

$$
0 < |\lambda(k)|^2 \leq c\,(\max\{k, 0\} + 1) \tag{3.2}
$$

for all $k \in \mathbb{Z}$. Let S be the unitary left shift on h defined by $Se_k = e_{k-1}$ and consider the self-adjoint operator N on h defined by

$$
D(N) = \left\{ x \in h \mid \sum_k k^2 |x_k|^2 < +\infty \right\}, \quad Ne_k = ke_k
$$

The q.s.d.e.

$$
\begin{cases} dU_t = U_t \left(\bar{\lambda}(N)S^* dA(t) + (S^* - 1)d\Lambda(t) - \lambda(N)dA^\dagger(t) - \dfrac{1}{2}|\lambda(N)|^2 dt \right) \\ U_0 = I \end{cases}
$$

$$
\tag{3.3}
$$

has a unique unitary solution in virtue of Theorem 3.1 in [2]. Moreover U^* is a unitary solution of the adjoint equation. The associated quantum flow on the *-algebra of bounded functions on $\mathbb{Z}$ describes a pure birth process with birth rates $(|\lambda(k)|^2)_{k\in\mathbb{Z}}$ (see Example 3.3 in [7]).

For all $n \in \mathbb{Z}$ let P_n be the orthogonal projection onto the subspace generated by e_n. Then, for all $k \geq n$, and all $t \geq 0$

$$p_{kn}(t) = \|P_n U_t^* e_k e(0)\|^2$$

represents the probability that the process jumps from the state k at time 0 to the state n at time t. Appling the Ito's formula we easily get the recursion relations

$$p_{kn}(t) = \delta_{kn} + \int_0^t \left(-|\lambda(n)|^2 p_{kn}(s) + |\lambda(n-1)|^2 p_{k\,n-1}(s)\right)\, ds.$$

Clearly $p_{kn}(t) = 0$ for all $k > n$ and all $t \geq 0$. Moreover we can ealsily solve an ordinary differential equation to get $p_{kk}(t) = \exp(-|\lambda(k)|^2 t)$. The following estimate can be easily obtained by induction using the above formula.

Lemma 3.2. *Suppose that there exists $\alpha \in [0,1]$ such that*

$$|\lambda(k)|^2 \leq (\max\{k,0\}+1)^\alpha$$

for all $k \in \mathbb{Z}$. Let $\varepsilon(k,n) = \min_{k\leq j\leq n} |\lambda(j)|^2$. The following inequality holds

$$p_{kn}(t) \leq \exp(-\varepsilon(k,n)t)(ct)^{n-k}\left((n-k)!\right)^{\alpha-1}.$$

Let η (resp. ζ) be a sequence of real (resp. complex) numbers satisfying the growth condition (3.2). Consider the selfadjoint operator on h

$$K = \eta(N) + \bar\zeta(N)S^* + S\zeta(N).$$

The q.s.d.e.

$$\begin{cases} dV_t = V_t\left(\bar\lambda(N)S^* dA(t) + (S^*-1)d\Lambda(t)\right. \\ \qquad\left. - \lambda(N)dA^\dagger(t) + \left(iK - \dfrac{1}{2}|\lambda(N)|^2 dt\right)\right) \\ V_0 = I \end{cases} \qquad (3.4)$$

has a unique unitary solution by Theorem 3.1 in [2].

Proposition 3.3. *Let U and V be as above. Then Ω_+ exists in each one of the following cases:*
i) $\alpha = 1$ and η, ζ have finite support,
ii) $0 \leq \alpha < 1$, $\inf_{n \geq k} \varepsilon(k, n) = \varepsilon_k > 0$ for all $k \in \mathbb{Z}$, and there exists positive constants c_1, β, with $\beta < (1 - \alpha)/2$ such that

$$|\zeta(n)| + |\eta(n)| \leq c_1^n (n!)^\beta.$$

Proof. Let $\mathcal{D}$ be the linear manifold generated by vectors e_k. The same proof of Propositions 4.3, 4.4 in [2] allows to show that the domain assumptions of Theorem 2.2 are satisfied. Clearly the condition (2.2) is also satisfied. To check the condition (2.3) it suffices to show that, for all $k \in \mathbb{Z}$ we have

$$\int_0^\infty \left(\|\eta(N)U_s^* e_k e(0)\| + \|\bar{\zeta}(N+1)U_s^* e_k e(0)\| \right.$$
$$\left. + \|\zeta(N-1)U_s^* e_k e(0)\| \right) ds < +\infty.$$

For all $s \geq 0$ we have

$$\|\eta(N)U_s^* e_k e(0)\| \leq \sum_{n \geq k} |\eta(n)| \, \|P_n U_s^* e_k e(0)\|.$$

The other terms can be estimated in a similar way and the proof can be completed combining the estimate of Lemma 3.2 and the growth conditions on η, ζ. $\square$

(iv) *Ornstein-Uhlenbeck processes and pure death processes*

Let h, S and $(e_k)_{k \in \mathbb{Z}}$ be as in iii) and let $a^\dagger$, a denote the creation and annihilation operators defined by

$$a^\dagger e_k = \begin{cases} (k+1)^{1/2} e_{k+1} & \text{if } k \geq 0 \\ 0 & \text{if } k < 0 \end{cases}, \qquad a e_k = \begin{cases} k^{1/2} e_{k-1} & \text{if } k \geq 0 \\ 0 & \text{if } k < 0 \end{cases}.$$

Let U be the unitary cocycle corresponding to a quantum Ornstein-Uhlenbeck process which satisfies the q.s.d.e.

$$\begin{cases} dU_t = U_t \left(a^\dagger dA(t) - a dA^\dagger(t) - \dfrac{1}{2} a^\dagger a \, dt \right) \\ U_0 = I \end{cases}$$

and let V be the unitary cocycle corresponding to a pure death process (see Example 3.3 in [7]) which satisfies the q.s.d.e.

$$\begin{cases} dV_t = V_t \left(a^\dagger dA(t) + (S^* - 1) d\Lambda(t) - S^* a dA^\dagger(t) - \dfrac{1}{2} a^\dagger a \, dt \right) \\ V_0 = I \end{cases}$$

The domain assumptions in Theorem 2.2 can be proved as in [2] Propositions 4.3, 4.4. In this case also Ω_+ exists because the left-hand side of (2.2) and (2.3) vanishes.

Aknowledgement. This paper was written while K.B.S. was visiting the Mathematics Department of the University of Trento in December 1991. He would like to express his gratitude to the host institution.

References

[1] Bhat, R., Sinha, K.B.: Unitarity of solutions of a class of quantum stochastic differential equations, Indian Statistical Institute, Delhi Centre. Preprint 1992.

[2] Fagnola, F.: On quantum stochastic differential equations with unbounded coefficients. *Probab. Th. Rel. Fields*, **86**, 501–516 (1990).

[3] Hudson, R.L., Parthasarathy, K.R.: Quantum Ito's formula and stochastic evolutions, *Comm. Math. Phys.* **93**, 301-323 (1984).

[4] Jauch, J.M., Amrein, W.O., Sinha, K.B.: *Scattering Theory in Quantum Mechanics*. W.A. Benjamin, Reading Mass., 1977.

[5] Kato, T.: *Perturbation Theory for Linear Operators*. Springer-Verlag, Berlin Heidelberg New York 1976.

[6] Parthasarathy, K.R.: *An introduction to Quantum stochastic calculus*. Birkhauser-Verlag, Basel 1992.

[7] Parthasarathy, K.R., Sinha K.B.: Markov chains as Evans-Hudson diffusion in Fock space. In: Azéma, J., Yor, M. (eds.) Séminaire de Probabilités XXIV 1988/89. (Lect. Notes Math., vol. 1426, pp. 362-369). Berlin, Heidelberg, New York: Springer 1989.

Franco Fagnola   Kalyan B. Sinha
University of Trento   Indian Statistical Institute
Department of Mathematics   Delhi Centre
I - 38050 Povo (TN) - Italy   New Delhi - 110016 - India

Sur les Variations des Fonctions Aléatoires Gaussiennes

X. FERNIQUE

Abstract : For vector-valued gaussian random functions, we develop, without separability assumptions, properties of the oscillations and we analyse their asymptotic behaviour.

0. Introduction.

0.1   Soit X une fonction aléatoire gaussienne sur R à valeurs dans R ou dans un espace de Banach séparable (E, ‖.‖) ; on note E' le dual de E et E'$_1$ la boule unité de ce dual. On se propose d'étudier le comportement des trajectoires de X dans les deux directions suivantes : (a) comportement local ou uniforme à partir de la notion d'oscillation, (b) comportement asymptotique.

La première partie vise à étendre au domaine vectoriel l'utilisation de la notion d'oscillation introduite dans le domaine réel par Ito et Nisio [8], développée ensuite par Jaïn et Kallianpur [9] ; une telle extension a été tentée précédemment dans [3], on montrera ici que cette tentative était maladroite et inefficace ; elle a été aussi réalisée dans [1] pour les seules fonctions aléatoires gaussiennes séparables ; cette hypothèse de séparabilité, sans conséquence dans le domaine réel, rend le résultat de l'étude à peu près inutilisable dans le domaine vectoriel (cf [7], proposition 5.3) ; on ne supposera pas ici une telle séparabilité. Dans l'étude réalisée, on montrera en particulier (corollaire 1.4) comment la continuité des trajectoires de X pour certaine topologie peut être analysée, indépendamment de toute hypothèse de continuité uniforme en probabilité, à partir de son oscillation associée à une métrique adaptée.

Dans la seconde partie de l'étude, supposant X stationnaire à trajectoires continues, on analyse le comportement de ses trajectoires à l'infini. L'étude s'applique en particulier à résoudre le problème suivant (théorème 2.3) : Analyser le comportement vectoriel de X(t), t →∞, à partir des différents comportements réels des ⟨X(t), y⟩, t →∞, quand y ∈ E'.

1.Sur les oscillations des fonctions aléatoires gaussiennes à valeurs vectorielles.

1.1   Dans un travail précédent ([3]) et en vue d'analyser les trajectoires des fonctions aléatoires gaussiennes à valeurs vectorielles, on a introduit des notions d'oscillation vectorielle : Soit (T, δ) un espace pseudo-métrique séparable et soit X une fonction aléatoire gaussienne sur T *à valeurs dans un espace de Banach séparable* E ; on suppose que X est (localement) uniformément continue en probabilité et on fixe une partie dénombrable dense S de T ; pour tout $\omega \in \Omega$ et tout $t \in T$, on pose

1.1.1 $V(\omega, X, t) = \bigcap_{u>0} \overline{\left\{ X(\omega, s) - X(\omega, t) ; \delta(t, s) \leq u ; s \in S \right\}}$;

on sait alors ([3], théorème 2.2) que V est non aléatoire au sens suivant :

1.1.2   Il existe une application (non aléatoire) v de T dans l'ensemble des parties fermées de E et pour tout élément t de T une partie presque sûre Ω_t de Ω telles que :
$$\forall \, t \in T, \forall \, \omega \in \Omega_t , V(\omega, X, t) = v(t).$$
   On sait aussi que si les trajectoires de X possèdent des propriétés locales de compacité, la fonctionnelle V est bien régulière ; mais dans ce cas, on a montré ultérieurement ([4], théorème 4.2.1) que V est un outil inutile pour l'analyse de ces trajectoires. La proposition ci-dessous montrera que dans la situation stationnaire et si X a une modification à trajectoires localement bornées, ce même outil est inefficace.

Proposition 1.1.3 : *Soit* X *une fonction aléatoire gaussienne stationnaire (localement) uniformément continue en probabilité sur* $\mathbf{R}^d$ *à valeurs dans un espace de Banach séparable* E *; on suppose que* X *possède une modification à trajectoires localement bornées ; alors pour tout* $t \in \mathbf{R}^d$, *on a :* $v(t) = \{0\}$.

Démonstration :   On vérifie immédiatement du fait de la stationnarité que $v(t) = v$ est indépendant de t et du fait de la continuité en probabilité que v contient l'origine ; v est d'ailleurs une partie bornée de E. Soit alors $x \in v$; il existe un ensemble presque sûr Ω_0 tel que pour tout $\omega \in \Omega_0$, tout $q \in S$ et tout $\varepsilon > 0$, il existe un élément s de S tel que :
$$\delta(s, q) < \varepsilon/2, \ \| X(s) - X(q) - x \| < \varepsilon/2,$$
et donc aussi un élément s' de S tel que :
$$\delta(s', s) < \varepsilon/2, \ \| X(s') - X(s) - x \| < \varepsilon/2 \ ;$$
on aura alors aussi :
$$\delta(s', q) < \varepsilon, \ \| X(s') - X(q) - 2 \, x \| < \varepsilon,$$
ceci signifie que $2x \in v$; puisque v est borné, on en déduit que $v = \{0\}$, c'est le résultat.

Remarque 1.1.4 :   La proposition s'appliquera par exemple dans $E = c_0$ à la fonction aléatoire très singulière X analysée dans l'exemple 3.1 de [3] ayant

presque toutes ses trajectoires bornées et non continues, même pour la topologie
affaiblie ; on peut d'ailleurs dans ce cas justifier directement sa conclusion à
partir de l'existence d'une modification X' de X continue dans l_∞ pour la
topologie affaiblie. On voit bien dans cette situation que le fait que l'oscillation
de X se réduise à {0}, signifie non pas que la trajectoire s'accumule en X(t) quand
s tend vers t, mais que l'espace E est assez grand pour que la trajectoire se
disperse dans une partie bornée sans s'y accumuler ailleurs qu'en X(t), ce qui rend
l'oscillation inefficace pour l'analyse.

1.2   On peut pourtant utiliser efficacement des notions d'oscillation numérique.
Soit (T, δ) un espace pseudo-métrique séparable et soit X une fonction aléatoire
gaussienne sur T à valeurs dans un espace vectoriel lusinien quasi-complet E ;
on suppose que X est uniformément continue en probabilité et on fixe une partie
dénombrable dense S de T ; soit enfin N une semi-norme continue sur E. Dans
ces conditions :

Théorème 1.2 :   *Il existe un ensemble presque sûr Ω_0 et une application non
aléatoire w de* T *dans* $\mathbf{R}^+$ *tels que pour tout:$\omega \in \Omega_0$ et tout* $t \in$ T,

$$\lim_{\varepsilon \to 0} [\sup \Big\{ N[X(\omega, s) - X(\omega, s')] ; \delta(s, t) \vee \delta(s', t) \leq \varepsilon, (s, s') \subset S \Big\}] = w(t) ,$$

$$w(t) = \lim_{\varepsilon \to 0} E \sup \Big\{ N[X(s) - X(s')] ; \delta(s, t) \vee \delta(s', t) \leq \varepsilon, (s, s') \subset S \Big\}.$$

Démonstration :   Puisque l'espace E^S est lusinien et quasi-complet, l'espace
auto-reproduisant H du vecteur gaussien $\{X(s), s \in S\}$ à valeurs dans E^S est séparable ; nous notons (h_n) une base orthonormale de H et (λ_n) la suite de variables
aléatoires gaussiennes associées. Puisque X est uniformément continue en probabilité, les h_n sont des applications uniformément continues de (S, δ) dans
(E, N). Pour toute fonction f définie sur S à valeurs dans E, tout $u > 0$ et tout
$t \in$ T, nous posons :

$$V(f,t,u) = \lim_{\varepsilon \to 0} \sup \Big\{ N[f(s) - f(s')], \delta(t,s) \vee \delta(t,s') \leq u, \delta(s,s') \leq \varepsilon, (s,s') \subset S \Big\},$$

$$V(f, t) = \lim_{\varepsilon \to 0} \sup \Big\{ N[f(s) - f(s')], \delta(t, s) \vee \delta(t, s') \leq \varepsilon, (s, s') \subset S \Big\};$$

on a donc : $V(f, t) = \lim_{u \to 0} V(f, t, u).$
Il existe alors un ensemble presque sûr Ω_0 tel que :

$$\forall \ \omega \in \Omega_0, \forall s \in S, X(\omega, s) = \sum_{n=0}^{\infty} \lambda_n(\omega) \ h_n(s) ;$$

on pose alors : $\forall \ \omega \in \Omega_0, \ \forall \ s \in S, \ \forall \ k \in \mathbf{N}, \ R_k(\omega, s) = \sum_{n=k}^{\infty} \lambda_n(\omega) \ h_n(s)$

et on a immédiatement du fait de l'uniforme continuité des h_n :
    $\forall \ \omega \in \Omega_0, \ \forall \ t \in T, \ \forall \ u > 0, \ \forall \ k \in \mathbf{N}, \ V(X(\omega), t, u) = V(R_k(\omega), t, u) ;$
il en résulte que la variable aléatoire : $\omega \to V(X(\omega), t, u)$ est une variable aléa-
toire terminale pour la suite indépendante (λ_n), c'est donc une variable aléatoire

dégénérée ; le théorème de convergence monotone fixe d'ailleurs sa valeur presque sûre : il existe un ensemble presque sûr Ω_1 tel que :
$$\forall \; \omega \in \Omega_1, \; \forall \; t \in S, \; \forall \; u \in Q^+, \; V(x(\omega), t, u) = v(t, u)$$
où v(t,u) est égal à :
$$\lim_{\varepsilon \to 0} E[\sup \left\{ N[X(s) - X(s')] \; ; \; \delta(s,t) \vee \delta(s',t) \leq u, \; \delta(s,s') \leq \varepsilon, \; (s,s') \subset S \right\}] .$$
L'implication évidente :
$$\delta(t, s) \leq h \;\; \Rightarrow \;\; V(f, t, u) \leq V(f, s, u+h), \; v(t, u) \leq v(s, u+h)$$
et la densité de S montrent alors que :
$$\forall \; \omega \in \Omega_1, \; \forall \; u > 0, \; \forall \; t \in T, \;\; v(t, u^-) \leq V(X(\omega), t, u) \leq v(t, u^+) \; ;$$
on a donc :
$$\forall \; \omega \in \Omega_1, \; \forall \; t \in T, \;\; V(X(\omega), t) = \lim_{u \to 0} v(t, u), \text{ c'est le résultat du théorème.}$$
Du théorème précédent, on déduit alors :

Corollaire 1.3 : *Dans les mêmes conditions, on suppose que* (T, δ) *est compact et que :*
$$\sup_{t \in T} \lim_{\varepsilon \to 0} E \sup \{N[X(t) - X(s)], \delta(s, t) \leq \varepsilon, s \in S \} = 0 \; ;$$
on a alors aussi :
$$\lim_{\varepsilon \to 0} E \sup \{N[X(t) - X(s)], \delta(s, t) \leq \varepsilon, (s, t) \subset S \} = 0 .$$

Corollaire 1.4 : *Soit X une fonction aléatoire gaussienne sur un espace topologique T contenant une partie dénombrable et dense S à valeurs dans un espace de Fréchet séparable E; on suppose que X est continue en probabilité ; on suppose aussi qu'elle est p.s. continue en tout point de T au sens suivant :*
$$\forall \; t \in T, \; P\{ \lim_{s \to t} \{ X(s) - X(t) , s \in S\} = 0\} = 1 .$$
Dans ces conditions, X a une modification à trajectoires continues.

Démonstration du corollaire 1.4 : Soit (N_k) une suite de semi-normes définissant la topologie de E ; nous définissons une distance bornée d sur E en posant : $d(x, y) = \sum_{k=0}^{\infty} [1 \wedge N_k(x-y)].2^{-k}$; nous lui associons la pseudo-distance δ sur T définie par : $\delta(s, t) = E \, d(X(s), X(t))$; on peut alors appliquer le théorème 1.2 à la fonction aléatoire X qui est uniformément continue en probabilité sur l'espace pseudo-métrique séparable (T, δ) ; la conclusion du corollaire en résulte.

2. Comportement asymptotique des fonctions aléatoires gaussiennes stationnaires.

On suppose dans ce paragraphe que X est stationnaire et à trajectoires continues et on étudie le comportement de X(t) , $t \to \infty$. On se base pour cela (a) sur les résultats de [2], § 3, qui caractérisent les différents comportements de X à partir de paramètres intrinsèques et (b) sur ceux de [5] qui permettent de relier les comportements de X à ceux des $\langle X, y \rangle$, $y \in E'$ et qu'on rappelle ci-dessous :

Lemme 2.0 : *Il existe une constante absolue K telle que* :

$\mathbf{E} \sup_{t \in [0, 1]} \| X(t) \| \leq K [\mathbf{E} \| X(0) \| + \sup_{y \in E'_1} \mathbf{E} \sup_{t \in [0, 1]} \langle X(t), y \rangle]$.

Dans toute la suite, on réservera la lettre K à la constante de ce lemme.

Les résultats de [2] fournissent ici directement :

Théorème 2.1 : *Les trajectoires de X sont p.s. bornées sur* $\mathbf{R}^+$ *si et seulement si la fonction* F = F_X *définie par* F(T) = $\mathbf{E} \sup_{t \in [0,T]} \|X(t)\|$ *est bornée ; dans le cas contraire, on a* :

$$\lim \sup_{t \to \infty} \frac{\|X(t)\|}{F(t)} = 1 \text{ p.s.}$$

Supposant F non bornée, on peut alors préciser :

Corollaire 2.2 : *Pour toute fonction positive croissante f sur* $\mathbf{R}^+$, *on a* :

2.2.1 $\qquad \lim \sup_{t \to \infty} \frac{\|X(t)\|}{f(t)} = \lim \sup_{t \to \infty} \frac{F(t)}{f(t)}$ p.s.

Remarque : L'égalité (2.2.1) est un peu inattendue si le rapport F/f n'a pas de limite à l'infini. Notons en effet L son second membre ; le fait que que $\lim \sup_{t \to \infty} \frac{\|X(t)\|}{f(t)}$ soit presque sûrement inférieur ou égal à L est immédiat et ceci suffit pour établir l'égalité si L est nulle. Mais la force du corollaire réside dans l'inégalité inverse quand L est positive, que F/f n'a pas de limite et donc que f n'est pas bornée.

Démonstration du corollaire : Nous supposons que F/f n'a pas de limite et donc que L est positive, f est alors non bornée ; fixons m < L, il existe donc une suite (t_n, n∈N), tendant en croissant vers l'infini, telle que pour tout n∈N, F(t_n) > m.f(t_n) ; les inégalités de Borell montrent alors que la médiane $\mu(t_n)$ = méd[$\sup_{[0,t_n]} \|X(t)\|$] vérifie pour tout n∈N:

$$\mu(t_n) > F(t_n) - \sup_{y \in E'_1} [E\langle X(0), y \rangle^2]^{1/2},$$

on a donc :

$$\mathbf{P}\{\sup_{[0,t_n]} \|X(t)\| > m.f(t_n) - \sup_{y \in E'_1} [E\langle X(0), y \rangle^2]^{1/2}\} \geq 1/2,$$

les lois zéro-un applicables ici puisque F et f sont non bornées montrent alors que :

$$\lim \sup_{T \to \infty} \frac{\sup_{[0,T]} \|X(t)\|}{f(T)} \geq m \text{ p.s.}$$

et le résultat s'ensuit par un schéma simple.

Théorème 2.3 : *Soit f une fonction croissante positive sur* $\mathbf{R}^+$ *; on suppose que* F = F_X *n'est pas bornée. Dans ces conditions, pour que X/f ait p.s. des* *trajectoires bornées, il faut et il suffit que pour tout* $y \in E'_1$, $\langle X, y \rangle / f$ *ait p.s. des trajectoires bornées. On a plus précisément :*

2.3.1 $\lim \sup_{t \to \infty} \dfrac{\|X(t)\|}{f(t)} = \lambda . \lim \sup_{t \to \infty} \left\{ \sup_{y \in E'_1} \dfrac{F\langle X, y \rangle(t)}{f(t)} \right\} < \infty$ p.s.

où $\lambda = \lambda(X, f)$ *est un élément non aléatoire de l'intervalle* $[1, K]$.

Démonstration : Nous démontrons la première partie de l'énoncé, la seconde utiliserait le même schéma de preuve. Pour que X/f ait presque toutes ses trajectoires bornées, il faut et il suffit (théorème 2.1) que la fonction F_X/f soit bornée ; il faut et il suffit pour cela (lemme 2.0) que $\sup_{y \in E'_1} F\langle X, y \rangle /f$ le soit, c'est-à-dire (théorème de Baire) pour tout $y \in E'_1$, que $F\langle X, y \rangle/f$ soit bornée et donc (théorème 2.1) que $\langle X, y \rangle/f$ ait presque toutes ses trajectoires bornées, c'est la première partie du théorème.

Remarque : L'égalité 2.3.1 n'est pas parfaitement satisfaisante du fait de l'indétermination de λ dans l'intervalle $[1, K]$; les techniques employées pour prouver le lemme 0.2 ne semblent pas permettre une meilleure évaluation qui pourrait par contre résulter de la conjecture suivante :

Conjecture 2.4 : Pour tout $\varepsilon > 0$, il existe une constante $K' = K'(\varepsilon)$ telle que pour tout entier n et toute fonction aléatoire gaussienne stationnaire X sur **R** à valeurs dans $\mathbf{R}^n$ à trajectoires continues, on ait :

$$E \sup_{t \in [0, 1]} \sup_{1 \leq k \leq n} |X_k(t)| \leq K'(\varepsilon) E \sup_{1 \leq k \leq n} |X_k(0)| +$$
$$+ (1+\varepsilon) \sup_{1 \leq k \leq n} E \sup_{t \in [0, 1]} X_k(t).$$

2.5   Exemple : Soient $p \geq 1$ et $X = (X_n, n \in N)$ une fonction aléatoire gaussienne stationnaire d'Ornstein-Uhlenbeck sur **R** à valeurs dans $E = l_p$, à composantes indépendantes et continues de sorte que pour tout entier n, $X_n(t)$ s'écrive $a_n x_n(b_n t)$, où les x_n sont des fonctions aléatoires gaussiennes stationnaires réelles indépendantes et de mêmes lois ayant toutes la covariance $E x_n(s) x_n(t) = \exp(-|s - t|/2)$ et où les a_n, b_n vérifient les propriétés caractéristiques énoncées dans [6]. Dans ces conditions, un emploi direct des inégalités de Borell fournit :

2.5.1 $\lim \sup_{t \to \infty} \dfrac{\|X(t)\|}{[\sup_{y \in E'_1} \Sigma \{ a_k^2 y_k^2 I_{b_k > 0} \} 2 \log t]^{1/2}} \leq 1$ p.s.

Par ailleurs dans la même situation, posant pour tout $\varepsilon > 0$, $\sigma^2(\varepsilon) = \Sigma \{ a_k^2 y_k^2 I_{b_k > \varepsilon} \}$, fixant $T > 0$ et l'entier $N = cT$ supérieur à T, l'application des inégalités de Slépian fournit pour tout $y \in E'$:

$$\text{méd}[\sup_{[0, T]} \langle X(t), y \rangle] \geq \sqrt{\Sigma \{ a_n^2 y_n^2 [1 - \exp(-b_n^2 T^2/2N^2)] \}} \ \Phi^{-1}[\frac{\log 2}{N+1}],$$

où Φ a la signification gaussienne usuelle.
Du théorème 2.3, on déduit alors :

$$1 = \lim \sup_{T \to \infty} \dfrac{\|X(T)\|}{E \sup_{[0, T]} \|X(t)\|} \leq$$

$$\leq \lim \sup_{t\to\infty} \frac{\|X(T)\|}{[\sigma^2(\varepsilon)[1-\exp(-\varepsilon^2/2c^2)]\, 2 \log c \, T]^{1/2}} \quad \text{p.s.}$$

En faisant tendre dans l'ordre T vers l'infini, c vers l'infini et ε vers zéro, on en déduit :

$$1 = \lim \sup_{T\to\infty} \frac{\|X(T)\|}{E \sup_{[0,\,T]} \|X(t)\|} \leq$$

$$\leq \lim \sup_{T\to\infty} \frac{\|X(T)\|}{[\sup_{y\in E'_1} \Sigma\{a_k^2 y_k^2 I_{b_k>0}\}\, 2 \log t]^{1/2}}$$

et donc finalement en comparant avec (2.5.1)

$$\lim \sup_{T\to\infty} \frac{\|X(T)\|}{[\sup_{y\in E'_1} \Sigma\{a_k^2 y_k^2 I_{b_k>0}\}\, 2 \log t]^{1/2}} = 1 \text{ p.s. ;}$$

en distinguant suivant les valeurs de p, on obtient :

$$\lim \sup_{T\to\infty} \frac{\|X(T)\|}{[\Sigma\{|a_k|^r I_{b_k>0}\}^{1/r}]\, [2 \log t]^{1/2}} = 1 \text{ p.s., } r = 2p/(2-p), \text{ si } p<2,$$

$$\lim \sup_{T\to\infty} \frac{\|X(T)\|}{\sup\{|a_k|\, I_{b_k>0}\}[2 \log t]^{1/2}} = 1 \text{ p.s., si } p \geq 2.$$

Ce résultat est compatible avec la conjecture 2.4.

Références.

[1]   Buldygin V.V. and Solnstev S.A. Equivalence of sample and sequential continuity of Gaussian processes and the continuity of Gaussian Markov processes, Theory Probab. Appl., 33, (1988), 624-637.

[2]   Fernique X. Gaussian random vectors and their reproducing kernel Hilbert spaces, Technical Report Series of the Laboratory for Research in Statistics and Probability, Université d'Ottawa, 34, (1985).

[3]   Fernique X. Oscillations de fonctions aléatoires gaussiennes à valeurs vectorielles, Math. Scand., 60, (1987), 96-108.

[4]   Fernique X. Fonctions aléatoires à valeurs dans les espaces lusiniens, Expositiones Math., 8, (1990), 289-364.

[5]   Fernique X. Régularité de fonctions aléatoires gaussiennes à valeurs vectorielles, Ann. Probab., 18, (1990), 1739-1745.

[6]   Fernique X. Sur la régularité des fonctions aléatoires d'Ornstein-Uhlenbeck à valeurs dans l_p, $p \in [1,\infty]$, Ann. Probab., à paraître.

[8]   Fernique X. Analyse de fonctions aléatoires gaussiennes stationnaires à valeurs vectorielles, Technical Report 331, Center of Stochastic Processes, University of N.C.at Chapel Hill, 1991.

[9]   Ito K. and Nisio M. On the oscillation functions of Gaussian processes, Math. Scand., 22, (1968), 209-223.

[10]   Jaïn N.C. and Kallianpur G. Norm convergent expansions for Gaussian processes in Banach spaces, Proc. Amer. Math. Soc., 25, (1970), 890-895.

Institut de Recherche Mathématique Avancée, Université Louis Pasteur,
7 rue René Descartes, 67084 Strasbourg Cédex, France.

Random Allocation Methods
in an Epidemic Model

J. GANI[1]

Abstract. A random allocation model is developed for the numbers of susceptibles in a group of n IVDU's who become infected by i HIV-infectives sharing needles.
A comparison of this model and the classical Reed-Frost model is outlined.

1. Introduction

An important method of HIV transmission is through needle sharing
among intravenous drug users (IVDUs). In the USA, the number of AIDS
cases due to IVDU transmission was 20% larger in 1989 than 1988, and these
represented (heterosexual IVDUs) 20% of all AIDS cases in 1989. In Europe
heterosexual IVDUs were 28% of all AIDS cases (see Cohen *et al.* (1990)).
In Australia, Wolk *et al.* (1990) have described the behaviour of IVDUs in
Sydney; they have recorded a seropositive rate of $\alpha \sim 9\%$, and estimated the
probability of needle sharing as $p = 0.8$. The probability of HIV transmission
through an infected needle is estimated approximately as $\beta = 0.35$. To make
predictions about the future spread of HIV among IVDUs, it is useful to
develop a model; we can then modify its parameters to see the effects of
various preventive measures.

One such model is the random allocation model in which i infectives
make contact with n susceptibles, as described in Gani, Heathcote and
Nicholls (1991), Gani (1991) and Gani and Yakowitz (1991).

We can visualize this as i balls (infectives) being placed in n boxes
(susceptibles) where a box may contain 0, 1, ..., i infectives, as in Figure 1.

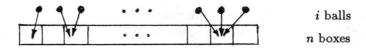

i balls

n boxes

Figure 1. Random allocation of i infectives to n susceptibles.

[1] Research carried out with the support of NIH grant R01 AI29426.

We then have

s_0 empty boxes,

s_1 boxes with 1 ball,

s_2 boxes with 2 balls,

$\cdots \qquad \cdots$

s_i boxes with i balls ($s_i = 0$ or 1).

We note that

$$\sum_{j=0}^{i} s_j = n,$$

and

$$\sum_{j=0}^{i} j s_j = i.$$

What we should like to find is the probability

$$p_{s_0 s_1 \cdots s_i}(i, n) = P\{s_0, s_1, \ldots, s_i \mid \sum_{j=0}^{i} s_j = n, \sum_{j=0}^{i} j s_j = i\}.$$

We shall show how it can be derived.

2. Sure infection by 1 or more contacts

Let us write

$$p_{s_0 s}(i, n) = P\{s_0, s_1 + s_2 + \cdots + s_i = s \mid \sum_{j=0}^{i} s_j = n, \sum_{j=0}^{i} j s_j = i\},$$

and assume that infection will occur for s susceptibles, each having 1 or more contacts with infectives. Note that $s_0 + s = n$, so that

$$p_s(i, n) = P\{s_1 + s_2 + \cdots + s_i = s \mid \sum_{j=0}^{i} s_j = n, \sum_{j=0}^{i} j s_j = i\}$$

will give us all the necessary information.

It is possible to find p_s by classical methods (see Johnson and Kotz (1977)). But it proves more interesting to consider the recurrence relation when the number of infectives increases from i to $i + 1$, as follows:

$$(2.1) \qquad p_s(i + 1, n) = p_{s-1}(i, n)\left(1 - \frac{s - 1}{n}\right) + p_s(i, n)\frac{s}{n}.$$

If we now define the probability generating function (p.g.f.)

$$\phi_{in}(\theta) = \sum_{s=0}^{\min(i,n)} p_s(i,n)\theta^s \qquad |\theta| < 1,$$

then we can derive the differential-difference equation

(2.2) $$\phi_{i+1,n}(\theta) = \frac{\theta}{n}(1-\theta)\frac{d\phi_{in}}{d\theta} + \theta\phi_{in},$$

from which, starting with $\phi_{0,n}(\theta) = 1$, we find

(2.3)
$$\phi_{1n}(\theta) = \theta$$
$$\phi_{2n}(\theta) = \theta^2(1-\frac{1}{n}) + \frac{\theta}{n}$$
$$\phi_{3n}(\theta) = \theta^3(1-\frac{1}{n})(1-\frac{2}{n}) + \frac{3\theta^2}{n}(1-\frac{1}{n}) + \frac{\theta}{n^2}$$
$$\cdots \qquad \cdots \qquad \cdots$$

Thus, for any given i, n, we can obtain the probabilities

$$p_s(i,n) = P\{s \text{ susceptibles are contacted by } i \text{ infectives}\}$$
$$s = 0, 1, \ldots, \min(i, n).$$

We can do a little more mathematics here by defining the generating function

$$\psi_n(u,\theta) = \sum_{i=0}^{\infty} \phi_{in}(\theta)u^i \qquad |u| < 1,$$

which can be shown from the equation (2.2) to satisfy the partial differential equation

(2.4) $$\frac{\theta}{n}(1-\theta)\frac{\partial\psi_n}{\partial\theta} + (\theta - \frac{1}{u})\psi_n + \frac{1}{u} = 0.$$

Using integrating factors, we can solve this to obtain

(2.5) $$\psi_n(u,\theta) = \sum_{j=0}^{n}\binom{n}{j}\frac{1}{1-(1-\frac{j}{n})u}(1-\theta)^j\theta^{n-j} + f(u)\theta^{\frac{n}{u}}(1-\theta)^{n(1-\frac{1}{u})},$$

where $f(u)$ is the integrating constant with respect to θ. Since for $0 < u < 1$, the coefficient of $f(u) \to \infty$ when $\theta \to 1$, then $f(u) \equiv 0$. We now have a technique for deriving the $p_s(i,n)$, and thus determining the number of new infectives after a needle sharing session, assuming infection follows 1 or more infectious contacts with probability 1. For further details, the reader is referred to Gani (1991).

3. Keeping track of the number of contacts

Suppose now that β (= 0.35, say) is the probability that HIV is transmitted through an infected needle making a single contact with a susceptible. Then if there are k infectious contacts, through k infected needles, the probability of HIV transmission is

$$\beta_k = 1 - (1 - \beta)^k.$$

Let us now denote by

$$\phi_{in}(\alpha_0, \alpha_1, \ldots, \alpha_i) = \sum_{s_0 \cdots s_i} p_{s_0 \cdots s_i}(i, n) \alpha_0^{s_0} \cdots \alpha_i^{s_i}$$

the p.g.f. of the probabilities

$$p_{s_0 \cdots s_i}(i, n) = P\{s_0, \ldots, s_i \mid \sum_{j=0}^{i} s_j = n, \sum_{j=0}^{i} j s_j = i\},$$

using the same notation ϕ_{in} as in Section 2, but indicating by the $\alpha_0, \ldots, \alpha_i$ that we are now keeping track of the numbers of susceptibles with $0, 1, \ldots, i$ infective contacts. Then we can readily see that the p.g.f. of the number of new infectives will be

$$\phi_{in}(1, \beta\theta + (1-\beta), \beta_2\theta + (1-\beta_2), \ldots, \beta_i\theta + (1-\beta_i))$$
$$= \sum_{s_0, \ldots, s_i} p_{s_0 \cdots s_i}(i, n)\{\beta\theta + (1-\beta)\}^{s_1} \cdots \{\beta_i\theta + (1-\beta_i)\}^{s_i} = \tilde{f}_{in}(\theta).$$

We now show how $\phi_{in}(\alpha_0, \alpha_1, \ldots, \alpha_i)$ can be found.

Our starting point is again a recurrence relation similar to (2.1); increasing the number of infectives from i to $i+1$,

$$p_{s_0 \cdots s_{i+1}}(i+1, n) = p_{s_0+1, s_1-1, \ldots, s_i}(i, n)\frac{s_0 + 1}{n}$$

(3.1)
$$+ p_{s_0, s_1+1, s_2-1, \ldots, s_i}(i, n)\frac{s_1 + 1}{n} + \cdots + p_{s_0, \ldots, s_i+1}(i, n)\frac{s_i + 1}{n},$$

which leads to

(3.2)
$$\phi_{i+1, n}(\alpha_0, \ldots, \alpha_{i+1}) = \frac{1}{n}\{\alpha_1 \frac{\partial \phi_{in}}{\partial \alpha_0} + \cdots + \alpha_{i+1} \frac{\partial \phi_{in}}{\partial \alpha_i}\}.$$

Starting from $\phi_{0n}(\alpha_0) = \alpha_0^n$, we can readily derive

$$(3.3) \quad \phi_{1n}(\alpha_0, \alpha_1) = \alpha_0^{n-1}\alpha_1$$

$$\phi_{2n}(\alpha_0, \alpha_1, \alpha_2) = \frac{1}{n}\{\alpha_0^{n-1}\alpha_2 + \alpha_0^{n-2}\alpha_1^2(n-1)\}$$

$$\phi_{3n}(\alpha_0, \alpha_1, \alpha_2, \alpha_3) = \frac{1}{n^2}\{\alpha_0^{n-1}\alpha_3 + 3\alpha_0^{n-2}\alpha_1\alpha_2(n-1) + \alpha_0^{n-3}\alpha_1^3(n-1)(n-2)\}$$

$$\cdots \quad \cdots \quad \cdots$$

Thus, the p.g.f. of the number of new infectives, where the probability of infection varies with the number of contacts, when for example there are $i = 3$ infectives and n susceptibles, will be

$$\tilde{f}_{3n}(\theta) = \frac{1}{n^2}\{\beta_3\theta + (1 - \beta_3) + 3(n-1)(\beta\theta + [1 - \beta])(\beta_2\theta + [1 - \beta_2])$$
$$+ (\beta\theta + [1 - \beta])^3(n-1)(n-2)\}$$
$$(3.4) \quad = q_0(3, n) + q_1(3, n)\theta + q_2(3, n)\theta^2 + q_3(3, n)\theta^3$$

with

$$q_0(3, n) = \frac{1}{n^2}\{1 - \beta_3 + 3(n-1)(1 - \beta)(1 - \beta_2) + (n-1)(n-2)(1 - \beta)^3\}$$

$$q_1(3, n) = \frac{1}{n^2}\{\beta_3 + 3(n-1)\beta(1 - \beta_2) + 3(n-1)\beta_2(1 - \beta)$$
$$+ 3(n-1)(n-2)(1 - \beta)^2\beta\}$$

$$q_2(3, n) = \frac{1}{n^2}\{3(n-1)\beta\beta_2 + 3(n-1)(n-2)\beta^2(1 - \beta)\}$$

$$q_3(3, n) = \frac{1}{n^2}\{\beta^3(n-1)(n-2)\}.$$

If one were recording the number of susceptibles $\{X_t\}$ at times t, $t + 1$, then assuming $n \leq i$ for simplicity, the transition probability matrix will be the lower triangular:

$$\begin{pmatrix} 1 & 0 & \cdots & 0 & 0 \\ q_1(i + n - 1, 1) & q_0(i + n - 1, 1) & \cdots & 0 & 0 \\ \vdots & \vdots & \ddots & \vdots & \vdots \\ q_{n-1}(i + 1, n - 1) & q_{n-2}(i + 1, n - 1) & \cdots & q_0(i + 1, n - 1) & 0 \\ q_n(i, n) & q_{n-1}(i, n) & \cdots & q_1(i, n) & q_0(i, n) \end{pmatrix}$$

$$= \begin{pmatrix} 1 & 0 \\ F & Q \end{pmatrix}.$$

It is now easy to work out the distribution of the time T until all IVDUs in the group are infected. This is

$$P\{T = t \mid X_0 = n, Y_0 = i \text{ infectives}\} = \underset{1 \times n}{\mathrm{E}_n} \quad \underset{n \times n}{Q^{t-1}} \quad \underset{n \times 1}{F}$$

where

$$\underset{1 \times n}{\mathrm{E}_n} = (0 \ 0 \ \cdots \ 0 \ 1).$$

It can readily be shown that the mean and variance of T are

$$\mathrm{E}(T) = \mathrm{E}_n(I - Q)^{-2}F$$
$$V(T) = \mathrm{E}_n(I + Q)(I - Q)^{-3}F - \{\mathrm{E}_n(I - Q)^{-2}F\}^2.$$

4. The Reed-Frost and the random allocation models

In the Reed-Frost model, we have i infectives and n susceptibles mixing homogeneously at times $t = 0, 1, 2, \cdots$. Here, q is defined as

$$q = P\{\text{no contact with 1 susceptible for a single infective}$$
$$\text{in the interval}(t, t+1)\},$$

and if each infective acts independently

$$q^i = P\{\text{no contact with 1 susceptible for } i \text{ infectives in } (t, t+1)\}.$$

Hence for a single susceptible,

$$1 - q^i =$$
$$P\{\text{at least 1 contact with 1 susceptible for } i \text{ infectives in}(t, t+1)\}.$$

Assuming.that a contact leads to certain infection, the p.g.f. of the number of new infectives among the n independent susceptibles at time $t+1$ will be the binomial

(4.1) $$\tilde{f}_{in}(\theta) = \{(1 - q^i)\theta + q^i\}^n,$$

so that

$$P\{k \text{ new infectives at } t+1 | i \text{ infectives, } n \text{ susceptibes at } t\} =$$
$$p_k(i, n) = \binom{n}{k}(1 - q^i)^k q^{i(n-k)}.$$

This is the result commonly used in the Reed-Frost model (see Bailey (1975) and also Martin-Löf (1986)).

If we now suppose that any contact will lead to infection with probability $0 \leq \beta \leq 1$ rather than $\beta = 1$ as before, then this slightly generalized Reed-Frost model will lead to the p.g.f.

$$(4.2) \qquad \tilde{f}_{in}(\theta) = \{(1 - q^i)\beta\theta + 1 - (1 - q^i)\beta\}^n.$$

We note that this allocates the same probability β of infection to any type of contact for a single susceptible, whether this contact is with 1 or more infectives. If one wished to take account of the number of contacts causing infection, one would require the multinomial p.g.f.

$$\psi_{in}(\alpha_0, \ldots, \alpha_i) = \{\alpha_0 q^i + \alpha_1 \binom{i}{1} q^{i-1} p + \cdots + \alpha_{i-1} \binom{i}{i-1} q p^{i-1} + \alpha_i p^i\}^n$$

$$(4.3) \qquad = \sum p_{s_0 s_1 \cdots s_i} \alpha_0^{s_0} \cdots \alpha_i^{s_i}, \qquad \sum_{j=0}^{i} s_j = n,$$

for the joint distribution $\{p_{s_0 s_1 \cdots s_i}\}$ of the numbers $\{s_0, s_1, \cdots, s_i\}$ of susceptibles contacted by $0, 1, \cdots, i$ infectives. The earlier $\tilde{f}_{in}(\theta)$ of (4.1) is simply $\psi_{in}(1, \theta, \cdots \theta)$.

Let us now assume that the transmission of infection depends on the number of infectious contacts for each susceptible, so that

$$\beta_k = P\{\text{transmission due to } k \text{ contacts}\}$$
$$= 1 - (1 - \beta)^k$$

where $\beta = P\{\text{transmission due to 1 contact}\}$.

Then from (4.3) the p.g.f. of the number of new infectives will be

$$\tilde{f}_{in}(\theta) = \psi_{in}(1, \beta\theta + 1 - \beta, \cdots, \beta_i\theta + 1 - \beta_i)$$
$$= \psi_{in}(1, \beta\theta + 1 - \beta, \cdots, [1 - (1 - \beta)^i]\theta + (1 - \beta)^i)$$
$$(4.4) \qquad = \{\theta(1 - [q + p(1 - \beta)]^i) + [q + p(1 - \beta)]^i\}^n$$

which differs from (4.2), though still binomial in form.

In the Reed-Frost model, there is homogeneous mixing. In the IVDU model, there is random allocation of each infective (or the number of contacts allowed to each infective) to a susceptible, with permanent attachment of this total number of contacts. Thus we now need to specify the number of contacts $c \geq 1$ allowed for each infective; this is equivalent to having c infectives, each of which becomes permanently attached to a susceptible on contact.

The total number of contacts for the random allocation model will now be ic, and the associated p.g.f. of the number of susceptibles with $0, 1, \cdots, ic$ infectious contacts will be

$$\phi_{ic,n}(\alpha_0, \cdots, \alpha_{ic}).$$

This leads to the p.g.f.

(4.6) $\tilde{f}_{ic,n}(\theta) = \phi_{ic,n}(1, \beta\theta + (1-\beta), \ldots, \beta_{ic}\theta + (1 - \beta_{ic}))$

for the total number of infectives at time $t + 1$.

The result is relatively simple to evaluate for small family sizes, and a small contact number c. An example where a family of 4 has 2 initial infectives together with $c = 2$ contacts each ($i = 2, n = 2, c = 2$), results in the p.g.f.

$$\phi_{4,2}(\alpha_0, \cdots \alpha_4) = \frac{1}{8}\{\alpha_0\alpha_4 + 4\alpha_1\alpha_3 + 3\alpha_2^2\}.$$

Thus, the number of new infectives will have the p.g.f.
(4.7)
$$\tilde{f}_{4,2}(\theta) = \frac{1}{8}\{\beta_4\theta + (1-\beta_4) + 4[\beta\theta + (1-\beta)][\beta_3\theta + (1-\theta_3)] + 3[\beta_2\theta + (1-\beta_2)]^2\}$$

where $\beta_i = \{1 - (1-\beta)^i\}$.

For an appropriate comparison of the Reed-Frost and random allocation models, one might wish to postulate a slightly different random allocation model with i infectives, each having $0, 1, \ldots, n$ possible contacts with a binomial distribution, so that the p.g.f. of contacts per infective is $(p\theta + q)^n$. For the i infectives, the distribution of contacts is then $(p\theta + q)^{in}$ with mean nip just as for the Reed-Frost model. The random allocation p.g.f. is now

(4.8) $\psi_{in}(\alpha_0, \ldots, \alpha_{in}) = \sum_{j=0}^{in} \binom{in}{j} p^j q^{in-j} \phi_{jn}(\alpha_0, \ldots, \alpha_j),$

which is somewhat different from the Reed-Frost result (4.3). We find that the p.g.f. of the number of new infectives is

(4.9) $\tilde{f}_{in}(\theta) = \sum_{j=0}^{in} \binom{in}{j} p^j q^{in-j} \phi_{jn}(1, \beta\theta + 1 - \beta, \ldots, \beta_j\theta + 1 - \beta_j).$

We note that there is a difference between the classical Reed-Frost model with homogeneous mixing and the random allocation model where each infective may have a binomial distribution of contacts, even though the mean number of contacts nip remains the same. For example, in the case of $i = 2, n = 2$, we find for $p = 0.8$, $q = 0.2$ in the ordinary Reed-Frost model (4.1)

$$\tilde{f}_{22}(\theta) = \{(1 - q^2)\theta + q^2\}^2$$
$$= \theta^2(1 - q^2)^2 + 2\theta(1 - q^2)q^2 + q^4$$
$$= \theta^2(0.9216) + \theta(0.0768) + 0.0016.$$

For the Reed-Frost model (4.4) with infectious transmission depending on the number of contacts, we have for $\beta = 0.35$

$$\tilde{f}_{22}(\theta) = \{\theta(1 - [q + p(1 - \beta)]^2) + [q + p(1 - \beta)]^2\}^2$$
$$= \theta^2(1 - [q + p(1 - \beta)]^2)^2 + 2\theta(1 - [q + p(1 - \beta)]^2)(q + p(1 - \beta))^2$$
$$+ (q + p(1 - \beta))^4$$
$$= \theta^2(0.2319) + \theta(0.4993) + 0.2687.$$

Finally for the random allocation model (4.9) with $p = 0.8$, $q = 0.2$ and $\beta = 0.35$, we find

$$\tilde{f}_{22}(\theta) = q^4 + 4q^3 p \phi_{12}(1, \beta\theta + 1 - \beta)$$
$$+ 6q^2 p^2 \phi_{22}(1, \beta\theta + 1 - \beta, \beta_2\theta + 1 - \beta_2)$$
$$+ 4qp^3 \phi_{32}(1, \beta\theta + 1 - \beta, \beta_2\theta + 1 - \beta_2, \beta_3\theta + 1 - \beta_3)$$
$$+ p^4 \phi_{42}(1, \beta\theta + 1 - \beta, \ldots, \beta_4\theta + 1 - \beta_4)$$
$$= \theta^2(0.174722239) + \theta(0.5565392) + 0.26873856.$$

Here the probabilities of the number of new infectives is seen to differ somewhat from the Reed-Frost model, although the average number of contacts $nip = 3.2$ remains the same for both. It is therefore important that the relevant model be used in cases where infection may be spread by different methods. Detailed numerical studies are currently being carried out on the various models outlined above.

References

Bailey, N.T.J. (1975) *The Mathematical Theory of Infectious Diseases.* Griffin, London.

Cohen, P.T., Sande, M.A. and Volberding, P.A. (1990) *The AIDS Knowledge Base.* Medical Publishing Group, Waltham, Mass.

Gani, J. (1991) Generating function methods in a random allocation problem of epidemics. *Bull. Inst. Combinatorics Applns.* **3**, 43–50.

Gani, J., Heathcote, C.R. and Nicholls, D.F. (1991) A model of HIV infection through needle sharing. To appear in *J. Math. Phys. Sciences.*

Gani, J. and Yakowitz, S. (1991) Modelling the spread of HIV among intravenous drug users. Technical Report 181, U. of California, Santa Barbara.

Johnson, N.L. and Kotz. (1977) *Urn Models and their Applications.* John Wiley and Sons, New York.

Martin-Löf, A. (1986) Symmetric sampling procedures, general epidemic processes and their limit threshold theorems. *J. Appl. Prob.* **23**, 265–282.

Wolk, J., Wodak, A., Morlet, A., Guinan, J.J. and Gold, J. (1990) HIV-related risk-taking behaviour, knowledge and serostatus of intravenous drug users in Sydney. *Med. J. Aust.* **152**, 453–458.

Department of Statistics
University of California
Santa Barbara, CA
USA 93106-3110

On Hellinger transforms for solutions of martingale problems

B. GRIGELIONIS

Abstract

Explicit Feynmann–Kac type formulas are obtained for Hellinger transforms of probability measures corresponding to solutions of martingale problems with jumps.

Introduction.

Let $(\Omega, \mathcal{F})$ be a measurable space endowed with a family $\mathcal{P} = \{P^j, j = 1, \ldots, k\}$ of probability measures. The Hellinger transform of $\mathcal{P}$ is defined as the function

$$H(\gamma; \mathcal{P}) = E_Q[\prod_{j=1}^{k} (\frac{dP^j}{dQ})^{\gamma_j}],$$

$$\gamma = (\gamma_1, \gamma_2, \ldots, \gamma_k), \quad 0 < \gamma_j < 1, \quad \sum_{j=1}^{k} \gamma_j = 1,$$

which does not depend on Q, where Q is any probability measure such that $P^j \ll Q$, $j = 1, \ldots, k$, and E_Q denotes the expectation with respect to Q. The problem of evaluating $H(\gamma; \mathcal{P})$ is important in many aspects. Recall only that in a binary case the formula

$$\rho^2(P^1, P^2) = 1 - H((\frac{1}{2}, \frac{1}{2}); \{P^1, P^2\})$$

defines the famous Hellinger-Kakutani distance $\rho(P^1, P^2)$, which plays a fundamental role in the theory of statistical inference (see, e.g., [1]-[4] and bibliography therein).

When the measures P^j, $j = 1, \ldots, k$, are solutions to martingale problems defined on a filtered probability space $(\Omega, \mathcal{F}, \mathbf{F})$ and are locally dominated by some fixed measure Q, enjoying the martingale representation property, we find the explicit Feynmann-Kac type formulas for $H(\gamma; \mathcal{P}_t)$, where $\mathcal{P}_t$ denotes the family $\{P_t^j, \quad j = 1, \ldots, k\}$ of restrictions $P_t^j :=$

$P^j_{|\mathcal{F}_t}$, $t \geq 0$, $j = 1, \ldots, k$. In the Markovian case they permit to derive the evolution equations for computation of $H(\gamma; \mathcal{P}_t)$ (cf. [5]-[9]). The techniques of Itô–Watanabe multiplicative decomposition of positive special semimartingales will be applied. As an example, Hellinger transforms for probability measures, corresponding to locally infinitely divisible Markov processes will be discussed in detail.

§1. Hellinger transforms for probability measures locally dominated by a measure having the martingale representation property

Let $(\mathcal{X}, \mathcal{B}(\mathcal{X}))$ be a Blackwell space, p be an integer-valued random measure on $(\mathbf{R}_+ \times \mathcal{X}, \mathcal{B}(\mathbf{R}_+) \otimes \mathcal{B}(\mathcal{X}))$ with the $(Q, \mathbf{F})$–compensator Π, satisfying, for the sake of simplicity, $\Pi(\{t\} \times \mathcal{X}) \equiv 0$, $q(dt, dx) = p(dt, dx) - \Pi(dt, dx)$, $\mathcal{P}(\mathbf{F})$ be the σ–algebra of $\mathbf{F}$–predictable subsets of $[0, \infty) \times \Omega$, $L^c = (L^{c1}, \ldots, L^{cm})$, $L^{cj} \in \mathcal{M}^c_{loc}(Q, \mathbf{F})$, $j = 1, \ldots, m$,

$$\beta^{ij}_t = <L^{ci}, L^{cj}>_t, \quad \beta_t = \sum_{j=1}^{m} \beta^{jj}_t, \quad \beta^{ij\prime}(t) = d\beta^{ij}_t / d\beta_t,$$

$$B_t = \| \beta^{ij}_t \|^m_{i,j=1}, \quad B'(t) = \| \beta^{ij\prime}(t) \|^m_{i,j=1}.$$

We shall assume that for each $t > 0$, $j = 1, \ldots, k$,

$$(1) \qquad\qquad P^j_t \sim Q_t := Q_{|\mathcal{F}_t}, \quad P^j_0 = Q_0,$$

and the local density processes are of the form

$$(2) \qquad\qquad Z_j(t) := \frac{dP^j_t}{dQ_t} = \mathcal{E}_t(Y_j), \quad t \geq 0, \quad j = 1, \ldots, k,$$

where $\mathcal{E}_t(\cdot)$ denotes the Doléans-Dade exponential,

$$(3) \qquad Y_j(t) = \int_0^t g_j(s) dL^c_s + \int_0^t \int_{\mathcal{X}} (V_j(s, x) - 1) q(ds, dx)$$

for some m–dimensional $\mathcal{P}(\mathbf{F})$–measurable functions g_j and $\mathcal{P}(\mathbf{F}) \times \mathcal{B}(\mathcal{X})$–measurable strictly positive functions V_j such that for each $t > 0$, Q–a.e.,

$$\int_0^t (g_j(s) B'(s), g_j(s)) d\beta_s < \infty$$

and

$$\int_0^t \int_{\mathcal{X}} \frac{(V_j(s, x) - 1)^2}{1 + |V_j(s, x) - 1|} \Pi(ds, dx) < \infty, \quad j = 1, \ldots, k.$$

Let for each $t > 0$, $\quad j = 1, \ldots, k$, Q–a.e.

(4)
$$\int_0^t \int_{\mathcal{X}} |V^{\gamma_j}(s, x) - 1 - \gamma_j(V_j(s, x) - 1)|\Pi(ds, dx) < \infty$$

and

(5)
$$\int_0^t \int_{\mathcal{X}} |\prod_{i=1}^{j-1} V^{\gamma_i}(s, x) - 1|^2 \Pi(ds, dx) < \infty.$$

Theorem 1. *Under assumptions (1)-(5) we have that*
$$H(\gamma; \mathcal{P}_t) = E_Q[e^{-h(\gamma; \mathcal{P})(t)}\mathcal{E}_t(Y(\gamma; \mathcal{P}))],$$

where

$$
\begin{aligned}
h(\gamma; \mathcal{P})(t) &= \frac{1}{2} \big[\sum_{j=1}^k \gamma_j \int_0^t (g_j(s)B'(s), g_j(s))d\beta_s \\
&\quad - \int_0^t (\sum_{j=1}^k \gamma_j g_j(s)B'(s), \sum_{j=1}^k \gamma_j g_j(s))d\beta_s\big] \\
&\quad + \int_0^t \int_{\mathcal{X}} (\sum_{j=1}^k \gamma_j V_j(s, x) - \prod_{j=1}^k V_j^{\gamma_j}(s, x))\Pi(ds, dx),
\end{aligned}
$$

is called the Hellinger process of order γ for $\mathcal{P}$, and

$$Y(\gamma; \mathcal{P})(t) = \int_0^t (\sum_{j=1}^k \gamma_j g_j(s))dL_s^c + \int_0^t \int_{\mathcal{X}} (\prod_{j=1}^k V_j^{\gamma_j}(s, x) - 1)q(ds, dx).$$

The proof of this theorem is divided in two steps. Let us recall first (see, e.g. [10]) that for a strictly positive special $(Q, \mathbf{F})$– semimartingale X with the canonical decomposition

$$X_t = X_0 + A_t + M_t, \quad t \geq 0,$$

where $A \in \mathcal{A}_{loc}^c(Q, \mathbf{F})$, $\quad M \in \mathcal{M}_{loc}(Q, \mathbf{F})$, the Itô–Watanabe multiplicative decomposition holds:

(6)
$$X_t = X_0 \mathcal{E}_t(\hat{A})\mathcal{E}_t(\hat{M}), \quad t \geq 0,$$

where

$$\hat{A}_t = \int_0^t X_{s_-}^{-1} dA_s, \quad \hat{X}_t = \int_0^t X_{s_-}^{-1} dM_s, \quad t \geq 0.$$

Step 1. Denote $\gamma^{(\ell)} = (\gamma_1, \dots, \gamma_\ell)$, $u^{(\ell)} = (u_1, \dots, u_\ell)$, $u_j \in \mathbf{R}_+$, $j = 1, \dots, \ell$,

$$\phi_\ell^{(\gamma^{(\ell)})}(u^{(\ell)}) = \sum_{j=1}^\ell \gamma_j u_j - \prod_{j=1}^\ell u_j^{\gamma_j} + 1 - \sum_{j=1}^\ell \gamma_j, \quad \ell = 1, \dots, k,$$

and remark that

$$\phi_{\ell+1}^{(\gamma^{(\ell+1)})}(u^{(\ell+1)})$$

(7)
$$= \phi_\ell^{(\gamma^{(\ell)})}(u^{(\ell)}) + \phi_1^{(\gamma_{\ell+1})}(u_{\ell+1}) - (\prod_{j=1}^\ell u_j^{\gamma_j} - 1)(u_{\ell+1}^{\gamma_{\ell+1}} - 1).$$

Let

$$Y_j^{(\gamma_j)}(t) = \int_0^t \gamma_j g_j(s) dL_s^c + \int_0^t \int_{\mathcal{X}} (V_j^{\gamma_j}(s, x) - 1) q(ds, dx),$$

and

$$
\begin{aligned}
h_j^{(\gamma_j)}(t) &= \frac{1}{2} \gamma_j (1 - \gamma_j) \int_0^t (g_j(s) B'(s), g_j(s)) d\beta_s \\
&\quad + \int_0^t \int_{\mathcal{X}} \phi_1^{(\gamma_j)}(V_j(s, x)) \Pi(ds, dx), \quad t \geq 0, \quad j = 1, \dots, k.
\end{aligned}
$$

Lemma 1. *Under assumptions (1) - (4) the following decomposition holds:*
$$\mathcal{E}_t^{\gamma_j}(Y_j) = \mathcal{E}_t(Y_j^{(\gamma_j)}) e^{-h_j^{(\gamma_j)}(t)}, \quad j = 1, \dots, k, \quad t \geq 0.$$

Proof. Under strict positivity of $\mathcal{E}_t(Y_j)$, $t \geq 0$, and applying Itô's formula (see, e.g. [10]), we find that

$$
\begin{aligned}
X_t^{(j)} &:= \mathcal{E}_t^{\gamma_j}(Y_j) = (1 + \int_0^t \mathcal{E}_{s-}(Y_j) dY_j(s))^{\gamma_j} \\
&= 1 + \gamma_j \int_0^t \mathcal{E}_{s-}^{\gamma_j}(Y_j) dY_j(s) - \frac{1}{2} \gamma_j (1 - \gamma_j) \int_0^t \mathcal{E}_{s-}^{\gamma_j}(Y_j) d < Y_j^c >_s \\
&\quad - \int_0^t \int_{\mathcal{X}} \mathcal{E}_{s-}^{\gamma_j}(Y_j) \phi_1^{(\gamma_j)}(V_j(s, x)) \Pi(ds, dx) \\
&\quad - \int_0^t \int_{\mathcal{X}} \mathcal{E}_{s-}^{\gamma_j}(Y_j) \phi_1^{(\gamma_j)}(V_j(s, x)) q(ds, dx)
\end{aligned}
$$

(8)
$$= 1 + A_t^{(j)} + M_t^{(j)}, \quad t \geq 0, \quad j = 1, \dots, k,$$

where for $t \geq 0$, $j = 1, \ldots, k$,

$$
(9) \qquad A_t^{(j)} = - \int_0^t X_{s-}^{(j)} \, dh_j^{(\gamma_j)}(s)
$$

and

$$
(10) \qquad M_t^{(j)} = \int_0^t X_{s-}^{(j)} \, dY_j^{(\gamma_j)}(s).
$$

Thus the processes $X^{(j)}$ are strictly positive special semimartingales. From (9) and (10) we find $\hat{A}_t^{(j)} = -h_j^{(\gamma_j)}(t)$, $\hat{M}_t^{(j)} = Y_j^{(\gamma_j)}(t)$, $t \geq 0$, and now Lemma 1 follows from formula (6).

Step 2. Introduce the following notation for $t \geq 0$, $\ell = 1, \ldots, k$.

$$
g_\ell^{(\gamma^{(\ell)})}(t) = \sum_{j=1}^{\ell} \gamma_j g_j(t),
$$

$$
V_\ell^{(\gamma^{(\ell)})}(t, x) = \prod_{j=1}^{\ell} V_j^{\gamma_j}(t, x), \quad V^{(\ell)}(t, x) = (V_1(t, x), \ldots, V_\ell(t, x)),
$$

$$
Y_\ell^{(\gamma^{(\ell)})}(t) = \int_0^t g_\ell^{(\gamma^{(\ell)})}(s) dL_s^c + \int_0^t \int_{\mathcal{X}} (V_\ell^{(\gamma^{(\ell)})}(s, x) - 1) q(ds, dx),
$$

$$
h_\ell^{(\gamma^{(\ell)})}(t) = \frac{1}{2} \, [\sum_{j=1}^{\ell} \gamma_j \int_0^t (g_j(s) B'(s), g_j(s)) d\beta_s
$$

$$
- \int_0^t (g_\ell^{(\gamma^{(\ell)})}(s) B'(s), g_\ell^{(\gamma^{(\ell)})}(s)) d\beta_s]
$$

$$
+ \int_0^t \int_{\mathcal{X}} \phi_\ell^{(\gamma^{(\ell)})}(V^{(\ell)}(s, x)) \Pi(ds, dx).
$$

Lemma 2. *Under assumptions (1)-(5) for each* $t \geq 0$, $\ell = 1, \ldots k$, Q–*a.e.*

$$
(11) \qquad \prod_{j=1}^{\ell} \mathcal{E}_t^{\gamma_j}(Y_j) = \mathcal{E}_t(Y_\ell^{(\gamma^{(\ell)})}) e^{-h_\ell^{(\gamma^{(\ell)})}(t)}.
$$

Proof. When $\ell = 1$ the equality (11) follows from Lemma 1. The proof of Lemma 2 will be complete if we check that for each $\ell \geq 1$ and $t \geq 0$ Q–a.e.

$$
\mathcal{E}_t^{\gamma_{\ell+1}}(Y_{\ell+1}) \mathcal{E}_t(Y_\ell^{(\gamma^{(\ell)})}) \exp[-h_\ell^{(\gamma^{(\ell)})}(t)] = \mathcal{E}_t(Y_{\ell+1}^{(\gamma^{(\ell+1)})}) \exp[-h_{\ell+1}^{(\gamma^{(\ell)})}(t)]
$$

or, according to Lemma 1,

$$
\mathcal{E}_t(Y_{\ell+1}^{(\gamma_{\ell+1})})\mathcal{E}_t(Y_\ell^{(\gamma^{(\ell)})})
$$

$$
(12) \qquad = \mathcal{E}_t(Y_{\ell+1}^{(\gamma^{(\ell+1)})}) \exp\{h_\ell^{(\gamma^{(\ell)})}(t) + h_{\ell+1}^{(\gamma_{\ell+1})}(t) - h_{\ell+1}^{(\gamma^{(\ell+1)})}(t)\}.
$$

Using the known properties of stochastic exponentials and stochastic integrals, we have that

$$
\mathcal{E}_t(Y_{\ell+1}^{(\gamma_{\ell+1})})\mathcal{E}_t(Y_\ell^{(\gamma^{(\ell)})}) = \mathcal{E}_t(Y_{\ell+1}^{(\gamma_{\ell+1})} + Y_\ell^{(\gamma^{(\ell)})} + [Y_{\ell+1}^{(\gamma_{\ell+1})}, Y_\ell^{(\gamma^{(\ell)})}]
$$

$$
(13) \qquad - < Y_{\ell+1}^{(\gamma_{\ell+1})}, Y_\ell^{(\gamma^{(\ell)})} >) \exp[< Y_{\ell+1}^{(\gamma_{\ell+1})}, Y_\ell^{(\gamma^{(\ell)})} >_t]
$$

and

$$
[Y_{\ell+1}^{(\gamma_{\ell+1})}, Y_\ell^{(\gamma^{(\ell)})}]_t - < Y_{\ell+1}^{(\gamma_{\ell+1})}, Y_\ell^{(\gamma^{(\ell)})} >_t
$$

$$
= \int_0^t \int_{\mathcal{X}} (V_{\ell+1}^{\gamma_{\ell+1}}(s,x) - 1)(V_\ell^{(\gamma^{(\ell)})}(s,x) - 1)q(ds,dx).
$$

Noting that

$$
(V_{\ell+1}^{\gamma_{\ell+1}}(s,x) - 1)(V_\ell^{(\gamma^{(\ell)})}(s,x) - 1) + (V_{\ell+1}^{\gamma_{\ell+1}}(s,x) - 1) + (V_\ell^{(\gamma^{(\ell)})}(s,x) - 1)
$$

$$
= V_{\ell+1}^{(\gamma^{(\ell+1)})}(s,x) - 1
$$

and

$$
\gamma_{\ell+1}g_{\ell+1}(s) + g_\ell^{(\gamma^{(\ell)})}(s) = g_{\ell+1}^{(\gamma^{(\ell+1)})}(s),
$$

from (13) we find that

$$
(14) \qquad \mathcal{E}_t(Y_{\ell+1}^{(\gamma_{\ell+1})})\mathcal{E}_t(Y_\ell^{(\gamma^{(\ell)})})
$$

$$
= \mathcal{E}_t(Y_{\ell+1}^{(\gamma^{(\ell+1)})}) \exp\{\int_0^t \int_{\mathcal{X}} (V_{\ell+1}^{\gamma_{\ell+1}}(s,x) - 1)(V_\ell^{(\gamma^{(\ell)})}(s,x) - 1)\Pi(ds,dx)\}.
$$

Applying (7), it is easy to check that

$$
h_\ell^{(\gamma^{(\ell)})}(t) + h_{\ell+1}^{(\gamma_{\ell+1})}(t) - h_{\ell+1}^{(\gamma^{(\ell+1)})}(t)
$$

$$
(15) \qquad = \int_0^t \int_{\mathcal{X}} (V_{\ell+1}^{\gamma_{\ell+1}}(s,x) - 1)(V_\ell^{(\gamma^{(\ell)})}(s,x) - 1)\Pi(ds,dx).
$$

Equalities (14)-(15) prove (12) and at the same time Lemma 2.

Taking $\ell = k$ and keeping in mind that $\sum_{j=1}^k \gamma_j = 1$, we obviously derive the statement of the Theorem 1 from Lemma 2.

Remark that Lemmas 1 and 2 are valid for any real numbers $\gamma_1, \ldots, \gamma_k$, satisfying (4) and (5).

§2. Hellinger transforms for probability measures corresponding to locally infinitely divisible Markov processes

For $f \in C_b^2(\mathbf{R}^m)$ consider the integro-differential operators

$$A_r f(x) = \frac{1}{2} \sum_{i,j=1}^m a_{ij}(x) \frac{\partial^2}{\partial x_i \partial x_j} f(x) + \sum_{j=1}^m b_{rj}(x) \frac{\partial}{\partial x_j} f(x)$$

$$+ \int_{E_m} \left(f(x+y) - f(x) - \sum_{j=1}^m \frac{\partial}{\partial x_j} f(x) y_j 1_{\{|y| \le 1\}} \right) V_r(x,y) \pi(x, dy),$$

where $E_m = \mathbf{R}^m \backslash \{0\}$, $A(x) = \| a_{ij}(x) \|_{i,j=1}^m$ is a non–negative definite matrix, $V_r(x, y)$ are strictly positive functions, $V_0(x, y) \equiv 1$,

$$\int_{E_m} (|y|^2 \wedge 1) V_r(x, y) \pi(x, dy) < \infty, \quad x \in \mathbf{R}^m, \quad r = 0, 1, \dots, k.$$

Assume that there exist m–dimensional functions $g_r(x) = (g_{r_1}(x), \dots, g_{rm}(x))$ such that for $b_r(x) = (b_{r_1}(x), \dots, b_{rm}(x))$ the following equalities are satisfied for $r = 1, 2, \dots, k$,

$$b_r(x) = b_0(x) + g_r(x) A(x) + \int_{|y| \le 1} y(V_r(x, y) - 1) \pi(x, dy).$$

We shall consider now a filtered measurable space $(\Omega, \mathcal{F}, \mathbf{F})$ assuming that $\mathbf{F}$ is generated by some m–dimensional càdlàg process X, $\mathcal{F} = \bigvee_{t \ge 0} \mathcal{F}_t$ and there exist probability measures P_x^j, $j = 0, 1, \dots, k$, $x \in \mathbf{R}^m$, uniquely characterized by the properties that $P_x^j(X_0 = x) = 1$ and for all $f \in C_b^2(\mathbf{R}^m)$,

$$M_t^j(f) = f(X_t) - \int_0^t A_j f(X_s) ds, \quad t \ge 0,$$

are $(P_x^j, \mathbf{F})$–local martingales, i.e., P_x^j is a solution to the (x, A_j)–martingale problem, $j = 0, 1, \dots, k$.

It is well known that X is a $(P_x^0, \mathbf{F})$–semimartingale having the canonical decomposition:

$$X_t = X_0 + \int_0^t b_0(X_s) ds + X_t^c + \int_0^t \int_{|y| \le 1} y q(ds, dy) + \int_0^t \int_{|y| > 1} y p(ds, dy),$$

where X^c is the continuous local martingale part of X,

$$< X^{ci}, X^{cj} >_t = \int_0^t a_{ij}(X_s) ds, \quad i, j = 1, \dots, m,$$

$p(dt, dy)$ is the jump measure of X with the $(P_x^0, \mathbf{F})$-compensator

$$\Pi(dt, dy) = \pi(X_t, dy)dt, \quad q(dt, dy) = p(dt, dy) - \pi(X_t, dy)dt.$$

Assuming that for each $t > 0, \quad j = 1, \ldots, k$

$$P_{x,t}^j := P_{x|\mathcal{F}_t}^j \sim P_{x,t}^0,$$

it is well known (see, e.g., [1]) that

$$Z_j(t) := \frac{dP_{x,t}^j}{dP_{x,t}^0} = \mathcal{E}_t(Y_j),$$

where

$$Y_j(t) = \int_0^t g_j(X_{s-})dX_s^c + \int_0^t \int_{E_m} (V_j(X_{s-}, y) - 1)q(ds, dy), \quad t \geq 0, j = 1, \ldots, k.$$

Introduce the following notation with $f \in C_b^2(\mathbf{R}^m), \quad \mathcal{P}_{x,t} = \{P_{x,t}^j, j = 1, \ldots, k\}$,

$$g_{\gamma;\mathcal{P}}(x) = \sum_{j=1}^k \gamma_j g_j(x), \quad V_{\gamma;\mathcal{P}}(x, y) = \prod_{j=1}^k V_j^{\gamma_j}(x, y),$$

$$b_{\gamma;\mathcal{P}}(x) = b_0(x) + g_{\gamma;\mathcal{P}}(x)A(x) + \int_{|y| \leq 1} y(V_{\gamma;\mathcal{P}}(x, y) - 1)\pi(x, dy),$$

$$Y_{\gamma;\mathcal{P}}(t) = \int_0^t g_{\gamma;\mathcal{P}}(X_{s-})dX_s^c + \int_0^t \int_{E_m} (V_{\gamma;\mathcal{P}}(X_{s-}, y) - 1)q(ds, dy),$$

$$F_{\gamma;\mathcal{P}}(x) = \frac{1}{2}[\sum_{j=1}^k \gamma_j (g_j(x)A(x), g_j(x)) - (g_{\gamma;\mathcal{P}}(x)A(x), g_{\gamma;\mathcal{P}}(x))]$$

$$+ \int_{E_m} (\sum_{j=1}^k \gamma_j V_j(x, y) - V_{\gamma;\mathcal{P}}(x, y))\pi(x, dy),$$

$$A_{\gamma;\mathcal{P}}f(x) = \frac{1}{2}\sum_{i,j=1}^m a_{ij}(x)\frac{\partial^2}{\partial x_i \partial x_j}f(x) + \sum_{j=1}^m b_{\gamma;\mathcal{P}_j}(x)\frac{\partial}{\partial x_j}f(x)$$

$$+ \int_{E_m} (f(x + y) - f(x) - \sum_{j=1}^m \frac{\partial}{\partial x_j}f(x)y_j 1_{\{|y| \leq 1\}})V_{\gamma;\mathcal{P}}(x, y)\pi(x, dy).$$

Theorem 2. *Under the above assumptions, if for each* $x \in \mathbf{R}^m$, $j = 1, \ldots, k$,

$$\int_{E_m} [\frac{(V_j(x,y) - 1)^2}{1 + |V_j(x,y) - 1|} \quad + \quad |V_j^{\gamma_j}(x,y) - 1 - \gamma_j(V_j(x,y) - 1)|$$
$$+ \quad |\prod_{i=1}^{j-1} V^{\gamma_i}(x,y) - 1|^2] \pi(x, dy) < \infty$$

and for each $t > 0$

(16) $$E_{P_x^0} \mathcal{E}_t(Y_{\gamma;\mathcal{P}}) = 1,$$

then

$$H(\gamma; \mathcal{P}_{x,t}) = E_{P_x(\gamma;\mathcal{P})} \exp[-\int_0^t F_{\gamma;\mathcal{P}}(X_s) ds],$$

where $P_x(\gamma; \mathcal{P})$ *is a solution to the* $(x, A(\gamma; \mathcal{P}))$ *–martingale problem.*

This statement follows obviously from Theorem 1 and the Girsanov type transformation formulas for semimartingales (see e.g., [1]), noting that

$$P_{x,t}(\gamma; \mathcal{P})(d\omega) = \mathcal{E}_t(Y_{\gamma;\mathcal{P}}) P_x^0(d\omega).$$

Finally remark that, if the function $u(t, x) = H(\gamma; \mathcal{P}_{x,t})$ is sufficiently smooth, then it solves the Cauchy problem:

$$\frac{\partial u(t, x)}{\partial t} = A(\gamma; \mathcal{P})u(t, x) - F_{\gamma;\mathcal{P}}(x)u(t, x), \quad t > 0, \quad x \in \mathbf{R}^m,$$
$$u(0, x) \equiv 1.$$

In the simplest case, when $F_{\gamma;\mathcal{P}}$ does not depend on x, we have that $H(\gamma; \mathcal{P}_{x,t}) = e^{-t F_{\gamma;\mathcal{P}}}$ (cf. [1], [2], [5]-[7]).

References

[1] J. Jacod, A.N. Shiryaev, *Limit Theorems for Stochastic Processes*, Springer, Berlin, 1987.

[2] F. Liese, I. Vajda, *Convex Statistical Distances*, Teubner, Leipzig, 1987.

[3] J. Jacod, Une application de la topologie d'Emery: le processus information d'un modèle statistique filtré, *Sém. Probab.* XXIII, *Lecture Notes in Math.*, **Vol. 1372**, Springer, Berlin, 1989, pp. 448-474.

[4] J. Jacod, Convergence of filtered statistical models and Hellinger processes, *Stochastic Proc. Appl.*, **32** (1989), 47–68.

[5] C.M. Newman, The inner product of path space measures corresponding to random processes with independent increments, *Bull. Amer. Math. Soc.*, **78** (1972), 268–272.

[6] J. Mémin, A.N. Shiryaev, Distance de Hellinger–Kakutani des lois correspondant à deux processus à accroissements indépendants, *Z. Wahrsch. verw. Geb.*, **70** (1985), 67–90.

[7] J. Jacod, Filtered statistical models and Hellinger processes, *Stochastic Proc. Appl.*, **32** (1989), 3–45.

[8] B. Grigelionis, Hellinger integrals and Hellinger processes for solutions of martingale problems, *Proc. 5th Vilnius Conference Probab. Theory and Math. Statist.*, **Vol. 1**, VSP/Mokslas, Vilnius, 1990.

[9] B. Grigelionis, On statistical inference for stochastic processes with boundaries, *Stochastic Processes and Related Topics*, ed. M. Dozzi, A.J. Engelbert, D. Nualart, Math. Research, **vol. 61**, Akademie-Verlag, Berlin, 1991.

[10] R.S. Liptser, A.N. Shiryaev, *Theory of Martingales*, Nauka, Moscow, 1986.

Institute of Mathematics and Informatics
Lithuanian Academy of Sciences
Vilnius, Lithuania

The Homogeneous Chaos over Compact Lie Groups

LEONARD GROSS

Abstract. For a compact simply connected Lie group G and related groups, an orthogonal decomposition of L^2 (G, heat kernel measure) is described which reduces, in case $G = R^d$, to the well known decomposition of L^2 (R^d, Gauss measure) into orthogonal subspaces determined by products of Hermite polynomials.

1. Introduction.

Professor Kallianpur has been concerned in recent years with the analysis of certain functionals of a Gaussian process in terms of the associated reproducing kernel Hilbert space H. In particular Johnson and Kallianpur [JK2] have exploited the well known isometry [W, Ka, I, S1,2] between the space of symmetric tensors over H and the space of square integrable functionals of the process in order to understand better the relationship between multiple Ito integrals and multiple Fisk–Stratonovich integrals. See also [Hu1,2,3, HM1,2,3, JK1, Su1,2]. Their paper was a significant influence on this author in his recent work on Schrödinger operators over loop groups [G], in which is derived an analog of the above isometry for noncommutative groups. The real Hilbert space H is replaced by a compact Lie group G, the Gaussian measure "on H" is replaced by the heat kernel measure on G, and the space of symmetric tensors over H is replaced by a specific completion of the universal enveloping algebra of the Lie algebra of G. In Section two we will describe this isometry in more detail after establishing notation. In Section three we describe the noncommutative analog of the "homogeneous chaos" decomposition. These are the analogs of the spaces of symmetric n-tensors over H. The main result of this paper is Theorem 3.4.

2. Notation and Background.

Let $\mathcal{G}$ be a finite dimensional real Lie algebra with an inner product $< , >$. Denote by $T = T(\mathcal{G})$ the space of algebraic tensors over $\mathcal{G}$. T is an

[1] This work was partially supported by NSF Grant DMS-89-22941.

associative algebra under the product $uv = u \otimes v$. Let $J = J(\mathcal{G})$ be the two sided ideal in T generated by $\{\xi \otimes \eta - \eta \otimes \xi - [\xi, \eta] : \xi, \eta \in \mathcal{G}\}$, where $[\xi, \eta]$ is the Lie algebra product in $\mathcal{G}$. Any element β in T has the form $\beta = \sum_{k=0}^{m} \beta_k$ with β_k in $\mathcal{G}^{\otimes k}$, $k = 0, 1, \ldots, m$. Define

$$(2.1) \qquad \|\beta\|^2 = \sum_{k=0}^{m} k! |\beta_k|^2$$

and denote by $\overline{T}$ the completion of T in this norm. In (2.1) $|\beta_k|$ denotes the usual cross norm on $\mathcal{G}^{\otimes k}$ induced by the given inner product on $\mathcal{G}$. $\overline{T}$ is a real Hilbert space in this norm. Denote by $J^{\perp}$ the orthogonal complement of J in $\overline{T}$. Each element ξ in $\mathcal{G}$ determines the right multiplication operator $R_\xi : T \to T$ given by $R_\xi u = u \otimes \xi$. Write $A_\xi = R_\xi^*$ for the adjoint of R_ξ in the Hilbert space $\overline{T}$. It is not hard to see that R_ξ is unbounded. But since R_ξ is densely defined A_ξ is well defined. Since J is a two sided ideal it is invariant under R_ξ. Hence $J^{\perp}$ is invariant under A_ξ. The definition of A_ξ gives easily the identity $A_\xi(\xi_1 \otimes \cdots \otimes \xi_n) = n < \xi, \xi_n > \xi_1 \otimes \cdots \otimes \xi_{n-1}$. Hence A_ξ is densely defined in $\overline{T}$. But it is a serious technical question as to whether A_ξ is densely defined in $J^{\perp}$. This has been answered affirmatively in [G] in case the given inner product on $\mathcal{G}$ is ad $\mathcal{G}$ invariant in the sense that $< [\xi, \eta], \zeta > = - < \eta, [\xi, \zeta] >$ for all ξ, η, ζ in $\mathcal{G}$. This is the only case of interest to us. We will simply assume that A_ξ is densely defined in $J^{\perp}$ in the following.

EXAMPLE 2.1.   Take $\mathcal{G} = R^d$ and define $[\xi, \eta] \equiv 0$ for ξ and η in $\mathcal{G}$. In this case the ideal J is clearly the direct sum of its homogeneous components. Consequently $J^{\perp}$ is also the direct sum of its homogeneous components. But if β is in $\mathcal{G}^{\otimes n} \cap J^{\perp}$ then for any elements ξ, η in $\mathcal{G}$ and u in $\mathcal{G}^{\otimes r}$, v in $\mathcal{G}^{\otimes s}$ with $r + s + 2 = n$ we have $< \beta, u \otimes \xi \otimes \eta \otimes v > = < \beta, u \otimes \eta \otimes \xi \otimes v >$. It follows that β is invariant under the action of the permutation $r + 1 \leftrightarrow r + 2$ in $\mathcal{G}^{\otimes n}$. Since r is arbitrary in $\{0, 1, \ldots, n - 2\}$, β is a symmetric tensor. Conversely the same computation shows that any symmetric tensor in $\mathcal{G}^{\otimes n}$ is also in $J^{\perp}$. Thus in this (commutative) case $J^{\perp}$ is exactly the space of symmetric tensors in $\overline{T}$.

The factor $k!$ in (2.1) is not often used in studying the space of symmetric tensors over R^d. But if $\mathcal{G}$ is not commutative then J is not the sum of its homogeneous components and neither is $J^{\perp}$. In this case omission of the factor $k!$ would greatly complicate the algebra involved in the next theorem — which is the theorem that drives our interest in the space $J^{\perp}$. Let us note that in the commutative case (i.e., Example 2.1) the lowering operator $A_\xi \,|\, J^{\perp}$ is precisely the usual annihilation operator, in the sense that the map $u \to (k!)^{1/2} u$ from $\mathcal{G}^{\otimes k}$ to $\mathcal{G}^{\otimes k}$ extends to an isometry from

$\overline{T}$ to the completion of T in the cross norm and interchanges A_ξ with the standard annihilation operator. The well known connection between this algebraic structure on the one hand and Gauss measure ν on R^d may be described by the statement that there is a unique isometry between $L^2(R^d, \nu)$ and $J^\perp$ which satisfies a) the constant function equal to one corresponds to the zero rank unit tensor ω in $J^\perp$ and b) $\partial/\partial x_j$ corresponds to the annihilation operator A_{e_j} where $e_1, \dots, e_d$ is the standard basis of R^d. It is the infinite dimensional version of this isometry which Professor Kallianpur has exploited in much of his recent work. The following theorem extends this isometry to the noncommutative case.

Let G be a Lie group which is a product, $G = G_c \times R^d$ where G_c is compact. Assume G is connected and simply connected. Let $\mathcal{G}$ be the tangent space to G at e and choose an Ad G invariant inner product on $\mathcal{G}$. Let $e_1, \dots, e_r$ be an O.N. basis of $\mathcal{G}$ and $\tilde{e}_j$ the corresponding right invariant vector fields on G. The self–adjoint version of the Laplacian $\Delta := \sum_{j=1}^r \tilde{e}_j^2$ in $L^2(G, dx)$, where dx denotes Haar measure, generates a semigroup given by convolution, $e^{t\Delta} = p_t *$, wherein the "heat kernel" p_t is a strictly positive C^∞ function on G for $t > 0$.

THEOREM 2.2. *There exists a unique isometry U from $L^2(G, p_1(x)dx)$ onto $J^\perp$ such that*

　　a) $U(1) = \omega$ *(the unit zero rank tensor)*
　　b) $U\tilde{\xi} = A_\xi U$　$\xi \in \mathcal{G}$.

REMARK 2.3.　In spite of the fact that the statement of Theorem 2.2 is not in itself probabilistic the only proof I have of it is a lengthy one [G] which uses G valued Brownian motion, the associated Ito expansion and the relation between Ito and Stratonovich multiple integrals, as discussed by Hu, Meyer, Johnson and Kallianpur and Sugita [Hu1,2,3, HM1,2,3, JK1,2, Su1,2]. In case $G = R^d$ then the heat kernel $p_1(x)$ is just a Gaussian measure on R^d and the theorem reduces to the classical one discussed above. The novelty of Theorem 2.2 lies primarily in the compact case. If G is a compact group then $J^\perp$ is no longer the direct sum of the spaces of symmetric n tensors over $\mathcal{G}$. The purpose of this note is to describe the noncommutative analog of this decomposition of $J^\perp$ and thereby of $L^2(G, p_1(x)dx)$.

3. Orthogonal Decomposition of $J^\perp$.

We assume $<, >$ is an inner product on a finite dimensional real Lie algebra $\mathcal{G}$. But we need not assume that it is ad $\mathcal{G}$ invariant. Denote by P the orthogonal projection in $\overline{T}$ onto $J^\perp$. Let T_n denote the space of tensors over $\mathcal{G}$ of rank $\leq n$. Thus $T_n = \sum_{k=0}^n \mathcal{G}^{\otimes k}$. Let $K_n = PT_n$. T_n is finite dimensional. So K_n is also finite dimensional, hence closed. Since J

contains no zero rank tensors it follows that $T_0 \subset J^\perp$. Hence $K_0 = T_0 = \mathbb{R}$. Let $Q_0 = K_0$ and let

$$(3.1) \qquad Q_n = K_n \ominus K_{n-1} \quad n = 1, 2, \ldots .$$

Since $T = \bigcup_{n=0}^{\infty} T_n$ is dense in $\overline{T}$ and P is bounded the union of the K_n is dense in $J^\perp$. Hence

$$(3.2) \qquad J^\perp = \oplus_{n=0}^{\infty} Q_n .$$

Of course if $\mathcal{G}$ is commutative then P is the projection on the space of symmetric tensors in $\overline{T}$ as noted in Example 2.1 and Q_n is therefore the space of symmetric n–tensors over $\mathcal{G}$. But if $\mathcal{G}$ is not commutative we can gain some understanding of the space Q_n as follows.

LEMMA 3.1. *Let*

$$(3.3) \qquad \beta = \sum_{k=0}^{\infty} \beta_k \quad \beta_k \in \mathcal{G}^{\otimes k}$$

be in $J^\perp$. If $\beta_k = 0$ for $k = 0, 1, \ldots, n-1$ then β_n is symmetric.

PROOF: We need only consider $n \geq 2$. Let ξ and η be in $\mathcal{G}$. Let u be in $\mathcal{G}^{\otimes r}$ and let v be in $\mathcal{G}^{\otimes s}$ with $r + s + 2 = n$ and $r \geq 0$ and $s \geq 0$. Then $w \equiv u \otimes (\xi \otimes \eta - \eta \otimes \xi - [\xi, \eta]) \otimes v$ is in J and is therefore orthogonal to β. But since $\beta_{n-1} = 0$ it follows that β_n is orthogonal to $u \times (\xi \otimes \eta - \eta \otimes \xi) \otimes v$, which is the component of w of rank n. Symmetry of β_n now follows by the same argument as in Example 2.1.

LEMMA 3.2. *Suppose that β is given by (3.3). If β is in Q_n then $\beta_k = 0$ for $k = 0, 1, \ldots, n-1$.*

PROOF: By assumption β is orthogonal to K_{n-1}. Hence for all v in T_{n-1} we have $0 = (\beta, Pv)_{\overline{T}} = (P\beta, v)_{\overline{T}} = (\beta, v)_{\overline{T}} = \sum_{k=0}^{n-1} k! < \beta_k, v_k >_{\mathcal{G}^{\otimes k}}$. Since the components v_k are arbitrary elements of $\mathcal{G}^{\otimes k}$ for $k = 0, 1, \ldots, n-1$ the lemma follows.

Denote by $\mathcal{S}_n$ the space of symmetric tensors in $\mathcal{G}^{\otimes n}$.

LEMMA 3.3. *The map $\beta \to \beta_n$ is a one to one map from Q_n into $\mathcal{S}_n$.*

PROOF: If β is in Q_n then $\beta_k = 0$ for $k = 0, 1, \ldots, n-1$ by Lemma 3.2 and therefore β_n is in $\mathcal{S}_n$ by Lemma 3.1. If moreover $\beta_n = 0$ then for all u in T_n we have $0 = (\beta, u)_{\overline{T}} = (P\beta, u)_{\overline{T}} = (\beta, Pu)_{\overline{T}}$. Hence $\beta \perp K_n$. But β is in K_n. So $\beta = 0$. Therefore the map $\beta \to \beta_n$ is one to one.

Next, denote by $\overline{J}$ the closure of J in $\overline{T}$. It seems likely to the author that there are no finite rank tensors in $\overline{J}$ that were not already in J. That is, the identity

$$(3.4) \qquad\qquad \overline{J} \cap T = J$$

seems like a reasonable conjecture. But I don't have a proof. We will assume the validity of (3.4) in the next theorem to complete our analysis of the subspaces Q_n.

THEOREM 3.4. *Assume that (3.4) holds. Then the map $\beta \to \beta_n$ (see (3.3)) is a linear isomorphism of Q_n onto S_n.*

PROOF: In view of Lemma 3.3 it suffices to show that Q_n and S_n have the same dimension. We will use the Poincaré–Birkhoff–Witt (PBW) theorem [B, Chapter I, §2.7] which asserts that in the quotient algebra T/J ($\equiv$ universal enveloping algebra) the subspace S_n/J is a supplement to T_{n-1}/J in T_n/J. That is, for any element t in T_n there is a unique element s in S_n such that $t-s$ is in $T_{n-1}+J$. For such t and s we have $Pt = Ps + P(t-s) \in PS_n + K_{n-1}$. Hence $K_n \subset PS_n + K_{n-1}$ and since both summands on the right are contained in K_n we have equality:

$$(3.5) \qquad\qquad K_n = PS_n + K_{n-1}.$$

We will show that $PS_n \cap K_{n-1} = (0)$ under the assumption (3.4). If PS_n and K_{n-1} have a non trivial intersection then there is an element s in S_n and t in T_{n-1} such that $Ps = Pt$. That is, $P(t-s) = 0$. So $t-s$ is in $\overline{J}$. But $t-s$ is in T. Therefore by the assumption (3.4) $t-s$ is in J. Hence s is in $T_{n-1}+J$. By the uniqueness portion of the PBW theorem $s = 0$. Hence $PS_n \cap K_{n-1} = (0)$. The same argument shows that if $Ps = 0$ for some s in S_n then $s = 0$. Thus P is one to one on S_n under the assumption (3.4). Consequently $\dim K_n = \dim PS_n + \dim K_{n-1} = \dim S_n + \dim K_{n-1}$. But $K_n = Q_n \oplus K_{n-1}$. Hence $\dim Q_n = \dim S_n$. This concludes the proof of Theorem 3.4.

REMARK 3.5. Whereas, in the commutative case the finite rank tensors in $J^\perp$ are dense in $J^\perp$, in the noncommutative case this is false. For example if $[\mathcal{G}, \mathcal{G}] = \mathcal{G}$ (as in the case of $su(2)$) then there are no finite rank tensors in $J^\perp$ at all except the zero rank tensors. Indeed, if $\zeta = [\xi, \eta]$ and β is in $J^\perp$ with $\beta_n = 0$ for some $n \geq 2$, then, in the notation of the proof of Lemma 3.1, $- < \beta_{n-1}, u \otimes \zeta \otimes v > = < \beta, w > = 0$. It follows that $\beta_{n-1} = 0$, and by induction $\beta_k = 0$ for $k = 1, \dots, n-1$. Hence if β is of finite rank then it is of rank zero. Although Theorem 3.4 together with (3.2) shows that $J^\perp$ can be "parametrized" (algebraically) by the algebraic symmetric tensors over $\mathcal{G}$ the norms on Q_n and $\mathcal{S}_n$ are probably poorly related. Clearly $\|\beta\|_{\overline{T}} \geq \|\beta_n\|_{\overline{T}}$. But there is probably no bound going the other way which is uniform in n. Topologically, therefore, $J^\perp$ should, in all likelihood, be regarded as distinct from any perspicuous space of symmetric tensors over $\mathcal{G}$, in spite of Theorem 3.4.

REMARK 3.6. By means of the isometry of Theorem 2.2 we may identify the subspaces Q_n of $J^\perp$ with mutually orthogonal subspaces $\tilde{Q}_n$ of $L^2(G, p_1(x)dx)$. If $G = R^d$ then of course $\tilde{Q}_n$ is the usual "nth homogeneous chaos" and is spanned by products of Hermite polynomials. If G is compact then the "nth homogeneous chaos" $\tilde{Q}_n$ is a finite dimensional subspace of $L^2(G, p_1(x)dx)$. To my knowledge nothing is known about these analogs of the Hermite polynomials.

References

[B]   N. Bourbaki, *Lie Groups and Lie Algebras*, Chapters I–III, Springer–Verlag, New York, 1989.

[G]   L. Gross, *Uniqueness of ground states for Schrödinger operators over loop groups*, Cornell preprint, October 1991.

[HM1]   Y. Z. Hu and P. A. Meyer, *Chaos de Wiener et integral de Feynman*, in Sem. de Prob. XXII; Lect. Notes in Math. No. 1321, Springer, 1988, p. 51–71.

[HM2]   ——, *Sur les integrales multiples de Stratonovich*, in Sem. de Prob. XXII; Lect. Notes in Math. No. 1321, Springer, 1988, p. 72-81.

[HM3]   ——, *On the approximation of Stratonovich multiple integrals*, Strassbourg, preprint, 1991.

[Hu1]   Y. Z. Hu, *Some notes on multiple Stratonovich integrals*, Acta Math. Scientia 9 (1989), 453–462.

[Hu2]   ——, *Calculs formels sur les E. D. S. de Stratonovich*, Sem. de Prob. XXIV Lect. Notes in Math. No. 1426, Springer 1990, 453–460.

[Hu3]   ——, *Symmetric integral and canonical extension for jump processes – some combinatorial results*, Acta Math. Scientia 10 (1990), 448–458.

[I]   K. Ito, *Multiple Wiener integral*, J. Math. Soc. of Japan 3 (1951), 157–169.

[JK1]   G. W. Johnson and G. Kallianpur, *Homogeneous chaos, p-forms, scaling and the Feynman integral*, Univ. of North Carolina Center for Stoch. Proc. Tech. Report No. 274 (Feb. 1990), Trans. A.M.S., to appear.

[JK2] _____, *Multiple Wiener integrals on abstract Wiener spaces and liftings of p-linear forms* (T. Hida, H. H. Kuo, J. Potthoff, L. Streit, ed.), in "White Noise Analysis", World Scientific, 1990, pp. 208–219.

[Ka] S. Kakutani, *Spectrum of the flow of Brownian motion*, Proc. Nat. Acad. Sci. U.S.A. **36** (1950), 319–323.

[S1] I. E. Segal, *Tensor algebras over Hilbert spaces*, Trans. Amer. Math. Soc. **81** (1956), 106–134.

[S2] _____, *Distributions in Hilbert space and canonical systems of operators*, Trans. Amer. Math. Soc. **88** (1958), 12–41.

[Su1] H. Sugita, *Hu–Meyer's multiple Stratonovich integral and essential continuity of multiple Wiener integral*, Bull. Sc. Math. 2^e serie **113** (1989), 463–474.

[Su2] _____, *Various topologies in the Wiener space and Levy's stochastic area*, Saga preprint, June, 1991.

[W] N. Wiener, *The homogeneous chaos*, Amer. J. Math. **60** (1938), 897–936.

Department of Mathematics
Cornell University
Ithaca, NY 14853

Asymptotics for Two-dimensional Anisotropic Random Walks

C.C. HEYDE

Abstract. Asymptotic normality results are provided for a two-dimensional random walk on an anisotropic lattice. The results, which are motivated by transport problems in statistical physics, involve only macroscopic properties of the medium through which the walk takes place.

Key words: random walks, two dimensions, anisotropic lattices, transport processes, conductivity, fluid flow, central limit theorem, martingale methods.

1. Introduction.

Substantial motivation for research on 2–dimensional anisotropic random walks has come from statistical physics, for example in connection with transport phenomena (e.g. [4], [5]).

One area of application involves conductivity of various organic salts, such as tetrathiofulvalene (TTF)–tetracyanoquinodimethane (TCNQ), which show signs of superconductivity. They conduct strongly in one direction but not others. Indeed, the conductivity parallel to the structural axis is 100 times or more that perpendicular to it. The molecular forms of TCQN and TTF are planar so that they can be easily stacked and, in fact, the structure of the TTF–TCQN crystal is generally thought to consist of parallel columns of separately stacked TTF and TCQN molecules.

Another related area of application concerns transport in physical systems which lack complete connectivity such as with fluid flow through packed columns, for example in gas absorption or distillation processes. In such situations gases or liquids predominantly flow through vertical cylinders filled with a random packing of inert objects such as ceramic rings.

The predominance of one dimension for the transport, say along rows, plus a possible lack of complete connectivity, can be modelled rather more generally by an anisotropic 2–dimensional random walk in which the transition mechanism depends only on the index of the column which is at present occupied. Thus, we consider a random walk which, if situated at a site on column j, moves with probability p_j to either horizontal neighbour and with probability $\frac{1}{2} - p_j$ to either vertical neighbour at the next step.

2. Results and Proofs.

Let X_n, Y_n denote the horizontal and vertical positions of the walk after n steps beginning from $X_0 = Y_0 = 0$. The transition probabilities are

$$P\{(X_{n+1}, Y_{n+1}) = (j-1, k) \mid (X_n, Y_n) = (j, k)\} = p_j$$
$$P\{(X_{n+1}, Y_{n+1}) = (j+1, k) \mid (X_n, Y_n) = (j, k)\} = p_j \qquad (1)$$
$$P\{(X_{n+1}, Y_{n+1}) = (j, k-1) \mid (X_n, Y_n) = (j, k)\} = \frac{1}{2} - p_j$$
$$P\{(X_{n+1}, Y_{n+1}) = (j, k+1) \mid (X_n, Y_n) = (j, k)\} = \frac{1}{2} - p_j$$

for each $j, k \in Z$, the set of integers, and $n = 0, 1, 2, \ldots$.

It turns out that the asymptotic behaviour of the random walk (X_n, Y_n) depends only on the macroscopic properties of the medium (i.e. of the $\{p_j, j \in Z\}$). This is of considerable practical significance since, although many materials are heterogeneous on a microscopic scale they are essentially homogeneous on a macroscopic or laboratory scale.

We shall obtain the following result which gives a complete description of the asymptotic behaviour of (X_n, Y_n). This extends the convergence in distribution result of Heyde [3] which only dealt with the horizontal component X_n of the walk and which used a slightly stronger asymptotic density condition than (2) below. We use $\xrightarrow{d}$ to denote convergence in distribution, $N(0, 1)$ for the unit normal law, and $MVN(0, C)$ for the multivariate normal law with zero mean vector and covariance matrix C.

Theorem. *If*

$$\lim_{n \to \infty} n^{-1} \sum_{j=0}^{n} p_j^{-1} = \lim_{n \to \infty} n^{-1} \sum_{j=0}^{n} p_{-j}^{-1} = 2\gamma \qquad (2)$$

for some constant γ, $1 < \gamma < \infty$, *then* $n^{-\frac{1}{2}} X_n \xrightarrow{d} N(0, \gamma^{-1})$, *while* $EX_n^2 \sim \gamma^{-1} n$ *and* $EY_n^2 \sim (1 - \gamma^{-1})n$ *as* $n \to \infty$.

If

$$\lim_{n \to \infty} n^{-1} \sum_{j=0}^{n} (1 - 2p_j)^{-1} = \lim_{n \to \infty} n^{-1} \sum_{j=0}^{n} (1 - 2p_{-j})^{-1}$$
$$= (1 - \gamma^{-1})^{-1}, \qquad (3)$$

then $n^{-\frac{1}{2}} Y_n \xrightarrow{d} N(0, 1 - \gamma^{-1})$ *as* $n \to \infty$.

If both (2) and (3) hold, then

$$n^{-\frac{1}{2}}(X_n, Y_n)' \xrightarrow{d} MVN(0, \mathrm{diag}(\gamma^{-1}, 1 - \gamma^{-1}))$$

as $n \to \infty$.

Proof. As mentioned above, the result $n^{-\frac{1}{2}}X_n \xrightarrow{d} N(0, \gamma^{-1})$ as $n \to \infty$ under a slightly stronger condition than (2) has been established in [3] and we indicate the modifications necessary to the proof therein.

Let $\sigma_0 = 0 < \sigma_1 < \sigma_2 < \dots$ be the successive times at which the values of the $X_i - X_{i-1}$, $i = 1, 2, \dots$ are nonzero and put $S_k = X_{\sigma_k}$. Then $\{S_k, k \geq 0\}$ is a simple random walk, Since $X_n = X_{\sigma_k}$ for $\sigma_k \leq n < \sigma_{k+1}$, the proof of Anscombe's theorem (e.g. [1], p. 317) shows that to establish the required asymptotic normality result for X_n it suffices to prove that

$$n^{-1}\sigma_n \xrightarrow{p} \gamma \tag{4}$$

as $n \to \infty$ under (2).

Now from the proof of Theorem 2 of [3] we see that

$$W_n = \sigma_n - \frac{1}{2} \sum_{j=0}^{n-1} p_{S_j}^{-1}$$

is a martingale and

$$n^{-1} \sum_{j=0}^{n-1} p_{S_j}^{-1} \xrightarrow{\text{a.s.}} 2\gamma$$

as $n \to \infty$, so to obtain (4) it suffices to show that $n^{-1}W_n \xrightarrow{p} 0$.

Next, for large $A > 0$ we put

$$W_n' = \sum_{j=0}^{n-1}(W_{j+1} - W_j)I(|S_j| \leq A\, n^{\frac{1}{2}}),$$

I denoting the indicator function, and note that

$$P(W_n \neq W_n') \leq P(\max_{j \leq n} |S_j| > A\, n^{\frac{1}{2}}) \to 0$$

as $n \to \infty$, so that $n^{-1}W_n \xrightarrow{p} 0$ if $n^{-1}W_n' \xrightarrow{p} 0$.

If $\mathcal{F}_k$ is the σ-field generated by $S_1, \ldots, S_k, \sigma_1, \ldots, \sigma_k$, $k \geq 1$ and $\mathcal{F}_0$ is the trivial σ-field we have, since the $W_{j+1} - W_j$ are martingale differences and W_j is $\mathcal{F}_j$-measurable,

$$n^{-2} E(W_n')^2 = n^{-2} E[\sum_{j=0}^{n-1} E\{(W_{j+1} - W_j)^2 \mid \mathcal{F}_j)\} I(|S_j| \leq A n^{\frac{1}{2}})]$$

$$\leq (2n)^{-2} E[\sum_{j=0}^{n-1} p_{S_j}^{-2} I(|S_j| \leq A n^{\frac{1}{2}})] \tag{5}$$

$$\leq (2n)^{-2} (\max_{|z| \leq A n^{\frac{1}{2}}} p_z^{-1}) E[\sum_{j=0}^{n-1} p_{S_j}^{-1} I(|S_j| \leq A n^{\frac{1}{2}})].$$

Furthermore, if $N_n(z)$ is the number of $j \leq n$ with $S_j = z$, then

$$\sum_{j=0}^{n-1} p_{S_j}^{-1} I(|S_j| \leq A n^{\frac{1}{2}}) = \sum_{|z| \leq A n^{\frac{1}{2}}} N_n(z) p_z^{-1}$$

and, since $P(S_j = z) \leq C|j|^{-\frac{1}{2}}$ for some $C > 0$ (e.g. [6], p. 72) then $\sup_z E N_n(z) \leq C n^{\frac{1}{2}}$ and

$$E[\sum_{j=0}^{n-1} p_{S_j}^{-1} I(|S_j| \leq A n^{\frac{1}{2}})] = O(n^{\frac{1}{2}} \sum_{|z| \leq A n^{\frac{1}{2}}} p_z^{-1}) = O(n). \tag{6}$$

Finally, from (2) we have

$$\max_{|z| \leq A n^{\frac{1}{2}}} p_z^{-1} = o(n^{\frac{1}{2}}) \tag{7}$$

and using (6) and (7) in (5) gives $n^{-2} E(W_n')^2 = o(1)$ and hence $n^{-1} W_n' \overset{P}{\longrightarrow} 0$ as $n \to \infty$. This completes the proof of $n^{-\frac{1}{2}} X_n \overset{d}{\longrightarrow} N(0, \gamma^{-1})$.

That $E X_n^2 \sim \gamma^{-1} n$ as $n \to \infty$ follows from the proof of Corollary 2 of [3].

To obtain $EY_n^2 \sim (1 - \gamma^{-1})n$ as $n \to \infty$ we first calculate the conditional moments of X_{n+1}^2 and Y_{n+1}^2. Indeed, if $\mathcal{G}_k$ is the σ-field generated by $X_j, Y_j, j \leq k$, it follows from (1) that

$$E(Y_{n+1}^2 \mid \mathcal{G}_n) = Y_n^2 + 1 - 2p_{X_n}, \quad E(X_{n+1}^2 \mid \mathcal{G}_n) = X_n^2 + 2p_{X_n},$$

so that

$$E(X_{n+1}^2 + Y_{n+1}^2 \mid \mathcal{G}_n) = X_n^2 + Y_n^2 + 1$$

from which we deduce that $E(X_n^2 + Y_n^2) = n$. Then, $EY_n^2 \sim (1 - \gamma^{-1})n$ follows since $EX_n^2 \sim \gamma^{-1}n$ as $n \to \infty$.

The result for the asymptotic distribution of Y_n follows readily from the corresponding result for X_n. We can take $\tau_0 = 0 < \tau_1 < \tau_2 < \ldots$ as the successive times at which the values of $Y_i - Y_{i-1}$, $i = 1, 2, \ldots$ are nonzero and put $T_k = Y_{\tau_k}$. Now $\{T_k, k \geq 0\}$ is a simple random walk, and $Y_n = Y_{\tau_k}$ for $\tau_k \leq n < \tau_{k+1}$ etc. The results for X_n apply with p_j replaced by $1 - 2p_j$ for each j and lead immediately to the required result.

To obtain the asymptotic joint distribution of $n^{\frac{1}{2}}(X_n, Y_n)'$ we use the Cramér– Wold device. It is easily checked that for any real α, β, $\{\alpha X_n + \beta Y_n, \mathcal{G}_n, n \geq 1\}$ is a martingale and

$$n^{-1} \sum_{j=1}^{n} [\alpha(X_j - X_{j-1}) + \beta(Y_j - Y_{j-1})]^2 = n^{-1}[\alpha^2 k(n) + \beta^2 \ell(n)], \quad (8)$$

where

$$\sigma_{k(n)} = \max [j \; : \; j \leq n, \; X_j \neq X_{j-1}]$$
$$\tau_{\ell(n)} = \max [j \; : \; j \leq n, \; Y_j \neq Y_{j-1}].$$

But, since $n^{-1}\sigma_n \xrightarrow{P} \gamma$ and, correspondingly, $n^{-1}\tau_n \xrightarrow{P} (1 - \gamma^{-1})^{-1}$, we find that $n^{-1}k(n) \xrightarrow{P} \gamma^{-1}, n^{-1}\ell(n) \xrightarrow{P} 1 - \gamma^{-1}$ as $n \to \infty$ and consequently, from (8), that

$$n^{-1} \sum_{j=1}^{n} [\alpha(X_j - X_{j-1}) + \beta(Y_j - Y_{j-1})]^2 \xrightarrow{P} \alpha^2 \gamma^{-1} + \beta^2(1 - \gamma^{-1}). \quad (9)$$

Next, we have from (1) and the asymptotic results for EX_n^2, EY_n^2 that

$$s_n^2 = E(\alpha X_n + \beta Y_n)^2 = \alpha^2 E X_n^2 + \beta^2 E Y_n^2$$
$$\sim [\alpha^2 \gamma^{-1} + \beta^2(1 - \gamma^{-1})]n \quad (10)$$

as $n \to \infty$.

Finally, since $|X_j - X_{j-1}| \leq 1$, $|Y_j - Y_{j-1}| \leq 1$, the martingale central limit result of Theorem 3.2, p. 58 of [2] applies and

$$s_n^{-1}(\alpha X_n + \beta Y_n) \xrightarrow{d} N(0,1)$$

i.e., in view of (10),

$$n^{-\frac{1}{2}}(\alpha X_n + \beta Y_n) \xrightarrow{d} N(0, \alpha^2 \gamma^{-1} + \beta^2(1 - \gamma^{-1})).$$

The required result then follows since α and β are arbitrary real numbers.

3. Final Remark.

The so called <u>dimensional anisotropy</u> is $1 - \gamma^{-1}/\gamma^{-1}$ and for a material like TTF–TCNQ we expect this to be, say 10^{-2}. It should be noted that this would be achieved, for example, if only every 100th column were connective. We would have $p_j = \frac{1}{2}$ for j a non–connective column and $p_j = \frac{1}{4}$ for j a connective column.

REFERENCES

[1] Chow, Y.S. and Teicher, H. (1978). *Probability Theory. Independence, Interchangeability, Martingales*, Springer, New York.

[2] Hall, P. and Heyde, C.C. (1980) *Martingale Limit Theory and its Application*, Academic Press, New York.

[3] Heyde, C.C. (1982). On the asymptotic behaviour of random walks on an anisotropic lattice. *J. Statist. Phys.* **27**, 721-730.

[4] Shuler, K.E. and Mohanty, U. (1981). Random walk properties from lattice bond enumeration: Anisotropic diffusion in lattices with periodic and randomly distributed scatterers. *Proc. Natl. Acad. Sci. USA.* **78**, 6576-6578.

[5] Silver, H., Shuler, K.E. and Lindenberg, K. (1988). Two dimensional anisotropic walks. In *Statistical Mechanics and Statistical Methods in Theory and Application*, U. Landman Ed., Plenum, New York, pp. 463- 505.

[6] Spitzer, F. (1964). *Principles of Random Walk*. Van Nostrand, Princeton, N.J.

Statistics Research Section, School of Mathematical Sciences, Australian National University, GPO Box 4, Canberra ACT 2601, Australia

A role of the Lévy Laplacian in the causal calculus of generalized white noise functionals

Takeyuki Hida

Abstract. The Lévy Laplacian Δ_L plays an important role in the white noise analysis when it is considered as an infinite dimensional harmonic analysis arising from the rotation group. The Δ_L acts on the space of generalized white noise functionals effectively and enjoys different characters from ∞-dimensional Laplace-Beltrami operator.

§0. Introduction

P. Lévy proposed a Laplacian in the study of functionals defined on a Hilbert space $H = L^2([0,1])$ and showed that its properties are useful for the calculus of functionals on H. The best reference to this approach is, of couse, his book [2] published in 1 (also see the monograph [4]).

We have been motivated by Lévy's work in establishing our white noise analysis, and we can now see a strong relationship between harmonic analysis (in particular, the theory of the rotation group and the Laplacian operator) and our theory of the *causal calculus*.

This paper, after giving a short review of white noise analysis, discusses the Lévy Laplacian within the framework of the theory of white noise analysis. Then, we proceed to the main topic which deals with functional differential equations involving the Lévy Laplacian. We hope that the solutions to those equations would have close connection with infinite-dimensional Dirichlet forms which have been investigated in the course of our analysis.

§1. Background

Let μ be a white noise measure introduced on the space E^* of generalized functions on R (see [1] for reference). It is given by the characteristic functional $C(\xi)$ of the form

$$(1) \qquad C(\xi) = \exp[-\frac{1}{2}\|\xi\|^2] = \int_{E^*} \exp[i < x,\xi >]d\mu(x).$$

Then, a complex Hilbert space $(L^2) = L^2(E^*, \mu)$ is formed. A member of (L^2) is a *white noise functional*.

The S-transform of a white noise functional $\varphi(x)$ is defined by

$$(2) \qquad (S\varphi)(\xi) = \exp[-\|\xi\|^2] \int_{E^*} \exp[< x,\xi >]\varphi(x)d\mu(x).$$

We often denote the S-transform by $U(\xi)$ and call it the U-*functional* associated with $\varphi(x)$. The S-transform gives an injective map from (L^2) to a space of continuous functionals of ξ.

The differential operator ∂_t is defined in such a way that

$$(3) \qquad \partial_t \varphi(x) = S^{-1}\{\frac{\delta}{\delta\xi(t)}(S\varphi)(\xi)\},$$

where $\frac{\delta}{\delta\xi(t)}$ is the Fréchet derivative. It is easy to see that the domain of the operator ∂_t is rich enough in (L^2). We can therefore determine the adjoint operator ∂_t^* of ∂_t:

$$< \partial_t\varphi, \psi > = < \varphi, \partial_t^*\psi >_\mu,$$

where $<, >_\mu$ is the inner product in (L^2).

The operator ∂_t is, in terms of quantum dynamics, the annihilation operator, while the ∂_t^* is the creation operator. Multiplication by $x(t)$, called multiplication oprerator and denoted by m_t, is given by the formula

$$(4) \qquad m_t = \partial_t^* + \partial_t.$$

The index t may be thought of as the time parameter, so that ∂_t and ∂_t^* as well as m_t are fitting to carry on the calculus of white noise functionals that describe random phenomena chaging as time t goes by. Such a calculus is called the *causal calculus* because the time development can explicitly be described by using those operators.

In order to develop the causal calculus and for other reasons, it is necessary to introduce a larger class of white noise functionals, namely a class of *generalized white noise functionals*. Like the Schwartz space **S** of test functionals on **R**, we can introduce a nuclear space (see e.g. [3]) (**S**), and its dual space (**S**)* so that we have a Gel'fand triple:

$$(\mathbf{S}) \subset (L^2) \subset (\mathbf{S})^*.$$

The space (**S**) is an algebra consisting of test functionals. Thus, on the space (**S**)* we can carry on the causal calculus without worrying about the domain problem.

§2. The Lévy Laplacian

As suggested in Lévy's book [2], we define the Lévy Laplacian Δ_L by the following formula.

$$(5) \qquad \Delta_L\varphi = \int S^{-1}\{\frac{\delta^2}{\delta\xi(t)^2}U(\xi)\}(x)dt,$$

$$U(\xi) = (S\varphi)(\xi),$$

where $\frac{\delta^2}{\delta\xi(t)^2}$ is the second order functional derivative.

It is known (see e.g. [8], [9], [10]) that Δ_L has a domain large enough in the space $(\mathbf{S})^*$ and it plays a very important role in the causal calculus.

Remark: There are two kinds of the second order functional derivatives; one is as is used above and the other is $\frac{\delta^2}{\delta\xi(t)\delta\xi(s)}$. The lévy Laplacian actually uses the trace of the former operator.

An auxiliary operator is now introduced. Let $\gamma_{s,t}$ be given by

$$(6) \qquad\qquad \gamma_{s,t} = \partial_s^* \partial_t - \partial_t^* \partial_s, s, t \in \mathbf{R}.$$

Recall that the infinitesimal generator of the rotation around the origin in the (x_j, x_k)-plane $\subset \mathbf{R}^n$ can be expressed in the form

$$(7) \qquad\qquad x_j \cdot \frac{\partial}{\partial x_k} - x_k \cdot \frac{\partial}{\partial x_j}.$$

From the view point of our causal calculus, where time-development is involved explicitly, the formula (6) seems to be a reasonable extension of (7). Indeed, a formal analogue of (7) may be taken to be

$$m_s \cdot \partial_t - m_t \cdot \partial_s,$$

where m_t is the multiplication operator defined before. Using the formula (4), the above operator turns out to be $\gamma_{s,t}$. Such operators are useful in the characterization of Laplacian operators.

There is another Laplacian operator which is denoted by Δ_∞ and is called the *infinite dimensional Laplace-Beltrami operator* or the *infinite dimensional Ornstein-Uhlenbeck operator*. It is expressible as

$$(8) \qquad\qquad \Delta_\infty = - \int \partial_t^* \partial_t dt.$$

Remark: $-\Delta_\infty$ is the *number operator* in the language of quantum mechanics.

It should be noted that the operator Δ_L can be completely characterized by the infinite dimensional rotation group. This has been discussed in Obata [10]. Such a property is an infinite dimensional analogue of the well known fact that the (finite-dimensional) Laplace-Beltrami operator is characterized by the rotation group.

One of the crucial difference between the Lévy Laplacian Δ_L and the Laplace-Beltrami operator Δ_∞ is that Δ_∞ acts on the space (L^2) while Δ_L annihilates (L^2) and acts effectively on the space $(\mathbf{S})^*$ of generalized white noise functionals. Some of the characteristic properties of Δ_∞ and Δ_L can be found in [8], [9] and [10] and others.

Before closing this section, a few remarks are now in order. H.-H.Kuo has shown that Δ_L may be expressed in the form

$$(9) \qquad \Delta_L = \int \partial_t^2 (dt)^2.$$

Although the above integral has only formal meaning, plausible interpretation can be given to (9). When the Lévy's original idea is emphasized, one is given by the expression

$$(10) \qquad \Delta_L = \lim_{N \to \infty} \frac{1}{N} \sum_{n=1}^{N} \partial_n^2,$$

where ∂_n stands for the functional derivative with respect to the variable $< x, \xi_n >$, $\{\xi_n\}$ being a complete orthonormal system in $L^2(\mathbf{R})$.

Many interesting results have been obtained in connection with the Lévy Laplacian that is acting on the space $(\mathbf{S})^*$ which is our favorite space.

§3. Eigenfunctionals of Δ_L

From this section onward, the time variable t is assumed to run over the unit interval $[0,1]$, and hence the basic Hilbert space is $L^2([0,1])$.

We start with the eigenvalue problem for Δ_L. Namely, we shall first consider the following equation

$$(11) \qquad \Delta_L \varphi_\lambda = -\lambda \varphi_\lambda, \varphi_\lambda \in (\mathbf{S})^*, \lambda \in \mathbf{R}.$$

Before we come to the actual computation, we need to introduce some notation. Set

$$(12) \qquad \tilde{\Delta}_L = S\Delta_L = \int_0^1 \frac{\delta^2}{\delta\xi(t)^2} dt.$$

It is the operator acting on the U-functionals of the test function ξ. Then we have

LEMMA 1 (OBATA [10]). *The operator* $\tilde{\Delta}_L$ *is a derivation:*

$$(13) \qquad \tilde{\Delta}_L(U_1(\xi) \cdot U_2(\xi)) = \tilde{\Delta}_L U_1(\xi) \cdot U_2(\xi) + U_1(\xi) \cdot \tilde{\Delta}_L U_2(\xi)$$

for any U-functionals U_1 and U_2 in the domain of $\tilde{\Delta}_L$.

PROOF: Take the variation of the product $U_1(\xi) \cdot U_2(\xi)$ to see

$$U_1(\xi + \delta\xi) \cdot U_2(\xi + \delta\xi) - U_1(\xi) \cdot U_2(\xi)$$

$$= U_1(\xi + \delta\xi) \cdot \{ \int U_{2\xi}(t)\delta\xi(t)dt + \frac{1}{2} \int U_{2\xi\xi}(t)(\delta\xi(t))^2 dt$$

$$+\frac{1}{2} \iint U_{2\xi\eta}(t,s)\delta\xi(t)\delta\xi(s)dtds \} + o(|\delta\xi|^2)$$

$$+U_2(\xi) \cdot \{ \int U_{1\xi}(t)\delta\xi(t)dt + \frac{1}{2} \int U_{1\xi\xi}(t)(\delta\xi(t))^2 dt$$

$$+\frac{1}{2} \iint U_{1\xi\eta}(t,s)\delta\xi(t)\delta\xi(s)dtds \} + o(|\delta\xi|^2).$$

(In all the above integrals, t and s extend over the unit interval.) The Lévy Laplacian comes only from the terms involving $(\delta\xi(t))^2$ and so we must have

$$\tilde{\Delta}_L(U_1 \cdot U_2)(\xi) = U_1(\xi) \cdot \tilde{\Delta}_L U_2(\xi) + \tilde{\Delta}_L U_1(\xi) \cdot U_2(\xi),$$

which was to be proved. ∎

We now introduce a notion of a product of generalised functionals. Let $U_1(\xi)$ and $U_2(\xi)$ be U-functionals associated with the functionals φ_1 and φ_2 in $(\mathbf{S})^*$, respectively. If the product $U_1(\xi) \cdot U_2(\xi)$ is again a U-functional (a necessary and sufficient condition for a functional of ξ to be a U-functional is given in [11]. Roughly speaking, a functional of ξ is a U-functional if it is ray analytic and is entire of exponential type of order at most 2), then we define the ◇-product by the following formula

$$(14) \qquad\qquad \varphi_1 \diamond \varphi_2 = S^{-1}\{U_1(\xi)U_2(\xi)\}.$$

PROPOSITION. *The equation (11) has a solution in* $(\mathbf{S})^*$, *which is expressed in the form*

$$(15) \qquad\qquad \varphi_\lambda(x) = \psi(x)\diamond : \exp[-\frac{1}{2}\lambda r(x)^2] :,$$

where ψ *is harmonic in the sense that* $\Delta_L\psi = 0$, *and where* r^2 *is the generalized white noise functional, the S-transform of which is* $\|\xi\|^2$.

Remark : : in (15) means the renormalized functional.

Before we come to the proof of this proposition, a short remark is in order. Using a complete orthonormal system $\{\xi_n\}$ in $L^2([0,1])$, the so-called *infinite dimensional quasi-metric* $r(x)$ is defined by

$$r(x)^2 = \sum_n \{< x,\xi >^2 -1\}.$$

In fact, $r(x)^2$ can be defined only as a *generalized* white noise functional, since the sum diverges in (L^2), but it does converge in the space $(\mathbf{S})^*$. The associated U-functional is

$$S(r^2)(\xi) = \sum_n < \xi, \xi_n >^2 = \|\xi\|^2$$

(see Saitô [12]). This $r(x)^2$ is in the domain of Δ_L and

(16) $$\Delta_L r(x)^2 = 2$$

holds. Or equivalently,

$$\tilde{\Delta}_L \|\xi\|^2 = 2,$$

which is thought of as an infinite dimensional analogue of the relation between the Laplacian operator and the Euclidean metric.

PROOF OF PROPOSITION: Let $U_\lambda(\xi)$ be the U-functional associated with $\varphi_\lambda(x)$. Then, the equation (11) is equivalent to

(11') $$\tilde{\Delta}_L U_\lambda(\xi) = -\lambda U_\lambda(\xi).$$

Since the equation (11') is linear and Δ_L has a property like a first order differential operator, as in Lemma 1, a special solution to the equation (11') is easily obtained by using the equality (13):

$$U_{\lambda 0}(\xi) = A \cdot \exp[-\tfrac{1}{2}\lambda\|\xi\|^2], \quad A \text{ constant.}$$

Suppose there are two solutions $U_{\lambda 1}$ and $U_{\lambda 2}$ to (11'). Then,

$$\tilde{\Delta}_L\{\frac{U_{\lambda 1}}{U_{\lambda 2}}\} = \tilde{\Delta}_L U_{\lambda 1} \cdot U_{\lambda 2}^{-1} + U_{\lambda 1} \cdot \tilde{\Delta}_L U_{\lambda 2}^{-1}$$

$$= -\lambda U_{\lambda 1} \cdot U_{\lambda 2}^{-1} + U_{\lambda 1} \cdot [\lambda U_{\lambda 2} \cdot U_{\lambda 2}^{-2}] = 0.$$

This shows that any solution to (11') is expressible as a product of $U_{\lambda 0}$ given above and a $\tilde{\Delta}_L$-harmonic functional $H(\xi)$. This $H(\xi)$ has to be a U-functional satisfying the condition given by Potthoff and Streit [11], because the $U_{\lambda 0}$ is of the following particular type

const.$\cdot$exp [quadratic function of ξ].

Set $(S^{-1}H)(x) = \psi(x)$, then $\psi(x)$ is Δ_L-harmonic. This completes the proof. ∎

§4. Functional equations involving Δ_L

This section is devoted to the functional differential equations of the following type

$$(17) \qquad\qquad (-\Delta_L + \alpha(x)\diamond)f_\lambda(x) = \lambda f_\lambda(x).$$

In terms of U-functionals, (17) can be expressed in the form

$$(17') \qquad\qquad (-\tilde{\Delta}_L + A(\xi))\Phi_\lambda(\xi) = \lambda\Phi_\lambda(\xi),$$

where $\Phi_\lambda(\xi)$ and $A(\xi)$ are the U-functionals associated with $f_\lambda(x)$ and $\alpha(x)$, respectively.

To concretize the solutions to these equations we assume that $A(\xi)$ is expressible as

$$A(\xi) = \int g(\xi(t))dt.$$

We now have

LEMMA 2. *The functional $W(\xi)$ be defined by*

$$(19) \qquad W(\xi) = H(\xi) \cdot \exp[\int G(\xi(t))dt] \ \ with \ \ G''(t) = g(t),$$

where $\tilde{\Delta}_L H(\xi) = 0$,. Then W is a solution to the equation

$$(20) \qquad\qquad \tilde{\Delta}_L W(\xi) = A(\xi) \cdot W(\xi).$$

PROOF: Again use Lemma 1. Then, we have

$$\tilde{\Delta}_L W(\xi) = (\tilde{\Delta}_L H(\xi)) \cdot W(\xi) + H(\xi) \cdot \tilde{\Delta}_L \exp[\int G(\xi(t))dt]$$

$$= 0 + H(\xi) \cdot \int G''(\xi(t))dt \cdot \exp[\int G(\xi(t))dt]$$

$$= H(\xi) \cdot A(\xi) \cdot \exp[\int G(\xi(t))dt]$$

$$= A(\xi) \cdot W(\xi).$$

to arrive at the conclusion. ∎

We are now ready to discuss the equation (17').

THEOREM. *Assume that* $W(\xi)$, *which is given by the formula (19), is a U-functional, and take the solution* $\varphi_\lambda(x)$ *to the equation (11), given by (15). Set* $S^{-1}W(\xi) = w(x)$. *Then,*

    i) $f_\lambda(x) = w(x) \diamond \varphi_\lambda(x)$ *is in* $(S)^*$

or equivalently,

    i') $\Phi_\lambda(\xi) = W(\xi) \cdot U_\lambda(\xi)$ *is a U-functional.*

    ii) $f_\lambda(x)$ *is a solution to the equation (17).*

PROOF: i) By the assumption on W, the Potthoff-Streit condition in [11] is guaranteed for $\Phi_\lambda(\xi)$: the functional has a ray entire extension and is exponential type at most of order 2. Hence, the assertion i') follows immediately.

    ii) We have

$$\tilde{\Delta}_L \Phi_\lambda(\xi) = (\tilde{\Delta}_L W(\xi)) \cdot U_\lambda(\xi) + W(\xi) \cdot (\tilde{\Delta}_L U_\lambda(\xi))$$

$$= A(\xi)W(\xi)U_\lambda(\xi) - \lambda W(\xi)U_\lambda(\xi)$$

$$= \{A(\xi) - \lambda\}\Phi_\lambda(\xi)$$

which implies (17'). ∎

    So far the eigenvalue problems have been discussed, and with minor assumptions the solutions have been obtained. Here one can see some similarity to the finite dimensional case, as was expected. However, there are many dissimilarities as well; some come from the unusual property that the Lévy Laplacian is a derivation. Still we hope that the present approach will find close connection with the results in [5], for instance admissible densities (living in $(S)^*$) could be obtained in the manner established above.

References

[1] G.Kallianpur and R.L.Karandikar, White noise theory of prediction, filtering and smoothing. Stochastic Monographs vol.3, 1988, Gordon and Breach Science Pub.

[2] P. Lévy, Problèmes concrets d'analyse fonctionnelle. 1951, Gauthier-Villars.

[3] T.Hida, Brownian motion. Applications of Mathematics vol.11, 1980, Springer-Verlag.

[4] T.Hida, H.-H.Kuo, J.Potthoff and L.Streit, White noise - An infinite dimensional calculus. Monograph to appear.

[5] T.Hida, J.Potthoff and L.Streit, White noise analysis and applications. Mathematics + Physics vol.3, ed. L.Streit, 1988, World Scientific, 143-178.

[6] T.Hida, J.Pothoff and L.Streit, Dirichlet forms and white noise analysis. Commun. Math. Physics. 116 (1988), 235-245.

[7] T.Hida, N.Obata and K.Saitô, Infinite dimensional rotations and Laplacians in terms of white noise calculus. preprint.

[8] H.-H.Kuo, On Laplacian operators of generalized Brownian functionals. Lecture Notes in Math. vol.1203, Springer-Verlag, 1986, 119-128.

[9] H.-H.Kuo, N.Obata and K.Saitô, Lévy Laplacian of generalized functions on a nuclear space. J. Funct. Anal. 94 (1990), 74-92.

[10] N.Obata, A characterization of the Lévy Laplacian in terms of infinite dimensional rotation groups. Nagoya Math. J. 118 (1990), 111-132.

[11] J.Potthoff and L.Streit, A characterization of Hida distributions. J. Funct. Anal. 101 (1991), 212-229.

(See, by the same authors, Generalized Radon-Nikodym derivatives and Cameron-Martin theory. Proceedings of the Conference on Gaussian Random Fields held at Nagoya 1990. ed. K.Itô and T.Hida.)

[12] K.Saitô, Itô's formula and Lévy's Laplacian. I and II, Nagoya Math. J. 108 (1987), 67-76; 123 (1991), 153-169.

[13] L.Streit and T.Hida, Generalized Brownian functionals and the Feynman integral. Stochastic Processes and their Applications 16 (1983) 55-69.

Department of Mathematics, Meijo University,
Nagoya 468, Japan.

On the Approximation of Multiple Stratonovich Integrals

by Y.Z. Hu and P.A. Meyer

ABSTRACT. We present in an unified way the results of several recent papers on the definition of multiple Stratonovich integrals.

1. Introduction. In a previous paper (Hu–Meyer [1]) we have shown that multiple Stratonovich integrals provide a natural way to "carry" a random variable F on the standard Wiener space, given by its Wiener–Ito expansion $F = \sum_n I_n(f_n)/n!$, to a Wiener space with a different variance parameter (possibly 0), while preserving the multiplicative structure. In this notation, $I_n(f)$ is the multiple Ito integral of the symmetric function f of n variables. One transforms this expansion into a Stratonovich expansion $F = \sum_n S_n(g_n)/n!$, and then the coefficients g_n are kept fixed. The Stratonovich integral $S_m(f)$, on the other hand, is expressed by the following formula using Ito integrals

$$(1) \qquad S_m(f) = \sum_{2k \leq n} \frac{m!}{2^k k! (m-2k)!} \, I_{m-2k}(\operatorname{Tr}^k f) \,.$$

This formula isn't completely rigorous : the *iterated traces* it contains are defined formally starting with

$$\operatorname{Tr} f(s_1, \ldots, s_{m-2}) = \int f(s_1, \ldots, s_{m-2}, s, s) \, ds$$

and their meaning is clear only when f is an "elementary function"; these results are recalled in Section 3. If one tries to extend them by means of a Hilbertian interpretion of traces, following Johnson-Kallianpur [2] [3], the theory seems a little too restrictive to include the case of solutions of stochastic differential equations. On the other hand, a topological definition of traces is not entirely natural, since it makes use of an additional structure — see however Sugita [6]. Up to now, the most natural definition for this problem seems to be that given by Solé–Utzet [5]. A similar, but simpler, definition has been suggested very recently (without details) in a note due to Russo and Vallois.

For another approach to formula (1) using Hida's theory of white noise, see Yan [7]. Working on $\mathbb{R}^m$ instead of $\mathbb{R}_+^m$ (the special role of 0 being troublesome in a theory of distributions) and assuming that f is a test–function, the integral $S_m(f)$ may indeed be rewritten as

$$\int f(s_1, \ldots, s_m) \, \dot{X}_{s_1}(\omega) \ldots \dot{X}_{s_m}(\omega) \, ds_1 \ldots ds_n = \, < f , \dot{X}^{\otimes n} >$$

where the derivative of brownian motion is understood in the distribution sense.

Our purpose in this note is the unification of the different definitions of the Stratonovich integral we have just mentioned. To simplify our setup, we remain on the standard Wiener space ($\sigma^2 = 1$), and we try to define, under reasonably general conditions, a Stratonovich integral

$$(2) \qquad S(f) = \sum_m \frac{1}{m!} \int_{(S)} f_m(s_1, \ldots, s_m) \, dX_{s_1}(\omega) \ldots dX_{s_m}(\omega)$$

where f is a finite sequence of coefficients $f_m \in L_s^2(\mathbb{R}^m)$ (the symmetric L^2 space).

It is a great pleasure for us to dedicate this work to Prof. G. Kallianpur, whose work on the Feynman integral has been a constant source of inspiration for us.

2. Approximation methods. Since f is a finite sequence, we can define the following function on the Cameron–Martin space

$$(3) \qquad F(h) = \sum_m \frac{1}{m!} \int f_m(s_1, \ldots, s_m) \, \dot{h}(s_1) \ldots \dot{h}(s_m) \, ds_1 \ldots ds_m \ ,$$

a (non homogeneous) polynomial of finite degree.

We define the Stratonovich integral $S(f)$ as the limit, when it exists in L^2, of the r.v.'s $F \circ \alpha_i(\omega)$, where $\alpha_i(\omega)$ denotes a family of approximations of the path ω by Cameron–Martin functions. To each approximation procedure corresponds in this way a method for the computation of the "traces" in formula (1). The previously described results correspond to the following approximation procedures.

— (C) (for convolution) is the classical regularization method using convolution, depending on a parameter $\varepsilon \downarrow 0$

$$X_t^\varepsilon = \frac{1}{\varepsilon} \int_{(t-\varepsilon)^+}^t X_s \, ds \ .$$

This procedure, used for instance by Malliavin, leads to a differentiable function, but not to a Cameron–Martin one: to improve this, one may replace X_s by $X_{s \wedge (1/\varepsilon)}$ inside the integral. Russo and Vallois suggested a definition of traces using this method.

— (H) (for Hilbert) is the Hilbertian regularization method of Johnson–Kallianpur. An orthonormal basis (e_n) of the Cameron-Martin space having been selected, the path ω is formally expanded using this basis, with coefficients

$$c_n(\omega) = \int \dot{e}_n(s) \, dX_s(\omega) \ ,$$

and the Cameron–Martin approximation then is the partial sum $\sum_{k=1}^n c_k(\omega) \, e_k$.

— (P) (for polygonal) is the procedure of Solé–Utzet: with each finite subdivision $\Delta = \{0 = t_0 < \ldots < t_m < \infty\}$ of the line, we associate a linear interpolation of the path X.

$$X_t^\Delta = X_{t_i} + \frac{X(t_{i+1}) - X(t_i)}{t_{i+1} - t_i} (t - t_i) \quad \text{for} \quad t_i < t \le t_{i+1} \ , \quad = X_{t_m} \quad \text{for} \quad t > t_m \ .$$

To get a convenient sequence of approximations, one may use the dyadic subdivision Δ_n of step 2^{-n} on the interval $[0, 2^n]$.

All three procedures have a common definition as follows: one constructs a differentiable approximation $\alpha_\varphi(\omega)$ by means of a formula

$$(4) \qquad \dot{X}_s(\alpha(\omega)) = \int_0^\infty \varphi(s, t) \, dX_t(\omega)$$

where $\varphi(s, \cdot)$ tends, for fixed s, towards the Dirac function at s. For the procedure (C), one takes

$$(5) \qquad \varphi(s, t) = \frac{1}{\varepsilon} I_{]\,(s-\varepsilon)^+, s\wedge(1/\varepsilon)\,]}(t) \, ,$$

for the procedure (H),

$$(6) \qquad \varphi(s, t) = \sum_{k \le n} \dot{e}_k(s) \, \dot{e}_k(t) \, .$$

To describe (P), we call Δ_i the i-th interval $]\,t_{i-1}, t_i\,]$ $(i = 1, \ldots, m)$ of the subdivision, its length is called δ_i, and its indicator function is χ_i. Then we have

$$(7) \qquad \varphi(s, t) = \sum_i \frac{1}{\delta_i} \chi_i(s) \chi_i(t) \, .$$

In all three cases we have

$$(8) \qquad \int_0^\infty \varphi(s, t)^2 \, ds\,dt < \infty \, ,$$

a condition implying that the approximating path a.s. belongs to the Cameron–Martin space. We also put

$$(9) \qquad \psi(u, v) = \int \varphi(s, u) \varphi(s, v) \, ds$$

which according to (8) belongs to $L^2(\mathbb{R}^2_+)$.

On sequences of chaotic coefficients, or on symmetric functions of several variables, let us define two operators. The first one, $R = R_\varphi$, is a *smoothing* operator which preserves the number of variables. It maps a function $f(s_1, \ldots, s_m)$ into the function

$$(10) \quad Rf(s_1, \ldots, s_m) = \int f(t_1, \ldots, t_m) \varphi(s_1, t_1) \ldots \varphi(s_m, t_m) \, dt_1 \ldots dt_m \, .$$

(If $m = 0$, one takes $Rf = f$). The operator R on functions of m variables is the m-th tensor power of the same operator R on functions of one variable, which in case (H) is a projection, in case (C) a convolution operator, and in case (P) a conditional expectation. *Thus R has, in all three cases, an operator* *norm bounded by* 1. On the other hand, these operators converge strongly to the identity for $n \to \infty$ (H), $\varepsilon \to 0$ (C) or $\sup_i \delta_i \to 0$ (P).

The role of the operator R may be explained as follows (without going into all details, since we will not use the result). According to (8), the function φ defines a Hilbert–Schmidt kernel. In the same way as the product of two Hilbert–Schmidt operators is known to belong to the trace class, the smoothing operation R transforms a function f_m, which is only square integrable, into a function possessing iterated traces of all orders in the Hilbert space sense, and to which therefore the theory developed by Johnson–Kallianpur may be applied.

The second operator, denoted by $C = C_\varphi$, is a *contracted product* (in the sense of tensor calculus) with the function ψ. Il decreases by two the number of variables and transforms the function $f(s_1, \ldots, s_m)$ into

$$(11) \qquad Cf(s_1, \ldots, s_{m-2}) = \int f(s_1, \ldots, s_{m-2}, u, v)\, \psi(u, v)\, du\, dv \ .$$

For $m = 0, 1$ we define it to be 0. It is a bounded operator, but its operator norm tends to infinity along the family of approximations.

3. Elementary functions. As we said before, we use the notation $I_m(f)$ for the Ito integral of a symmetric function f of m variables. For $m = 1$ we also write $I_1(f) = \widetilde{f}$. Let us recall the main results of [1].

We say that a sequence f of symmetric coefficients is *elementary* if it is finite, and if each coefficient f_m is a finite linear combination of symmetric products (*i.e.* symmetrized tensor products) $h_1 \circ \ldots \circ h_m$ or, amounting to the same thing by polarization, of functions of the form $h^{\otimes m}$. Given an elementary function, there is no difficulty in defining its trace by linearity, starting from the formula $\mathrm{Tr}\,(h^{\otimes m})) = (h, h)\, h^{\otimes(m-2)}$ for $m \geq 2$, $= 0$ for $m < 2$. Since the trace of an elementary function is again elementary, iterated traces are well defined too. On the other hand, the Stratonovich integral of an elementary function can be defined by polarization starting from the formula

$$S_m(h^{\otimes m}) = (\widetilde{h})^m \quad ; \quad \sum_m \frac{\lambda^m}{m!}\, S_m(h^{\otimes m}) = e^{\lambda \widetilde{h}} \ .$$

In particular, we have

$$S_m(h_1 \circ \ldots \circ h_m) = \widetilde{h}_1 \ldots \widetilde{h}_m \ .$$

Formula (1) is easily justified for elementary functions: one starts from the relation

$$\sum_m \frac{\lambda^m}{m!}\, S_m(h^{\otimes m}) = e^{\lambda \widetilde{h}} = e^{\lambda^2 (h,h)/2} \sum_m \frac{\lambda^m}{m!}\, I_m(h^{\otimes m}) \ .$$

Identifying the coefficients of λ^m on both sides, one gets formula (1) for $f = h^{\otimes m}$, and the case of elementary functions follows by polarization.

When f is an elementary function, Rf and Cf are again elementary. More precisely, $R(h^{\otimes m}) = (Rh)^{\otimes m}$, and $Ch^{\otimes m} = (h \otimes h, \psi) \, h^{\otimes(m-2)}$. Then an immediate verification shows that

$$S(R_\varphi(f)) = F(\alpha_\varphi(\omega)) \,, \qquad \mathrm{Tr}^k Rf = R(C^k f) \,,$$

whenever f is elementary.

4. Convergence of the approximations.

We return to the equality given by formula (1), which has been proved for elementary functions

$$(12) \qquad F \circ \alpha_\varphi = S(R_\varphi(f)) = \sum_m \sum_{2p \le m} \frac{1}{2^p p! (m - 2p)!} \, I_{m-2p}(C_\varphi^p (R_\varphi(f))) \,.$$

We can extend it without difficulty to all finite sequences of square integrable coefficients. In order to see this, we approximate in L^2 each coefficient by elementary functions (φ remaining unchanged). The left side converges in the Cameron–Martin space, and on the other hand the operators R and C^k are continous. The passage to the limit then is obvious.

Explicitly, for $f \in L^2(\mathbb{R}_+^m)$ $S_m(\alpha_\varphi(f))$ has in the chaos of order $m - 2p$ a coefficient equal to $(\, m!/2^p p! (m - 2p)! \,) \, RC^p f$, and the value of this function is

$$\int \varphi(s_1, t_1) \ldots \varphi(s_{m-2p}, t_{m-2p}) \, f(t_1, \ldots t_{m-2p}, u_1, v_1, \ldots, u_p, v_p)$$

$$\psi(u_1, v_1) \ldots \psi(u_p, v_p) \, dt_1 \ldots dt_{m-2p} \, du_1 \, dv_1 \ldots du_p \, dv_p \,.$$

We now consider the whole family of approximations depending on the parameter i. The convergence of the sequence $F \circ \alpha_i$ *is equivalent* to the convergence in $L_s^2(\mathbb{R}_+^m)$, for each m and $p \le m/2$, of the functions $R_i(C_i^p f_m)$. On the other hand, it is *sufficient* to this order that the functions $C_i^p f_m$ should converge in L^2 to a function g_{m-2p}. Indeed, we have $\| R_i(C_i^p f_m - g_{m-2p}) \| \to 0$ since the operator norm of R_i remains bounded, and besides that $R_i g_{m-2p}$ converges strongly to g_{m-2p}. In conclusion

THEOREM . Let $f \in L_s^2(\mathbb{R}_+^m)$. If for every $p \le m/2$ the following limit exists in the strong sense in L^2, along the family α_i of approximations

$$\mathrm{Tr}_\alpha^p f = \lim_i C_i^p f = g_{m-2p} \,,$$

then the Stratonovich integral $S_\alpha(f)$ exists, and its value is

$$\sum_{2p \le m} \frac{m!}{2^p p! (m - 2p)!} \, I_{m-2p}(g_{m-2p}) \,.$$

5. Comments.

Let us describe explicitly the definition of traces that corresponds to the three approximation procedures. We consider for simplicity a symmetric function $f(s,t)$ of two variables, but the general case offers only difficulties of notation.

When procedure (H) is used, the approximate trace Cf is $\sum_{k=1}^{n}(\dot{e}_k, F\dot{e}_k)$, F denoting the Hilbert–Schmidt operator with kernel f. Since there is no orthonormal basis playing a specific role in our problem, one wonders naturally in which case this series converges for *every* orthonormal basis; it is well known that such an unrestricted convergence takes place iff F is a trace class operator in the usual sense (and the series converges to the standard trace). This is too restrictive for some applications.

For procedure (C), defining the Stratonovich double integral (on the square $[0,1]^2$ for definiteness) requires the convergence of the following approximate trace

$$\lim_{\varepsilon \to 0} \frac{1}{\varepsilon} \int_{[0,1]^2} I_{\{(u-\varepsilon)^+ < v < u\}} f(u,v)\, du\, dv \ .$$

The approximate trace corresponding to procedure (P) is

$$\sum_i \frac{1}{\delta_i} \int_{\Delta_i \times \Delta_i} f(u,v)\, du\, dv \ .$$

It is interesting to compare these two traces, taking $\varepsilon = 1/n$ and the subdivision Δ consisting of the points k/n, $0 \le k \le n$. Taking into account the symmetry of $f(u,v)$, the difference between the approximate traces is $\int f(u,v) j_n(u,v)\, du\, dv$, the function j assuming the values $\pm n$, and its L^1 norm being equal to 1 ; The functions j_n are not uniformly integrable, however, and the convergence of either one of the approximate traces (which in our case are scalars, not functions) does not imply that of the other one, even if f is bounded.

On the other hand, the conditions for convergence of the approximate traces do not seem to be comparable to those that would lead to the existence, in the sense of stochastic calculus, of an *iterated* Stratonovich integral $\int_{t<s<1} f(s,t)\, dX_s\, dX_t$, and they seem in fact more natural.

Let us conclude with a remark. If the function $f(s_1,\ldots,s_n)$ is continuous on the closed simplex $\{0 \le s_1 \ldots < s_n \le t\}$ and equal to 0 for $s_n > t$, it possesses iterated traces for procedures (C) and (P), which can be computed by integration on diagonals, and share the same continuity properties as f. Thus the Stratonovich integral of such a function is well defined, and can be expressed by means of Ito integrals as indicated in formula (1). The inverse formula (expressing Ito multiple integrals by means of Stratonovich integrals) follows from it in a purely algebraic way. Otherwise stated, the formal statements in [1] are now justified for random variables which are $\mathcal{F}_t$–measurable for some finite t, and have a chaos expansion with continuous coefficients *and only finitely many terms*. Thus the interesting case, that of solutions of stochastic differential equations, is still open, since the coefficients are known to be continuous but their number is infinite.

REFERENCES

[1] HU (Y.Z.) and MEYER (P.A.). Sur les intégrales multiples de Stratonovitch, *Sém. Prob. XXII*, Lect. Notes in M. **1321**, 1987, p. 72-81.

[2] JOHNSON (G.W.) and KALLIANPUR (G.). Some remarks on Hu and Meyer's paper and infinite dimensional calculus on finitely additive canonical Hilbert space, *Teorija Verojat.*, **34**, 1989, p. 742-752.

[3] JOHNSON (G.W.) et KALLIANPUR (G.). Homogeneous chaos, p-forms, scaling and the Feynman integral, Technical Report n° 274, Center of Stoch. Processes, Univ. of North Carolina, 1989. To appear in *Trans. Amer. Math. Soc.*.

[4] RUSSO (F.) et VALLOIS (P.). Intégrales progressive, rétrograde et symétrique de processus non adaptés, *C.R. Acad. Sci. Paris*, **312**, 1991, p. 615–618.

[5] SOLÉ (J.Ll.) and UTZET (F.). Stratonovich integral and trace, *Stochastics*, **29**, 1990, p. 203-220.

[6] SUGITA (H.). Hu–Meyer's multiple Stratonovich integral and essential continuity of multiple Wiener integrals, *Bull. Sc. Math.*, **113**, 1989, p. 463–474.

[7] YAN (J.A.). Notes on the Wiener semigroup and renormalization, *Sem. Prob. XXV*, Lect. Notes in M. **1485**, 1991, p. 79–94.

[8] ZAKAI (M.). Stochastic integration, trace, and the skeleton of Wiener functionals, *Stochastics*, **32**, 1990, p. 93-108.

Institut de Recherche Mathématique Avancée
Université Louis Pasteur
7 rue René Descartes
F–67084 Strasbourg Cedex
and (Y.Z.Hu)
Institute of Math. Research
Academia Sinica
Wuhan , Hubei, P.R. of China

Two Examples of Parameter Estimation for Stochastic Partial Differential Equations

M. HÜBNER, R. KHASMINSKII, B.L. ROZOVSKII[1]

Abstract

We study parameter estimation for two types of parabolic stochastic PDE's. Examples considered in this article suggest that asymptotic properties of maximum likelihood estimators (MLE's) for Galerkin approximations to these SPDE's depend critically on certain properties of the distributions of solutions to the original equations. In particular, singularity of the distributions for different values of the parameter provides for consistency of the MLE as the dimension of the approximation approaches infinity.

1 Introduction

The aim of this work is to consider some interesting phenomena arising in parameter estimation problems for stochastic partial differential equations (SPDE's). We consider the Dirichlet problem with zero boundary conditions for the equation

$$du_\epsilon(t, x) = \mathcal{L}^\theta u_\epsilon(t, x)dt + \epsilon dW_Q(t, x)$$

where $W_Q(t, x)$ is a Wiener process in $L_2(0, 1)$ with the nuclear covariance operator Q. This problem is studied in two cases:

$$(1.1) \qquad \mathcal{L}^\theta u_\epsilon(t, x) = \frac{\partial^2}{\partial x^2} u_\epsilon(t, x) + \theta u_\epsilon(t, x)$$

and

$$(1.2) \qquad \mathcal{L}^\theta u_\epsilon(t, x) = \theta \frac{\partial^2}{\partial x^2} u_\epsilon(t, x) .$$

For $\theta \in R_+$, both equations belong to the same class and their fundamental analytical properties (existence, uniqueness, smoothness of solutions, asymptotic properties, etc.) are identical. However, as we demonstrate below, when it comes to parameter estimation, these two cases are very

[1]Work partially supported by NSF Grant No. DMS-9002997 and ONR Grant No. N00014-91-J-1526

different. The first example is in some sense routine, because the measures P_θ^ϵ generated by the solutions of these equations become singular only if the intensity parameter ϵ for the white Gaussian noise is equal to 0. The results for this case are analogous to results obtained by [Ibragimov, Khasminskii (1981)] and [Kutoyants (1984)] where this problem is studied for ordinary SDEs. Here it is possible to consider a suitable N-dimensional projection of the observation and to prove that the solution of this finite dimensional estimation problem converges uniformly in ϵ to the solution of the infinite dimensional problem. The family of measures have the LAN property and the MLE $\hat{\theta}_{N,\epsilon}$ is asymptotically normal.

A completely different situation occurs in the second example. Here even for any positive ϵ the measures P_θ^ϵ are singular for different θ's, although the measures corresponding to the finite dimensional projections are absolutely continuous. The LAN property and asymptotic normality of the MLE as $\epsilon \to 0$ for these projections hold here, too, but not uniformly in N. We will show that the accuracy of the MLE will grow very quickly in N even for fixed $\epsilon = \epsilon_0$ and the problem is to study properties of best estimators for $N \to \infty$. The MLE $\hat{\theta}_{N,\epsilon_0}$ is asymptotically normal and asymptotically efficient for any bounded loss function.

2   The Case of Absolutely Continuous Distributions

Let us fix a probability space $(\Omega, \mathcal{F}, P)$ and consider the process $u^\epsilon(t, x)$ $0 < x < 1$, $0 \le t \le T$ governed by

$$(2.1) \qquad du_\epsilon(t, x) = (\Delta u_\epsilon(t, x) + \theta u_\epsilon(t, x))\, dt + \epsilon dW_Q(t, x)$$

where $\Delta = \frac{\partial^2}{\partial x^2}$. We assume that ϵ is a small parameter, $\epsilon \to 0$ and $\theta \in \Theta \subset \mathbf{R}$.

Equation (2.1) is considered together with initial and boundary conditions

$$u_\epsilon(0, x) = f(x), \quad f \in L_2(0, 1)$$

(2.2)

$$u_\epsilon(t, 0) = u_\epsilon(t, 1) = 0\,, \quad 0 \le t \le T$$

Q is the covariance operator for the Wiener process $W_Q(t, x)$, so that

$$W_Q(t, x) = Q^{1/2}W(t, x)$$

where $W(t, x)$ is a cylindrical Brownian motion in $L_2(0, 1)$ (a Wiener process in $L_2(0, 1)$ with unity covariance operator). It is a standard fact (see e.g. [Rozovskii(1990)] that given Q is nuclear,

$$(2.3) \qquad dW_Q(t,z) = \sum_{i=1}^{\infty} q_i^{1/2} e_i(x) dW_i(t)$$

where $W_1(t), W_2(t), \ldots$ are independent one-dimensional Wiener processes and $\{e_i\}, i = 1, 2, \ldots$, is a complete orthonormal system in $L_2(0,1)$ which consists of eigenvectors of Q. We denote q_i as the eigenvalue corresponding to e_i. For the sake of simplicity we choose a special covariance operator Q and CONS $\{e_i\}$. Write $e_k := \sin k\pi x$ and $\lambda_k = (\pi k)^2$. Obviously $\{e_k\}, k = 1, 2, \ldots$ is a CONS in $L_2(0,1)$ and $\{e_k, \lambda_k\}$, $k = 1, 2, \ldots$, is a solution to the spectral problem

$$(2.4) \qquad \begin{aligned} (\Delta + \lambda I)u &= 0 \\ u(0) = u(1) &= 0 \end{aligned}$$

Now we define Q by the formula

$$Qe_i = (1 + \lambda_i)^{-1} e_i \quad , \quad i = 1, 2 \ldots .$$

In other words, we choose $Q = (1 - \Delta)^{-1}$ for the operator Δ with zero boundary conditions and $q_i := (1 + \lambda_i)^{-1}$.

We understand problem (2.2) in a generalized sense (see [Rozovskii (1990)]). Specifically we define a solution $u_\epsilon(t,x)$ to this problem as a formal sum

$$u_\epsilon(t,x) = \sum_{i=1}^{\infty} u_{i\epsilon}(t) e_i(x)$$

where the Fourier coefficients $u_{i\epsilon}(t)$ of the above expansion satisfy the equations

$$(2.5) \qquad du_{i\epsilon}(t) = (\theta - \lambda_i) u_{i\epsilon}(t) dt + \frac{\epsilon}{\sqrt{\lambda_i + 1}} dW_i(t)$$

with initial conditions

$$(2.6) \qquad u_{i\epsilon}(0) = v_i$$

Here the v_i are determined by

$$(2.7) \qquad f(x) = \sum_{i=0}^{\infty} v_i e_i(x) \quad , \quad v_i = \int_0^1 f(x) e_i(x) dx$$

It can be shown (see [Rozovskii(1990)] that $u_\epsilon(t,x)$ belongs to $L_2([0,T] \times \Omega; L_2(0,1))$ together with its derivative in x. It vanishes at $0, 1$ and its norm in $L_2(0,1)$ is continuous in t. In addition, $u_\epsilon(t,x)$ is the only solution to (2.1),(2.2) with the above properties. We get an expression for the likelihood ratio of the projection of the solution $u_\epsilon(t, x)$ onto the subspace Π^N spanned by $\{e_1, \ldots, e_N\}$ (see[Liptser, Shiryayev (1978)])

$$u_\epsilon^N(t, x) = \sum_{i=1}^N u_{i\epsilon}(t) e_i(x)$$

as follows

$$(2.8) \qquad \frac{dP_{\theta_0+\theta}^{\epsilon,N}}{dP_{\theta_0}^{\epsilon,N}}(u_\epsilon^N) =$$

$$= \exp\left\{ \sum_{i=1}^N \frac{\lambda_i+1}{\epsilon^2} \left[\theta \int_0^T u_{i\epsilon}(t) du_{i\epsilon}(t) - \frac{1}{2}(\theta^2 + 2\theta(\theta_0 - \lambda_i)) \int_0^T u_{i\epsilon}^2(t) dt \right] \right\}$$

Fisher's information is given by

$$(2.9) \qquad I_{N,\epsilon}(\theta) = \sum_{i=1}^N \frac{(\lambda_i+1)}{\epsilon^2} E \int_0^T u_{i\epsilon}^2(t) dt .$$

This can be rewritten as

$$(2.10) \qquad \epsilon^2 I_{N,\epsilon}(\theta) =$$

$$\frac{1}{2} \sum_{i=1}^N \frac{\lambda_i+1}{\lambda_i-\theta} v_i^2 \left(1 - e^{2(\theta-\lambda_i)}\right) + \frac{\epsilon^2}{2} \sum_{i=1}^N \frac{1}{\lambda_i-\theta} \left(T + \frac{1-e^{2(\theta-\lambda_i)T}}{2(\theta-\lambda_i)}\right)$$

From this expression it is clear that $\epsilon^2 I_{N,\epsilon}(\theta)$ converges uniformly in ϵ as $N \to \infty$, and uniformly in N as $\epsilon \to 0$.

$$(2.11) \qquad \lim_{N\to\infty} \lim_{\epsilon\to 0} \epsilon^2 I_{N,\epsilon}(\theta) = \lim_{\epsilon\to 0} \lim_{N\to\infty} \epsilon^2 I_{N,\epsilon}(\theta) =$$

$$= \frac{1}{2} \sum_{i=1}^\infty \frac{\lambda_i+1}{\lambda_i-\theta} v_i^2 \left(1 - e^{2(\theta-\lambda_i)T}\right) =: I(\theta) .$$

Since $f \in L_2$, we have that $I(\theta) < \infty$, and $I(\theta) > 0$ if $f \not\equiv 0$.

Hence,

$$\frac{dP_{\theta_0+\theta}^{\epsilon,N}}{dP_{\theta_0}^{\epsilon,N}}(u_\epsilon^N) \to \frac{dP_{\theta_0+\theta}^\epsilon}{dP_{\theta_0}^\epsilon}(u_\epsilon) \qquad \text{(a.s.)}$$

as $N \to \infty$, where

$$(2.12) \qquad \frac{dP_{\theta_0+\theta}^\epsilon}{dP_{\theta_0}^\epsilon}(u_\epsilon) :=$$

$$:= \exp \left\{ \sum_{i=1}^{\infty} \frac{\lambda_i + 1}{\epsilon^2} \left[\theta \int_0^T u_{i\epsilon}(t) du_{i\epsilon}(t) - \frac{1}{2}(\theta^2 + 2\theta(\theta_0 - \lambda_i)) \int_0^T u_{i\epsilon}^2(t) dt \right] \right\} .$$

Adjusting slightly the proof of Lemma 4 from [Skorohod(1965) Ch. 4, §2], one can prove that the measures generated by solutions to (2.1), (2.2) corresponding to different θ are absolutely continuous with the likelihood ratio given by (2.12) (see also [Kozlov(1978)]), where this absolute continuity was obtained by another method).

It is easy to verify that the family of measures $\{P_\theta^{\epsilon,N}\}$ have the LAN (Local Asymptotic Normality) property (see [Ibragimov, Khasminskii(1981)]) with normalizing factor ϵ. Since

$$(2.13) \quad \sum_{i=1}^{N} (\lambda_i + 1) E \int_0^T u_{i\epsilon}^2(t) dt \longrightarrow \sum_{i=1}^{N} (\lambda_i + 1) \int_0^T u_{i0}^2(t) dt \quad \text{as} \quad \epsilon \to 0$$

with $u_{i0}(t) = v_i e^{(\theta - \lambda_i)t}$, the solution of (2.5)(2.6) with $\epsilon = 0$, the normalized likelihood ratio

$$(2.14) \qquad Z_{\epsilon,N}(z) = \frac{dP_{\theta + \epsilon z}^{\epsilon,N}}{dP_{\theta}^{\epsilon,N}}(u_\epsilon^N)$$

has the limiting representation ($\epsilon \to 0$)

$$(2.15)_N \qquad Z_{\epsilon,N}(z) = \exp \left\{ z \sum_{i=1}^{N} \sqrt{\lambda_i + 1} \int_0^T u_{i0}(t) dW_i(t) - \right.$$

$$\left. - \frac{1}{2} z^2 \sum_{i=1}^{N} (\lambda_i + 1) \int_0^T u_{i0}^2(t) dt + o(1) \right\}$$

Of course, (2.13)—$(2.15)_N$ also hold if we replace the finite sums by infinite sums and have thus obtained

Theorem 1: *The measures P_θ^ϵ generated by the solution of the problem (2.1)(2.2) are absolutely continuous, have the LAN property $(2.15)_\infty$ with Fisher's information (2.11).*

Consequently, for any estimator $\theta_{\epsilon,N}^*$ based on $u_\epsilon^N(t, x)$, or θ_ϵ^* based on $u_\epsilon(t, x)$ we get by the Hajek-LeCam inequality (see [Ibragimov, Khasminskii(1981)]), a lower bound for the variance

$$(2.16) \qquad \varliminf_{\epsilon \to 0} \sup_{|\theta - \theta_0| < \delta} E_{\theta,\epsilon} w\left(\epsilon^{-1}(\theta_\epsilon^* - \theta) \right) \geq E w(\xi) ,$$

where $\mathcal{L}(\xi) = \mathcal{N}(0, I_o(\theta)^{-1})$, and an analogous inequality holds for $\theta_{\epsilon,N}^*$, where $w \in W$, the class of loss functions $w(x)$, which are symmetric, $w(0) =$

0 and monotone for $x > 0$. Denote by $W_b = \{w \in W : w \text{ is bounded}\}$ the class of bounded loss functions. The Maximum Likelihood Estimator (MLE) for the projection onto Π^N has the form

$$(2.17) \qquad \hat{\theta}_{N,\epsilon} = \left(\sum_{i=1}^{N} (\lambda_i + 1) \int_0^T u_{i\epsilon}^2(t) dt \right)^{-1} \cdot$$

$$\cdot \left(\sum_{i=1}^{N} (\lambda_i + 1) \int_0^T u_{i\epsilon}(t)(du_{i\epsilon}(t) + \lambda_i u_{i\epsilon}(t)dt) \right)$$

which we can rewrite as follows

$$(2.18) \qquad \epsilon^{-1}(\hat{\theta}_{N,\epsilon} - \theta) = \left(\sum_{i=1}^{N} (\lambda_i + 1) \int_0^T u_{i\epsilon}^2(t) dt \right)^{-1} \times$$

$$\times \left(\sum_{i=1}^{N} \sqrt{\lambda_i + 1} \int_0^T u_{i\epsilon}(t) dW_i(t) \right)$$

In view of (2.13), by a central limit theorem for stochastic integrals (see Kutoyants(1977)])

$$(2.19) \qquad \mathcal{L}(\epsilon^{-1}(\hat{\theta}_{N,\epsilon} - \theta)) \to N(0, I_{N,0}(\theta)^{-1}) \quad \text{as} \quad \epsilon \to 0$$

It follows from (2.19) that $\hat{\theta}_{N,\epsilon}$ is asymptotically efficient for any $w \in W_b$, i.e. equality holds in (2.16). Here again, we can replace the finite sums by infinite sums and obtain the asymptotic normality of $\hat{\theta}_\epsilon$, the MLE of the estimation problem (2.1)(2.2).

Theorem 2: *The Maximum Likelihood Estimator $\hat{\theta}_{N,\epsilon}$ for the projection onto Π^N is consistent, asymptotically normal with parameters 0 and $I_{N,0}(\theta)^{-1}$, and asymptotically efficient for any loss function $w \in W_b$ as $\epsilon \to 0$.*

Also, the MLE $\hat{\theta}_\epsilon$ for the fully observed process $u_\epsilon(t, x)$ is consistent, asymptotically normal with parameters $0, I(\theta)^{-1}$ and asymptotically efficient for any $w \in W_b$ as $\epsilon \to 0$.

Remark: We intend to prove the asymptotic efficiency for more general $w \in W$ in another paper.

Remark: What happens if $T \to \infty$?

If $\theta < \lambda_1$ then Fisher's information $I(\theta, T) = \frac{1}{2} \sum_{i=1}^{\infty} v_i^2 \frac{(\lambda_i+1)}{\lambda_i - \theta}(1 - e^{-2(\lambda_i - \theta)T})$ will not grow much when T gets large, but stays bounded. So, observations over a long period of time does not increase the information considerably.

If $\theta > \lambda_1$ and $v_1 \neq 0$ then the normalized difference $\epsilon^{-1}(\hat{\theta}_{N,\epsilon} - \theta)$ has Gaussian distribution with variance $\left(\frac{v_1^2(\lambda_1+1)}{2(\theta-\lambda_1)}e^{2(\theta-\lambda_1)T}\right)^{-1}(1 + o(1))$ as $T \to \infty$, and this is the same variance for the normalized MLE $\hat{\theta}_{1,\epsilon}$ based on component $u_{1\epsilon}(t)$ only. We see that almost all information comes from $u_{1\epsilon}(t)$, and the variance decreases exponentially fast in T.

In the case where $\theta = \lambda_1$ the term $\frac{e^{2(\theta-\lambda_1)T}-1}{\theta-\lambda_1}$ is to be understood as a limit. Thus $\frac{e^{2(\theta-\lambda_1)T}-1}{\theta-\lambda_1} \to 2T$ as $\theta \to \lambda_1$, so here the improvement would only be linear in T.

3 The Case of Singular Distributions

We can use the same approach for the estimation of $\theta > 0$ in the case where $u_\epsilon(t, x)$ is a solution to

$$(3.1) \qquad du_\epsilon(t, x) = \theta \Delta u_\epsilon(t, x)dt + \epsilon(I - \Delta)^{-\frac{1}{2}}dW(t, x)$$

satisfying initial and boundary conditions

$$(3.2) \qquad \begin{aligned} u_\epsilon(0, x) &= f(x) & 0 < x < 1 \\ u_\epsilon(t, 0) &= u_\epsilon(t, 1) = 0 & 0 \leq t \leq T \end{aligned}$$

Analogously to (2.5)(2.6) we have in this case the following equations for the Fourier coefficients $u_{i\epsilon}(t)$

$$(3.3) \qquad du_{i\epsilon}(t) = -\theta\lambda_i u_{i\epsilon}(t)dt + \frac{\epsilon}{\sqrt{\lambda_i + 1}}dW_i(t)$$

$$(3.4) \qquad u_{i\epsilon}(0) = v_i .$$

As in the first example we assume that $f \in L_2(0, 1)$ and $v_i = \int_0^1 f(x)e_i(x)dx$. We have now a slightly different expression for the likelihood ratio for the projection $u_\epsilon^N(t, x)$

$$(3.5) \qquad \ln\frac{dP_{\theta+\theta_0}^{\epsilon,N}}{dP_{\theta_0}^{\epsilon,N}}(u_\epsilon^N) =$$

$$-\frac{1}{\epsilon^2}\sum_{i=1}^N \lambda_i(\lambda_i+1)\left[\theta\int_0^T u_{i\epsilon}(t)(du_{i\epsilon}(t) + \theta_0\lambda_i u_{i\epsilon}(t)dt) + \frac{1}{2}\theta^2\lambda_i\int_0^T u_{i\epsilon}^2(t)dt\right].$$

It can be shown that the measures P_θ^ϵ corresponding to the solutions of equations (3.1)(3.2) with different θ's are *not* absolutely continuous. This fact corresponds to the one that the series

$$\sum_{i=1}^\infty \lambda_i^2(\lambda_i + 1) \int_0^T E u_{i\epsilon}^2(t) dt$$

diverges (if $\epsilon \neq 0$) which can be verified by straightforward computations. (This relation will be discussed in more detail elsewhere.) So the problem of estimating θ is more delicate in this case.

In applications the following two situations could be of interest:

Assuming that the statistician can observe the process $u_\epsilon(t, x)$ for all t, x, so we know the functions $u_{i\epsilon}(t)$ for all $i = 1, 2, \ldots$. The problem is to find an "estimator" which is equal to θ if that is possible. More common is the situation where the statistician knows only some approximation for $u_\epsilon(t, x)$ (e.g. Galerkin's approximation). In such cases some natural questions arise:

1. What is the rate of convergence of the best estimator $\theta_{\epsilon,N}^*$ based on $u_\epsilon^N(t, x)$ if N is fixed and $\epsilon \to 0$?

2. What is the improvement of the risk if the statistician can observe better approximations of the solution $u_\epsilon^{N'}(t, x)$ for $N' > N$?

Part 1) is more similar to the previous example, and we will study it first.

Analogously to (2.15) it is clear from (3.3)(3.4) that the limit of the normalized likelihood ratio $Z_{\epsilon,N}(z)$ exists as $\epsilon \to 0$ and is equal to

(3.6)
$$\lim_{\epsilon \to 0} \frac{dP_{\theta+\epsilon z}^{\epsilon,N}}{dP_\theta^{\epsilon,N}}(u_\epsilon^N) =$$

$$\exp\left\{ z \sum_{i=1}^N \lambda_i \sqrt{\lambda_i + 1} \int_0^T u_{i0}(t) dW_i(t) - \frac{1}{2}z^2 \sum_{i=1}^N \lambda_i^2(\lambda_i + 1) \int_0^T u_{i0}^2(t) dt \right\}$$

So the family of measures $\{P_\theta^{\epsilon,N}\}$ satisfies the LAN condition with normalizing factor $\epsilon \left(\sum_1^N \lambda_i^2(\lambda_i + 1) \int_0^T u_{i0}^2(t) dt \right)^{-\frac{1}{2}}$ where $u_{i0}(t)$ is the solution of (3.3)(3.4) with $\epsilon = 0$,

$$u_{i0}(t) = v_i e^{-\theta \lambda_i t}$$

and Fisher's information

(3.7) $$I_N(\theta) := \lim_{\epsilon \to 0} \epsilon^2 I_{N,\epsilon}(\theta) = \frac{1}{2\theta} \sum_{i=1}^N \lambda_i(\lambda_i + 1) v_i^2 (1 - e^{-2\theta \lambda_i T})$$

$I_N(\theta)$ has a finite limit as $N \to \infty$ if and only if $f'' \in L_2(0,1)$ and, again, $I_N(\theta) > 0$ if $f \not\equiv 0$.

By the Hajek-LeCam inequality we get a lower bound for any estimator $\theta^*_{\epsilon,N}$ and any $w \in W$

$$(3.8) \qquad \varliminf_{\epsilon \to 0} \sup_{|\theta - \theta_0| < \delta} E^{\epsilon,N}_\theta w\left(\epsilon^{-1}\left(\theta^*_{\epsilon,N} - \theta \right) \right) \geq Ew(\xi)$$

$$\text{where} \quad \mathcal{L}(\xi) = \mathcal{N}(0, I_N(\theta)^{-1}) \ .$$

It is clear from (3.5) that the MLE has the form

$$(3.9) \qquad \hat{\theta}_{N,\epsilon} = -\frac{\sum_1^N \lambda_i(\lambda_i + 1) \int_0^T u_{i\epsilon}(t) du_{i\epsilon}(t)}{\sum_1^N \lambda_i^2(\lambda_i + 1) \int_0^T u_{i\epsilon}^2(t) dt}$$

and using (3.3) we get

$$(3.10) \qquad \epsilon^{-1}(\hat{\theta}_{N,\epsilon} - \theta) = -\frac{\sum_1^N \lambda_i(\sqrt{\lambda_i + 1}) \int_0^T u_{i\epsilon}(t) dW_i(t)}{\sum_1^N \lambda_i^2(\lambda_i + 1) \int_0^T u_{i\epsilon}^2(t) dt}$$

Again, by the central limit theorem for stochastic integrals (cf.(2.19))

$$(3.11) \qquad \mathcal{L}\left(\epsilon^{-1}(\hat{\theta}_{N,\epsilon} - \theta) \right) \to \mathcal{N}\left(0, I_N(\theta)^{-1} \right) \quad \text{as} \quad \epsilon \to 0$$

and we have obtained the result

Theorem 3: *The family of measures $\{P^{\epsilon,N}_\theta\}$ generated by the projections of the observations (3.1)(3.2) onto Π^N has the LAN property (3.6) with normalizing factor $\epsilon(I_N(\theta))^{-\frac{1}{2}}$ defined by (3.7) as $\epsilon \to 0$.*

The MLE (3.9) based on this projection is consistent, asymptotically normal with parameters $0, I_N(\theta)^{-1}$, and asymptotically efficient for $w \in W_b$ as $\epsilon \to 0$.

Consider now the problem of estimating θ in (3.1) (3.2) when the statistician can observe $u_{i\epsilon}(t)$ for all $i = 1, 2, \ldots$.

The MLE (3.9) converges to the "true" parameter θ in probability as $N \to \infty$, for any fixed $\epsilon > 0$.

Theorem 4: *For any fixed $\epsilon > 0$*

$$(3.12) \qquad \hat{\theta}_{N,\epsilon} \to \theta \quad \text{in probability} \quad P^\epsilon_\theta \quad \text{as} \quad N \to \infty$$

Proof: Rewriting (3.10) we get

$$(\hat{\theta}_{N,\epsilon} - \theta) = \frac{-\epsilon \sum_1^N \lambda t_i \sqrt{\lambda_i + 1} \int_0^T u_{i\epsilon}(t) dW_i(t)}{E Z_1^N \lambda_i^2(\lambda_i + 1) \int_0^T u_{i\epsilon}^2(t) dt} \cdot \frac{E \sum_1^N \lambda_i^2(\lambda_i + 1) \int_0^T u_{i\epsilon}^2(t) dt}{\sum_1^N \lambda_i^2(\lambda_i + 1) \int_0^T u_{i\epsilon}^2(t) dt}$$

$$= -J_1 \cdot J_2$$

Straightforward computations give that for $N \to \infty$ $EJ_1^2 \to 0$ and $E(J_2^{-1} - 1) \to 0$. Assertion (3.12) follows from these facts.

So we see that the MLE $\hat{\theta}_{N,\epsilon}$ has unusual good properties if $N \to \infty$. This is possible because the measures P_θ^ϵ corresponding to different θ's are singular. The next question is about the rate of convergence.

The following lemma is well known (see e.g. Jacod, Shiryayev [1987]).
Lemma: *Let $W_i(t), i = 1, 2, \ldots$ be independent Wiener processes, $A_{i,N}(t)$ nonanticipating processes such that $\sum_1^N \int_0^T A_{iN}^2(t)dt < \infty$ a.s. and*

$$\sum_1^N \int_0^T A_{i,N}^2(t)dt \to 1 \quad \text{in probability as} \quad N \to \infty$$

then

$$\mathcal{L}\left(\sum_1^N \int_0^T A_{iN}(t)dW_i(t)\right) \to \mathcal{N}(0,1) \quad \text{as} \quad N \to \infty$$

To study the LAN property as $N \to \infty$ we fix ϵ and write $u_i(t)$ instead of $u_{i\epsilon}(t)$.

The natural normalizing factor is

$$(3.13) \qquad \varphi_N(\theta) = \left(\sum_1^N \lambda_i^2(\lambda_i + 1)E \int_0^T u_i^2(t)dt\right)^{-1/2} =$$

$$= \sqrt{2\theta}\left(\sum_1^N \lambda_i(\lambda_i + 1)v_i^2(1 - e^{-2\theta\lambda_i T}) + T\sum_1^N \lambda_i\right)^{-1/2}(1 + o(1))$$

Theorem 5: *The family of measures generated by (3.3)(3.4) for $i = 1, \ldots, N$ is LAN with normalizing factor $\varphi_N(\theta)$:*

$$(3.14) \qquad Z_{N,\theta}(z) = \frac{dP_{\theta+\varphi_N(\theta)z}^N}{dP_\theta^N} = \exp\{\Delta_N z - \frac{1}{2}z^2 + o(1)\}$$

where $\mathcal{L}(\Delta_N \mid P_\theta^N) \to \mathcal{N}(0,1)$ as $N \to \infty$.

Proof: Analogously to (2.8) we have

$$Z_{N,\theta}(z) = \exp\left\{-\varphi_N(\theta)z\sum_1^N \lambda_i\sqrt{\lambda_i + 1}\int_0^T u_i(t)dW_i(t) - \right.$$

$$\left. -\frac{1}{2}z^2\varphi_N(\theta)^2\sum_1^N \lambda_i^2(\lambda_i + 1)\int_0^T u_i^2(t)dt\right\}$$

The second term in the exponent tends to $-\frac{z^2}{2}$ in probability P_θ^N as in the proof of Theorem 3.

By the lemma, the first term is asymptotically normal where $A_{iN} = \varphi_N(\theta)\lambda_i\sqrt{\lambda_i + 1}u_i(t)$ and the theorem holds.

Consequently, by the Hajek-LeCam inequality for any estimator θ_N^* and loss function $w \in W$

$$(3.15) \quad \varliminf_{N\to\infty} \sup_{|\theta-\theta_0|<\delta} E_\theta^N w\left(\varphi_N^{-1}(\theta)(\theta_N^* - \theta)\right) \geq Ew(\xi), \mathcal{L}(\xi) = \mathcal{N}(0,1).$$

Theorem 6: *The MLE (3.9) (with fixed ϵ) is asymptotically normal,*

$$(3.16) \qquad \mathcal{L}\left(\varphi_N^{-1}(\theta)(\hat{\theta}_N - \theta)\right) \to \mathcal{N}(0,1) \quad as \quad N \to \infty$$

and equality holds in (3.15) for $w \in W_b$ and $\theta_N^* = \hat{\theta}_N$

Proof: We can write

$$\varphi_N^{-1}(\theta)(\hat{\theta}_N - \theta) =$$

$$= \frac{-\epsilon\left(\sum_1^N \lambda_i^2(\lambda_i + 1)E\int_0^T u_i^2(t)dt\right)^{-1/2}\left(\sum_1^N \lambda_i\sqrt{\lambda_i+1}\int_0^T u_i(t)dW_i(t)\right)}{\left(\sum_1^N \lambda_i^2(\lambda_i + 1)E\int_0^T u_i^2(t)dt\right)^{-1}\left(\sum_1^N \lambda_i^2(\lambda_i + 1)\int_0^T u_i^2(t)dt\right)}$$

Then the denominator converges to 1 as $N \to \infty$ which was proved in Theorem 3 while the numerator converges in distribution to a normal random variable by the lemma. Hence (3.16) follows by a standard result of weak convergence.

References

1. I.A. Ibragimov, R.Z. Khasminskii: Statistical estimation. *Asymptotic Theory*, Springer Verlag (1981).

2. J. Jacod, A.N. Shiryayev: *Limit Theorems for Stochastic Processes*, Springer-Verlag (1987)

3. Yu.A. Kutoyants: On a hypothesis testing problem and asymptotic normality of stochastic integrals. *Theor. Prob. Appl.* **20** (1975), 376-389.

4. Yu.A. Kutoyants: Parameter Estimation for stochastic processes. Heldermann Verlag (1984).

5. S.M. Kozlov: Some problems for stochastic partial differential equations. *Proceedings of Petrovski's Seminar*, **8** (1978) (in Russian).

6. R.S. Liptser, A.N. Shiryayev: Statistics of random processes, Springer Verlag (1978).

7. W. Loges: Girsanov's Theorem in Hilbert space and an application to the statistics of Hilbert space valued stochastic differential equations. *Stoch. Proc. Appl.* **17** (1984), 243-263.

8. B.L. Rozovskii: Stochastic evolution systems. *Linear Theory and Applications to Non-Linear Filtering*, Kluwer Academic Publishers (1990).

9. A.V. Skorohod: *Studies in the Theory of Random Processes*, Addison-Wesley (1965)

M. Hübner: Department of Mathematics, University of Southern California, Los Angeles, CA 90089-1113

R. Khasminskii: Institute for Problems of Information Transmission, Moscow 101447, ul. Ermolova 19, Russia

B.L. Rozovskii: Center for Applied Mathematical Sciences, University of Southern California, Los Angeles, CA 90089-1113

Computer Simulation of α–stable Ornstein–Uhlenbeck Processes

A. JANICKI, K. PODGÓRSKI and A. WERON

Abstract

We present a method of numerical approximation and computer simulation of stable Ornstein–Uhlenbeck processes derived as solutions of linear stochastic differential equations driven by a stable Lévy motion and some results on the convergence of this method. Making use of some statistical methods of construction of density estimators and applying computer graphics we get additional interesting quantitative and visual information on the family of stable Ornstein–Uhlenbeck processes that satisfy these equations.

1 Introduction

One of the most important examples of α–stable Ornstein–Uhlenbeck process $\{X(t,\omega);\ t \geq 0,\ \omega \in \Omega\}$ (or in shortened notation $\{X(t);\ t \geq 0\}$) based on a given α–stable Lévy motion process $\{L_\alpha(t,\omega);\ t \geq 0,\ \omega \in \Omega\}$ ($\{L_\alpha(t);\ t \geq 0\}$), can be described as follows

$$X(t,\omega) = e^{-\lambda t}\, X(0,\omega) + \mu \int_0^t e^{-\lambda(t-s)}\ dL_\alpha(s,\omega),\quad \lambda > 0,\quad \mu > 0.$$

(For exact definitions of α–stable stochastic measures on the real line $\mathcal{R}$ and of α–stable stochastic integrals of deterministic functions we refer the reader to Hardin [8], Samorodnitsky–Taqqu [13], Weron [15].)

In this paper by an α–stable (or simply stable) Ornstein–Uhlenbeck process we mean a solution of a linear Itô–type stochastic differential equation driven by α–stable Lévy motion, i.e. an equation in the following form

$$X(t) = X_0 + \int_0^t \left(a(s) + b(s)X(s)\right)\ ds + \int_0^t c(s)\ dL_\alpha(s),\tag{1}$$

where $t \in [0, T]$. (The idea of construction of such stochastic processes goes back to the pioneering work of Doob [6].)

Note that the general solution of (1) belongs to the class of α–stable processes and can be expressed in the following form

$$X(t) = \Phi(t,0)X_0 + \int_0^t \Phi(t,s)\, a(s)\ ds + \int_0^t \Phi(t,s)\, c(s)\ dL_\alpha(s),\tag{2}$$

where $\Phi(t,s) = \exp\left\{\int_s^t b(u)\ du\right\}$.

It is not so commonly understood that a vast class of diffusion processes $\{X(t);\ t \geq 0\}$ with given drift and dispersion coefficients can be described by the following stochastic differential equation

$$X(t) = X_0 + \int_0^t a(s, X(s))\ ds + \int_0^t c(s)\ dL_\alpha(s), \quad t > 0, \quad X(0) = X_0. \quad (3)$$

Note that diffusion processes $\{X(t);\ t \geq 0\}$ defined by (3) are not in general α–stable processes (for any $\alpha \in (0, 2]$).

In contrast, thanks to Itô's theory, it is commonly understood that any continuous diffusion process $\{X(t);\ t \geq 0\}$ with given drift and dispersion coefficients can be obtained as a solution of the following stochastic differential equation

$$X(t) = X_0 + \int_0^t a(s, X(s))\ ds + \int_0^t c(s, X(s))\ dB(s), \quad (4)$$

where $\{B(t)\}$ stands for Brownian motion process ($B(t) = 2^{-1/2} L_2(t)$). The theory of such stochastic differential equations has been developed for a long time (see Arnold [2] or Kallianpur [10]).

To our knowledge, up to now the numerical analysis of stochastic differential systems driven by Brownian motion has essentially focused on such problems as mean–square approximation, pathwise approximation or approximation of expectations of the solution, etc. (see e.g. Pardoux–Talay [11], Talay [14], Yamada [17]).

Our aim is to adopt some of these constructive computer techniques based on discretization of the time parameter t to the case of equation (1). We describe some results on the convergence of approximate numerical solutions.

Our idea is to represent the discrete time process solving a stochastic finite difference system (6)–(7) approximating (1) by appropriately constructed finite set of random samples, so we can obtain kernel estimators of densities of the stable Ornstein–Uhlenbeck process solving (1), for a finite set of values of t.

We also present an example of computer simulations of stable Ornstein–Uhlenbeck processes. An original application of computer graphics yields interesting visualization of such processes, providing useful quantitative information on their behavior. To the best of our knowledge the enclosed figures present the first visual representation of α–stable Ornstein–Uhlenbeck diffusions.

2   Existence results and convergence of approximate methods

Now our aim is to recall briefly the most important facts on the existence of solutions of stochastic differential equations in question, to describe the simplest method of approximation of equation (1) and to prove its convergence.

It is well known that in the case of equation (4), a Lipschitz continuity condition on the drift and dispersion functions a, c implies the existence and uniqueness of a solution in an appropriate process space. Lipschitz continuity condition can be weakened and replaced by the so called Yamada–Watanabe condition. It is even sufficient to ensure convergence of some numerical schemes approximating (4) (see Yamada [17]).

We are not aware of any similar results concerning equation (3).

However we can make use of theorems on the existence of solutions of stochastic differential equations driven by semimartingales (we refer the interested reader to Protter [12]). It is enough to notice that α–stable Lévy motion can serve as an example of a semimartingale.

Unfortunately, as far as we know, research on practically useful numerical approximate methods of solution of such equations, applicable in computer calculations, only now begins to attract the interest of mathematicians.

We think that the class of stochastic processes solving equation (1) is rich enough to be considered interesting in stochastic modeling. In the case of $\alpha = 2$ a system response is Gaussian and a fluctuation–dissipation relation can be satisfied (see Gardner [7]). When $\{L_\alpha(t); \ t \geq 0\}$ is a non–Gaussian stable process ($\alpha \in (1,2)$), then the response of the system is an α–stable process. In this case the system response has infinite variance, which corresponds to the situation when the particle with a given velocity and subject to linear damping has infinite kinetic energy. Thus the fluctuations supply an infinite amount of energy, which cannot be balanced by the linear dissipation. This means that the fluctuation must be regarded as external and no fluctuation–dissipation relation can be imposed on the system. (See West–Seshardi [16] for more details.)

Looking for an approximation of the process $\{X(t); \ t \in [0,T]\}$ solving equation (1) we have to approximate this equation by a time discretized explicit system of the form

$$X_{t_i}^\tau = \mathcal{F}\left(X_{t_{i-1}}, L_\alpha([t_{i-1}, t_i))\right), \tag{5}$$

where the set $\{t_i = i\tau, \ i = 0, 1, ..., I\}$, $\tau = T/I$, describes a fixed mesh on the interval $[0, T]$.

The simplest, but from our point of view sufficient, example of (5) provides the well known Euler scheme. The method consists in the construction of a sequence of stable random variables $\{X_{t_i}^\tau\}_{i=0}^I$ defined in the following way:

fix $X_0^\tau = X_0 \sim S_\alpha(\sigma_0, 0, \mu_0)$ and compute

$$X_{t_i}^\tau = X_{t_{i-1}}^\tau + Y_{t_i}^\tau, \tag{6}$$

$$Y_{t_i}^\tau = (a(t_{i-1}) + b(t_{i-1})X(t_{i-1})) \, \tau + c(t_{i-1})\Delta L_{\alpha,i}^\tau, \tag{7}$$

for $i = 1, 2, ..., I$, where the finite sequence of i.i.d. stable measures $\Delta L_{\alpha,i}^\tau$ of intervals $[t_{i-1}, t_i)$ is defined by $\Delta L_{\alpha,i}^\tau = L_\alpha([t_{i-1}, t_i)) \sim S_\alpha(\tau, 0, 0)$.

(Let us recall that writing $X \sim S_\alpha(\sigma, 0, \mu)$ we mean that X is a stable random variable with characteristic function of the form $\phi(\theta) = \exp(-\sigma^\alpha |\theta|^\alpha + i\mu)$.)

In order to formulate a result on the convergence of the method described above we define the continuous time interpolation process $\{X^\tau(t) : \ t \in [0, T]\}$

$$
\begin{aligned}
X^\tau(0) &= X_0, \\
X^\tau(t) &= X_{i-1}^\tau + \left(a(t_{i-1}) + b(t_{i-1})X_{i-1}^\tau\right)(t - t_{i-1}) + c(t_{i-1})\,\Delta L_{\alpha,i}^\tau,
\end{aligned} \tag{8}
$$

for $t \in [t_{i-1}, t_i)$, $i = 1, 2, ..., I$.

We prove the following result by simultaneous application of some techniques commonly used in studies of numerical schemes approximating stochastic differential equations driven by Brownian motion (see for example Pardoux–Talay [11]) and some techniques developed for the investigation of stable measures and integrals (see Samorodnitsky–Taqqu [13] or Weron [15]).

Proposition 1 *The family $\{X^\tau(t); \ t \in [0,T]\}$, defined by (8), of approximate solutions of the stochastic equation (1) with coefficient functions $a=a(t)$, $b=b(t)$ and $c=c(t)$ continuous on $[0,T]$, converges uniformly in probability to the exact solution $\{X(t); \ t \in [0,T]\}$ of (1) on $[0,T]$, when $\tau \to 0$.*

Proof. In order to simplify the notation let us define

$$\sigma(t, X(t)) = a(t) + b(t)X(t),$$

$$\sigma_\tau(t, X^\tau(t)) = \sum_{i=0}^{I-1} \sigma(t_i, X^\tau(t_i)) \mathbf{1}_{[t_i, t_{i+1})}(t),$$

$$c_\tau(t) = \sum_{i=0}^{I-1} c(t_i) \mathbf{1}_{[t_i, t_{i+1})}(t).$$

Then formulas (1) and (8) can be rewritten as follows

$$X(t) = X_0 + \int_0^t \sigma(s, X(s))ds + \int_0^t c(s)dL_\alpha(s),$$

$$X^\tau(t) = X_0 + \int_0^t \sigma_\tau(s, X_\tau(s))ds + \int_0^t c_\tau(s)dL_\alpha(s).$$

By the triangle inequality we have

$$|X(t) - X^\tau(t)| \le \int_0^t |\sigma(s, X(s)) - \sigma_\tau(s, X^\tau(s))|ds + \int_0^T |c(s) - c_\tau(s)|dL_\alpha(s).$$

Observe that $c_\tau(s)$ uniformly approximates $c(s)$, so the second term converges in probability to 0.

From the continuity of $a(\cdot)$ and $b(\cdot)$ it follows that

$$\int_0^t |\sigma(s, X(s)) - \sigma_\tau(s, X^\tau(s))|ds$$
$$\le \int_0^t |\sigma(s, X(s)) - \sigma(s, X^\tau(s))|ds + \int_0^t |\sigma(s, X^\tau(s)) - \sigma_\tau(s, X^\tau(s))|ds$$
$$\le K \int_0^t |X(s) - X^\tau(s)|dt + T \sup_{0 \le s \le T} |\sigma(s, X^\tau(s)) - \sigma_\tau(s, X^\tau(s))|$$

with some positive constant K.

Since $X^\tau(t)$ is a continuous process and $\sigma_\tau(\cdot, \cdot)$ uniformly approximates $\sigma(\cdot, \cdot)$, thus the last term converges to 0 with probability 1.

Hence we obtain the following estimate

$$|X(t) - X^\tau(t)| \leq K \int_0^t |X(s) - X^\tau(s)| ds + \epsilon_\tau,$$

where $\epsilon_\tau \to 0$ in probability. Thus from the Gronwall lemma we obtain

$$|X(t) - X^\tau(t)| \leq \int_0^t e^{K(t-s)} \epsilon_\tau ds + \epsilon_\tau \leq \epsilon_\tau e^{KT},$$

so we have uniform convergence in probability to 0 for $t \in [0, T]$ □.

3 Simulation and visualization of α–stable Ornstein–Uhlenbeck processes

Computer methods of construction of stochastic processes with stationary independent increments involve at least two kinds of discretization techniques: discretization of the time parameter and approximate representation of appropriate random variables with the aid of artificially produced finite time series data sets or random samples.

In order to describe fully our simulation algorithm we have to explain how to generate α–stable random variables $X \sim S_\alpha(1, 0, 0)$. We recommend the following method (cf. Chambers et al. [4]):

- generate a random variable V uniformly distributed on $(-\pi/2, \pi/2)$ and an exponential random variable W with mean 1;

- compute $\quad X = \frac{\sin(\alpha V)}{\{\cos(V)\}^{1/\alpha}} \times \left\{ \frac{\cos(V - \alpha V)}{W} \right\}^{(1-\alpha)/\alpha}.$

(It is obvious how to produce $X \sim S_\alpha(\sigma, 0, \mu)$.)

Now we are in a position to present the computer algorithm simulating approximate solutions of equation (1).

In computer calculations each random variable $X_{t_i}^\tau$ defined by (6) is represented by its N independent realizations, i.e. a random sample $\{X_i^\tau(n)\}_{n=1}^N$. So let us fix $N \in \mathcal{N}$ large enough. The algorithm consists in the following:

1. simulate a random sample $\{X_0^\tau(n)\}_{n=1}^N$ for X_0^τ;

2. for $i = 1, 2, ..., I$ simulate a random sample $\{\Delta L_{\alpha,i}^\tau(n)\}_{n=1}^N$ for α–stable random variable $\Delta L_{\alpha,i}^\tau$;

3. for $i = 1, 2, ..., I$ compute the random sample $\{Y_i^\tau(n)\}_{n=1}^N$ of Y_i^τ defined by (7), and $X_i^\tau(n) = X_{i-1}^\tau(n) + Y_i^\tau(n)$, $n = 1, 2, ..., N$;

4. construct kernel density estimators $f_i = f_i^{I,N} = f_i^{I,N}(x)$ of the densities of $X(t_i)$, using for example the optimal version of the Rosenblatt–Parzen method, and their distribution functions $F_i = F_i^{I,N} = F_i^{I,N}(x)$.

Observe that we have produced N finite time series of the form $\{X_i^\tau(n)\}_{i=0}^I$ for $n = 1, 2, ..., N$. We regard them as "good" approximations of the trajectories of the process $\{X(t); \ t \in [0, T]\}$, e.g. the effect of jumps of trajectories is visualized.

In order to obtain graphical approximation of the process $\{X(t); \ t \in [0, T]\}$ solving equation (1) we propose the following:

1. fix a rectangle $[0, T] \times [c, d]$, that should include the trajectories of $\{X(t)\}$;

2. for each $n = 1, 2, ..., n_{max}$ (with fixed $n_{max} \ll N$) draw the line segments determined by the points $(t_{i-1}, X_{i-1}^\tau(n))$ and $(t_i, X_i^\tau(n))$ for $i = 1, 2, ..., I$, constructing n_{max} approximate trajectories of the process X (thin lines on all figures);

3. fix a few values of a "probability parameter" p_j from $(0, 1/2)$ for $j = 1, 2, ..., J$ and for each of them compute 2 quantiles: $q_{min}^{i,j} = F_i^{-1}(p_j)$ and $q_{max}^{i,j} = F_i^{-1}(1 - p_j)$ for all $i = 1, 2, ..., I$; then draw the line segments determined by the points $(t_{i-1}, q_{min}^{i-1,j})$, $(t_i, q_{min}^{i,j})$ and $(t_{i-1}, q_{max}^{i-1,j})$, $(t_i, q_{max}^{i,j})$ for $i = 1, 2, ..., I$ and $j = 1, 2, ..., J$, constructing J (varying in time) prediction intervals (thick lines on all figures), that determine subdomains of $\mathcal{R}^2$ to which the trajectories of the approximated process should belong with probabilities $1 - 2p_j$ at any fixed moment of time $t = t_i$.

Note that the algorithm described above can be easily adjusted to the case of formula (2), which makes it more efficient in some situations (see an example of the stable Lévy bridge process below).

Combining some methods yielding convergence of numerical schemes approximating stochastic differential equations presented above with some techniques developed in the theory of nonparametric statistical estimation (see e.g. Devroye–Györfi [5]) one can prove several properties, e.g. the following one.

Proposition 2 *Let $f(x, T)$ be the nondegenerate density of the solution $\{X(t)\}$ of the equation (1) at $t = T$ and let $f_I^{I,N}(x)$ be its kernel density estimator obtained by the application of the method described above. Then*

$$\lim_{I,N \to \infty} \int | f_I^{I,N}(x) - f(x, T) | \ dx = 0 \quad \text{in probability.}$$

Note that for $\{X(t)\}$ described by (2) we have $X(T) \sim S_\alpha(\sigma_T, 0, \mu_T)$, with the parameters σ_T, μ_T given by

$$\mu_T = \Phi(T, 0)\mu_0 + \int_0^T \Phi(T, s) \ a(s) \ ds, \quad \sigma_T^\alpha = |\Phi(T, 0)|^\alpha \sigma_0^\alpha + \int_0^T |\Phi(T, s) \ c(s)|^\alpha \ ds.$$

So, computing these parameters and constructing Fourier transform of the characteristic function of $X(T)$ (with the use of FFT method) one can obtain another computer approximation of the density of $X(T)$.

4 α–stable Lévy Bridge

As an example illustrating the computer simulation techniques and the usefulness of computer graphics we propose the α-stable Lévy bridge process. This process can be defined as the solution of the following linear stochastic equation

$$B_\alpha(t) = \int_0^t \frac{B_\alpha(s)}{s-1}\, ds + \int_0^t dL_\alpha(s)$$

or, according to (2), in the following explicit form

$$B_\alpha(t) = (t-1)\int_0^t \frac{dL_\alpha(s)}{s-1}.$$

We believe that the figures at the end of the paper demonstrate in a surprisingly interesting manner how the behavior of the process $\{B_\alpha(t);\ t \in [0,1]\}$ depends on α.

Figures 1–3 show 10 approximate trajectories of the process $B_\alpha(t)$ for three different values of α: $\alpha = 2$, $\alpha = 1.1$ and $\alpha = 0.7$. In all cases trajectories are included in the same rectangle $(t, B_\alpha(t)) \in [0,1] \times [-2,2]$. Trajectories are represented by thin lines. Vertical lines, in our convention, illustrate the effect of jumps of the process $B_\alpha(t)$.

Three pairs of thick lines represent prediction intervals; for any fixed moment of time $t = t_i$ they show the lengths of intervals including trajectories with probabilities 0.5, 0.7 and 0.9, respectively. They were produced on the basis of random samples of $N = 2000$ realizations of $B_\alpha(t_i)$. In each case the time step τ was equal to 0.001 (I=1000).

In order to illustrate the exactness of our computer simulations we present in Fig. 4 the histogram and the kernel density estimator of $B_{1.1}(1)$. Observe that the error of this simulation can be derived as a difference between Dirac's δ and its approximation given by the numerically constructed density estimators.

For more details concerning our approach and for other examples we refer the reader to Janicki–Weron [9], where also general nonlinear case (3) is treated.

References

[1] Adler, R.J., Cambanis, S. and Samorodnitsky, G., (1990) *On stable Markov processes*, Stoch. Proc. Appl. **34**, 1–17.

[2] Arnold, L., (1974) *Stochastic Differential Equations*, Wiley, New York.

[3] Cambanis, S., Samorodnitsky, G. and Taqqu, M., eds. (1991) *Stable Processes and Related Topics*, Birkhäuser, Boston.

[4] Chambers, J.M., Mallows, C.L. and Stuck, B.W., (1976) *A method for simulating stable random variables*, J. Amer. Stat. Assoc., **71**, 340–344.

[5] Devroye, L. and Györfi, L., (1985) *Nonparametric Density Estimation: The L_1 View*, Wiley, New York.

[6] Doob, J.L., (1942) *The Brownian movement and stochastic equations*, Ann. Math. **43**, 351–369.

[7] Gardner, W.A., (1985) *Introduction to Random Processes with Applications to Signals and Systems*, MacMillan, London.

[8] Hardin jr.,C.D., (1982) *On the spectral representation of symmetric stable processes*, J. Multivariate Anal. **12**, 385–401.

[9] Janicki, A. and Weron, A., *Simulation and Ergodic Behavior of Stable Stochastic Processes*, (book in preparation).

[10] Kallianpur, G., (1980) *Stochastic Filtering Theory*, Springer, New York.

[11] Pardoux, E. and Talay, D., (1985) *Discretization and simulation of stochastic differential equations*, Acta Applicandae Mathematicae **3**, 23–47.

[12] Protter, P., (1990) *Stochastic Integration and Differential Equations: A New Approach*, Springer, New York.

[13] Samorodnitsky, G. and Taqqu, M.S., *Non–Gausssian Stable Processes* (book in preparation).

[14] Talay, D., (1983) *Résolution trajectorielle et analyse numérique des équations différentielles stochastiques*, Stochastics **9**, 275–306.

[15] Weron, A., (1984) *Stable processes and measures: A survey*, pp. 306–364 in *Probability Theory on Vector Spaces III*, Szynal, D. and Weron, A., eds. *Lecture Notes in Mathematics* **1080**, Springer, New York.

[16] West, B.J. and Seshadri, V., (1982) *Linear systems with Lévy fluctuations*, Physica **113A**, 203-216.

[17] Yamada, T., (1976) *Sur l'approximation des equations différentielles stochastiques*, Zeit. Wahrsch. verw. Geb. **36**, 133-140.

Hugo Steinhaus Center, Institute of Mathematics
Technical University of Wrocław
50–370 Wrocław, Poland

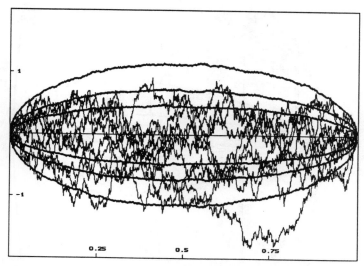

Figure 1: Visualization of 2.0–stable Lévy bridge.

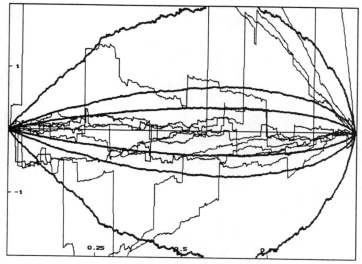

Figure 2: Visualization of 1.1–stable Lévy bridge.

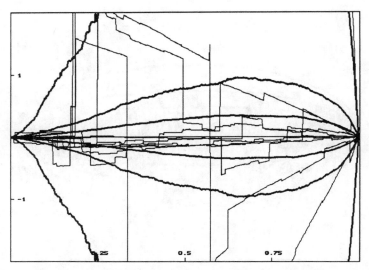

Figure 3: Visualization of 0.7–stable Lévy bridge.

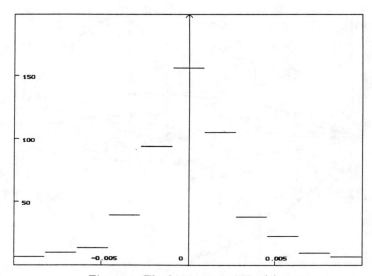

Figure 4: The histogram of $B_{1.1}(1)$.

Some Linear Random Functionals Characterized by L^p-Symmetries

OLAV KALLENBERG

Abstract. Consider a linear random functional ξ on some real or complex linear space $\mathcal{L}$, along with a linear operator A from $\mathcal{L}$ into a space $L^p(S, \mu)$. Under suitable regularity conditions, it is shown that if $P(\xi f)^{-1}$ depends only on $\|Af\|_p$, then ξ must be of the form $\sigma \int Af \, dX$, where X denotes symmetric p-stable noise on S with control measure μ, while $\sigma \geq 0$ is an independent r.v. The result generalizes Schoenberg's classical theorem and has many interesting applications.

1. Introduction.

Schoenberg's theorem [18] gives a rather surprising connection between positive definite and completely monotone functions. The result has also an interesting probabilistic interpretation, apparently first noted by Freedman [6], though later rediscovered by many authors. In the latter form it can be deduced most easily from de Finetti's theorem (cf. Aldous [2], p. 23).

The probabilistic version concerns an arbitrary infinite sequence of random variables (r.v.'s), $\xi = (\xi_1, \xi_2, \ldots)$, and asserts that every finite subsequence $(\xi_1, \ldots, \xi_n)$ has a spherically symmetric distribution, iff ξ is a mixture of centered Gaussian i.i.d. sequences. (As usual, the term *mixture* refers to the distributions rather than to the sequences themselves.) An equivalent condition is that, given a suitable r.v. $\sigma \geq 0$, the elements ξ_k are conditionally independent and $N(0, \sigma^2)$.

The stated symmetry condition may be recast into an equivalent functional form, where we regard ξ as a random functional $\xi f = \sum \xi_k f_k$ on the space $\mathcal{L}$ of all terminating sequences $f = (f_1, \ldots, f_n, 0, \ldots, 0)$ in R, and assume that $P(\xi f)^{-1}$, the distribution of ξf, should only depend on

[1]Research supported in part by NSF Grant No. DMS-9002732

$\|f\|_2$. Using the ℓ^p-norm instead, for some $p \in (0,2)$, leads to a similar characterization involving symmetric p-stable distributions, first noted by Bretagnolle, Dacunha-Castelle, and Krivine [3].

There exists a vast literature discussing different approaches to and extensions of the mentioned results, including some recent work by Diaconis and Freedman [4, 5], Lauritzen [12], and Ressel [15]. The continuous parameter theory was pioneered by Freedman [6], and further publications relating to spherical symmetry in continuous time or in higher dimensions may be traced from the papers by Aldous [1], Kallenberg [10, 11], and Ressel [16]. For a related property of *pseudo-isotropy*, see Hardin [7] and Misiewicz [14].

In Section 2 below, we shall prove an abstract version of the mentioned functional characterization, involving a linear random functional ξ on some real or complex linear space $\mathcal{L}$, and a linear mapping A of $\mathcal{L}$ into some space $L^p(S, \mu)$. Thus it will be shown under suitable regularity conditions that, for f restricted to $\mathcal{L}$, the distribution $P(\xi f)^{-1}$ depends only on $\|Af\|_p$, iff

$$(1) \qquad\qquad \xi f = \sigma \int Af\, dX, \quad f \in \mathcal{L},$$

for some symmetric p-stable noise X on S with control measure μ, and some independent r.v. $\sigma \geq 0$. Here the integral may be thought of as being of Wiener type, and equality will only hold in distribution, in general, unless additional assumptions are made.

To recover Schoenberg's theorem from the quoted result, it suffices to take $\mathcal{L}$ as the class of terminating sequences in R, let S be the natural numbers with associated counting measure μ, and choose A to be the natural embedding of $\mathcal{L}$ into ℓ^2. For a more interesting example with $p = 2$, we may take $\mathcal{L}$ to be the space of all signed measures on R with finite support, let A be the Fourier-Stieltjes transform on $\mathcal{L}$, and let S be R endowed with an arbitrary finite measure μ. Then (1) reduces, apart from a scale mixture, to the familiar spectral representation of a stationary Gaussian process. Other choices lead to various integral representations of measurable stable processes. Those and further applications will be disussed in Section 3.

2. Main result.

Fix a linear space $\mathcal{L}$ over the real or complex number field F. By a *linear random functional* on $\mathcal{L}$ we shall mean an F-valued process ξ on $\mathcal{L}$, such that

$$\xi(af + bg) = a\xi f + b\xi g \text{ a.s.}, \quad a, b \in F, \; f, g \in \mathcal{L}.$$

Fixing a linear mapping A of $\mathcal{L}$ into some space $L^p(S, \mu)$, we shall say that $P(\xi f)^{-1}$ depends only on $\|Af\|_p$, if $\xi f \overset{d}{=} \xi g$ for all pairs $f, g \in \mathcal{L}$ with $\|Af\|_p = \|Ag\|_p$. Note that ξ must then be *symmetric*, in the sense that $c\xi \overset{d}{=} \xi$ for all $c \in F$ with $|c| = 1$, where $\overset{d}{=}$ denotes equality of all finite-dimensional distributions.

The *symmetric p-stable noise* X with *control measure* μ occurring in (1) may be regarded most conveniently as a linear random functional on $L^p(S, \mu)$, such that the r.v. Xf is symmetric p-stable with scale parameter $\|f\|_p$ for every $f \in L^p(S, \mu)$. Thus

$$(2) \qquad E \exp(i \operatorname{Re} Xf) = \exp(-\|f\|_p^p), \quad f \in L^p(S, \mu).$$

Note that (2) determines all finite-dimensional distributions of X and yields independence between the r.v.'s $Xf_1, \ldots, Xf_n$ whenever $f_1, \ldots, f_n$ have disjoint supports. Note also that X must be symmetric. The existence of a random functional with the stated properties is well-known and follows easily from the Daniell-Kolmogorov theorem. The integral $\int f dX$ in (1) may now be defined simply as Xf. Thus (1) is equivalent to

$$(3) \qquad \xi f = \sigma X(Af), \quad f \in \mathcal{L}.$$

A crucial hypothesis below is to assume the closure $\overline{A\mathcal{L}}$ of $A\mathcal{L}$ in $L^p(S, \mu)$ to contain some functions $f_1, f_2, \ldots$ with $\|f_k\|_p \equiv 1$ and disjoint supports. The f_k will then be said to form a *separating sequence* in $\overline{A\mathcal{L}}$.

In order for ξ to have an a.s. representation on the original probability space (Ω, P), the latter should be rich enough to support a r.v. ϑ independent of ξ and $U(0, 1)$ (uniformly distributed on $[0, 1]$). Alternatively, we may define ϑ on the *extended probability space* $(\Omega \times [0, 1], P \times \lambda)$, where λ denotes Lebesgue measure on $[0, 1]$.

We may now state our main result.

THEOREM 2.1 *Let η be a linear random functional on some real or complex linear space $\mathcal{L}$, and let A be a linear mapping of $\mathcal{L}$ into some space $L^p(S, \mu)$ with $p \in (0, 2]$, such that $\overline{A\mathcal{L}}$ contains a separating sequence. Further assume for f in $\mathcal{L}$ that $P(\eta f)^{-1}$ depends only on $\|Af\|_p$. Then $\eta \overset{d}{=} \xi$, with ξ given by (3) in terms of some symmetric p-stable noise X on S with control*

measure μ and an independent r.v. $\sigma \geq 0$. If in addition S is Polish while
μ is σ-finite, then X and σ may be defined on an extension of the original
probability space, such that $\eta f = \xi f$ a.s. for all $f \in \mathcal{L}$.

We shall need three lemmas, where the first one is equivalent to the
result in Bretagnolle et al. [3], the second is a version of the coupling
Lemma 1.1 in Kallenberg [9], and the third is essentially a separability
property of p-stable noise.

LEMMA 2.2 *Fix $p \in (0, 2]$ and a random sequence $\xi = (\xi_1, \xi_2, \dots)$ in $F = R$*
or C. Assume for terminating sequences $f = (f_1, f_2, \dots)$ in F that $P(\xi f)^{-1}$
depends only on $\|f\|_p$. Then there exists a unique probability measure ν on
R_+, such that

$$(4) \qquad E \, \exp(i \, \mathrm{Re} \, \xi f) = \int \exp(-r\|f\|_p^p)\nu(dr), \quad f \in \ell^p.$$

Proof: For fixed $n \in N$, consider sequences f and g in F with $\|f\|_p = \|g\|_p$
and supported by $\{1, \dots, n\}$, and let h be a terminating sequence supported
by $\{n + 1, n + 2, \dots\}$. Then $\|rf + h\|_p = \|rg + h\|_p$ for any $r \in F$, so
$\xi(rf + h) \overset{d}{=} \xi(rg + h)$, and by the Cramér-Wold theorem,

$$(\xi f, \xi_{n+1}, \xi_{n+2}, \dots) \overset{d}{=} (\xi g, \xi_{n+1}, \xi_{n+2}, \dots).$$

Thus $\xi f \overset{d}{=} \xi g$ remains conditionally true, given $(\xi_{n+1}, \xi_{+2}, \dots)$. Letting
$n \to \infty$, we get the same result given the tail σ-field $\mathcal{T}$. Extending from
sequences f and g with rational coordinates, it is seen that our symmetry
condition remains conditionally true, given $\mathcal{T}$. But given $\mathcal{T}$, the ξ_k are
also conditionally i.i.d., by de Finetti's theorem. It follows easily that each
ξ_k is conditionally symmetric p-stable. The scale parameter is then $\mathcal{T}$-
measurable, and (4) follows. The uniqueness assertion follows from the
theory of Laplace transforms. $\square$

LEMMA 2.3 *Fix a set T, some Polish spaces S and S', and some measurable*
functions $f_t : S' \to S$, $t \in T$. Consider an S-valued process X on T and
a random element η' in S', such that the process $X'_t = f_t(\eta'), t \in T$, has
the same finite-dimensional distributions as X. Then there exists, on an
extension of the original probability space, some random element $\eta \overset{d}{=} \eta'$ in
S' such that $X_t = f_t(\eta)$ a.s., $t \in T$.

Proof: First reduce by a Borel isomorphism to the case when $S' = R$. Consider X and $X' = (f_t(\eta'), t \in T)$ as random elements in the product space S^T, endowed with the σ-field $\mathcal{S}^T$ induced by all coordinate projections $\pi_t : S^T \to S$, $t \in T$, where $\mathcal{S}$ denotes the Borel σ-field in S. By the theory of regular conditional distributions, there exists a probability kernel μ from S^T to R, such that

(5)
$$P[\eta' \in \cdot | X'] = \mu(X'; \cdot) \text{ a.s.},$$

and we may define a process Y on $[0, 1]$ by

$$Y_s = \inf\{x \in R;\ \mu(X; (-\infty, x]) > s\}, \quad s \in [0, 1].$$

Letting ϑ be $U(0, 1)$ and independent of X, we may finally put $\eta = Y_\vartheta$, which is a r.v. since Y is product measurable. By Fubini's theorem, $P[\eta \in \cdot | X] = \mu(X; \cdot)$ a.s. Comparing with (5) and recalling that $X \overset{d}{=} X'$, we get for any measurable function $g : S^T \times R \to R_+$,

$$Eg(X, \eta) = E \int g(X, y)\mu(X, dy) = E \int g(X', y)\mu(X', dy) = Eg(X', \eta'),$$

so $(X, \eta) \overset{d}{=} (X', \eta')$. In particular, $X_t - f_t(\eta) \overset{d}{=} X'_t - f_t(\eta') = 0$, $t \in T$, and the assertion follows.   □

LEMMA 2.4 *Let μ be a σ-finite measure on some Polish space S. Then there exist some measurable functions $G_f : [0, 1] \to R$ or C, $f \in L^p(S, \mu)$, such that if ϑ is $U(0, 1)$, then $Xf = G_f(\vartheta)$, $f \in L^p(S, \mu)$, is a symmetric p-stable noise on S with control measure μ.*

Proof: Since S is Polish while μ is σ-finite, the space $L^p = L^p(S, \mu)$ is separable, and we may choose $f_1, f_2, \ldots$ dense in L^p. Let Y be any symmetric p-stable noise on S with control measure μ. By a standard construction, we may choose measurable functions $g_1, g_2, \ldots, : [0, 1] \to R$ or C, such that if ϑ is $U(0, 1)$, the sequences $\xi \equiv (\xi_n) \equiv (g_n(\vartheta))$ and $\eta \equiv (\eta_n) \equiv (Yf_n)$ have the same distribution. Now define for arbitrary $f \in L^p$ the sequence

$$n_k = \inf\{n \in N;\ \|f - f_n\|_p \leq 2^{-k}\}, \quad k \in N.$$

Then for some constant $c > 0$,

$$E \sum_k |Yf - Yf_{n_k}|^{p/2} = c \sum_k \|f - f_{n_k}\|_p^{p/2} \leq c \sum_k 2^{-kp/2} < \infty,$$

so $\eta_{n_k} = Y f_{n_k} \to Y f$ a.s. Put

$$X f = \limsup_{k \to \infty} \xi_{n_k} = h_f(\xi) = G_f(\vartheta).$$

Since also $Y f = h_f(\eta)$ a.s., while $\xi \overset{d}{=} \eta$, it is clear that $X \overset{d}{=} Y$. □

Proof of Theorem 2.1 If $f, g \in \mathcal{L}$ with $Af = Ag$, then $\|A(f - g)\|_p = 0 = \|A0\|_p$, so $\eta f - \eta g = \eta(f - g) \overset{d}{=} \eta 0 = 0$, i.e. $\eta f = \eta g$ a.s. Hence there exists a linear random functional ζ on $A\mathcal{L}$, such that $\zeta A f = \eta f$ a.s. for all $f \in \mathcal{L}$. For $f \in A\mathcal{L}$, it is clear that $P(\zeta f)^{-1}$ depends only on $\|f\|_p$.

For any $f \in \overline{A\mathcal{L}}$, we may choose $f_1, f_2, \ldots, \in A\mathcal{L}$ with $\|f_1\|_p = 1$ and $f_n \to f$ in L^p. Then $\zeta(f_m - f_n) \overset{d}{=} \|f_m - f_n\|_p \zeta f_1$, so the sequence (ζf_n) is Cauchy in probability, say with limit ζf. The latter r.v. is a.s. independent of the choice of sequence (f_n), and the construction yields a continuous extension of ζ to $\overline{A\mathcal{L}}$. Note that $P(\zeta f)^{-1}$ still depends only on $\|f\|_p$.

Since $\overline{A\mathcal{L}}$ contains a separating sequence, Lemma 2.2 shows that

$$E \exp(i \operatorname{Re} \zeta f) = \int \exp(-r\|f\|_p^p)\nu(dr), \quad f \in \overline{A\mathcal{L}},$$

for some probability measure ν on R_+. For X as stated and with $\sigma \geq 0$ an independent r.v. with $P(\sigma^p)^{-1} = \nu$, we get by Fubini's theorem,

$$E \exp(i \operatorname{Re} \sigma X f) = E \exp(-\|\sigma f\|_p^p) = \int \exp(-r\|f\|_p^p)\nu(dr), \quad f \in L^p(S, \mu).$$

Hence $\zeta \overset{d}{=} \sigma X$ on $\overline{A\mathcal{L}}$ by the Cramér-Wold theorem, so $\eta \overset{d}{=} \sigma X(A \cdot) = \xi$.

If X is Polish while μ is σ-finite, there exist by Lemma 2.4 some measurable functions G_f, $f \in L^p$, such that we can choose $X f = G_f(\vartheta), f \in L^p$, with $\vartheta \ U(0, 1)$ and independent of σ. Then $\xi f = \sigma G_{Af}(\vartheta)$, $f \in \mathcal{L}$, so by Lemma 2.3 there exists, on some extension of the probability space, a pair $(\tilde{\sigma}, \tilde{\vartheta}) \overset{d}{=} (\sigma, \vartheta)$ with $\eta f = \tilde{\sigma} G_{Af}(\tilde{\vartheta})$ a.s., $f \in \mathcal{L}$. It is further clear that the functional $\tilde{X} f = G_f(\tilde{\vartheta})$, $f \in L^p$, is independent of $\tilde{\sigma}$ with $\tilde{X} \overset{d}{=} X$. □

3. Applications.

Starting with a simple example, we may take $\mathcal{L} = C[0, 1]$, and let A be the identity mapping on $\mathcal{L}$. Then the representation in Theorem 2.1 becomes $\xi f = \sigma \int f \, dX$, where X is a symmetric p-stable Lévy process on

$[0, 1]$, and the integral is of Wiener type. The representation clearly defines an extension of ξ to a linear random functional on $L^p([0, 1])$.

For a more interesting example, take $\mathcal{L} = C[0, 1]$ and $(S, \mu) = ([0, 1], \lambda)$ as before, and let $Af = f - \overline{f}$, where $\overline{f} = \int f d\lambda$. Then for $p = 2$ the representation becomes

$$\xi f = \sigma \int (f - \overline{f}) dX = \sigma \int f_t (dX_t - X_1 dt) = \sigma \int f dB,$$

where X is a Brownian motion and B a Brownian bridge independent of σ. A simple modification yields instead an integral with respect to a Kiefer process.

Next consider a bounded measurable function $g : [0, 1]^2 \to R$, and define an operator A on the space $\mathcal{L}$ of bounded measurable functions $f : [0, 1] \to R$ by $(Af)_t = \int f_s g_{st} ds$. Assuming g to be such that the conditions in Theorem 2.1 are fulfilled, we get a representation

$$\xi f = \sigma \int dX_t \int f_s g_{st} ds = \sigma \int f_s ds \int g_{st} dX_t,$$

for some symmetric p-stable noise X on $[0, 1]$, where the last relation holds by an easily established Fubini-type formula for Wiener integrals. To appreciate this example, note that every measurable symmetric p-stable process on $[0, 1]$ is of the form $\int g_{st} dX_t$ (cf. Rosiński and Woyczyński [17]).

We conclude with a more subtle application.

PROPOSITION 3.1 *Fix a locally compact Abelian group G with separable dual $\hat{G}$, a bounded positive measure μ on $\hat{G}$ with infinite support, and a number $p \in (0, 2]$. Let ξ be a linear random functional on some linear space $\mathcal{L}$ of bounded real or complex measures on G, such that $\mathcal{L}$ is weakly dense in the space $\overline{\mathcal{L}}$ of all such measures. Assume for $f \in \mathcal{L}$ that $P(\xi f)^{-1}$ depends only on $\|\hat{f}\|_p$. Then $\xi f = \sigma \int \hat{f} dX$ a.s., $f \in \mathcal{L}$, for some r.v. $\sigma \geq 0$ and some independent symmetric p-stable noise X on $\hat{G}$ with control measure μ.*

To appreciate this result, we may e.g. take $G = R$ or Z. Then the formula for the process $Y_t = \xi \delta_t$ becomes $Y_t = \sigma \int e^{itu} dX_u$, which for $p = 2$ is essentially the familiar spectral representation of a stationary Gaussian process with energy spectrum μ. For $p < 2$, the integral defines instead a so called strongly stationary stable process (cf. Marcus and Pisier [13]).

Proof: Writing $\mathcal{A}$ for the class of absolutely continuous measures in $\overline{\mathcal{L}}$, and letting $(\cdot)^-$ denote closure in $L^p(G, \mu)$, we have in the complex case

$$(6) \qquad C_0(\hat{G}) \subset (\mathcal{A})^- \subset ((\overline{\mathcal{L}})^\wedge)^- \subset (\hat{\mathcal{L}})^-,$$

where the first relation comes from the fact that $\mathcal{A}$ is dense in $C_0(\hat{G})$ in the uniform topology (cf. (31.5) in [8]), while the third one holds by continuity of the characters and dominated convergence. By Theorem 2.1, it is hence enough to construct a separating sequence $f_1, f_2, \ldots$ in $C_0(\hat{G})$. Then recall from (8.13) and (23.15) in [8] that $\hat{G}$ is again locally compact Hausdorff and hence completely regular. It is now easy to construct the f_n recursively, making sure that $(\text{supp } \mu) \cap \{|f_1| + \cdots + |f_n| = 0\}$ remains infinite for each n.

In the real case, put $\mu^s(A) = \frac{1}{2}(\mu(A) + \mu(-A))$, let $\|\cdot\|_p^s$ be the norm in $L^p(\mu^s)$, and let $C_0^s(\hat{G})$ denote the class of even functions in $C_0(\hat{G})$. For any $f \in C_0^s(\hat{G})$, there exist by (6) some bounded complex measures $\nu_1, \nu_2, \ldots$ on G, such that $\|\hat{\nu}_n - f\|_p^s \to 0$. Writing $h_n = \text{Re} \, \nu_n$, we get

$$\|\hat{h}_n - f\|_p = \|\hat{h}_n - f\|_p^s \le \|\hat{\nu}_n - f\|_p^s \to 0.$$

Thus $C_0^s(\hat{G}) \subset ((\overline{\mathcal{L}})^\wedge)^- \subset (\hat{\mathcal{L}})^-$, and the proof may be completed as before.
□

References.

[1] Aldous, D.J.: Representations for partially exchangeable arrays of random variables. J. Multivariate Anal. **11** (1981), 581-598.

[2] Aldous, D.J.: Exchangeability and Related Topics. In: École d'Été de Probabilités de Saint-Flour XIII-1983 (ed. P.L. Hennequin), pp. 1-198. Lecture Notes in Mathematics **1117**. Springer, Berlin 1985.

[3] Bretagnolle, J., Dacunha-Castelle, D., Krivine, J.L.: Lois stables et espaces L^p. Ann. Inst. H. Poincaré B **2** (1966), 231-259.

[4] Diaconis, P., Freedman, D.: Partial exchangeability and sufficiency. In: Statistics: Applications and New Directions (eds. J.K. Gosh and J. Roy), pp. 205-236. Indian Statistical Institute, Calcutta 1984.

[5] Diaconis, P., Freedman, D.: Cauchy's equation and de Finetti's theorem. Scand. J. Statist **17** (1990), 235-250.

[6] Freedman, D.: Invariants under mixing which generalize de Finetti's theorem. Ann. Math. Statist. **33** (1962), 916-923, **34** (1963), 1194-1216.

[7] Hardin, C.D.: On the linearity of regression. Z. Wahrscheinlichkeitstheorie verw. Gebiete **61** (1982), 293-302.

[8] Hewitt, E., Ross, K.A.: Abstract Harmonic Analysis, I (2nd ed.) and II. Springer, Berlin 1979, 1970.

[9] Kallenberg, O.: Spreading and predictable sampling in exchangeable sequences and processes. Ann. Probab. **16** (1988), 508-534.

[10] Kallenberg, O.: Some new representations in bivariate exchangeability. Probab. Th. Rel. Fields **77** (1988), 415-455.

[11] Kallenberg, O.: Some time change representations of stable integrals, via predictable transformations of local martingales. Stoch. Proc. Appl. (to appear).

[12] Lauritzen, S.L.: Extremal Families and Systems of Sufficient Statistics. Lecture Notes in Statistics **49**. Springer, Berlin 1989.

[13] Marcus, M.B., Pisier, G.: Characterizations of almost surely continuous p-stable random Fourier series and strongly stationary processes. Acta Math. **152** (1984), 245-301.

[14] Misiewicz, J.: Pseudo isotropic measures. Nieuw Arch. Wisk. **8** (1990), 111-152.

[15] Ressel, P.: de Finetti-type theorems; an analytical approach. Ann. Probab. **13** (1985), 898-922.

[16] Ressel, P.: Integral representations for distributions of symmetric stochastic processes. Probab. Th. Rel. Fields **79** (1988), 451-467.

[17] Rosiński, J., Woyczyński, W.A.: On Itô stochastic integration with respect to p-stable motion: inner clock, integrability of sample paths, double and multiple integrals. Ann. Probab. **14** (1986), 271-286.

[18] Schoenberg, I.J.: Metric spaces and positive definite functions. Trans. Amer. Math. Soc. **44** (1938), 522-536.

Author's Address:
    Departments of Mathematics
    Auburn University
    120 Math Annex
    Auburn, AL 36849-5307
    U.S.A.

Higher Order Approximate Markov Chain Filters

P.E. KLOEDEN, E. PLATEN and H. SCHURZ

Abstract. The aim of this paper is to construct higher order approximate discrete time filters for continuous time finite-state Markov chains with observations that are perturbed by the noise of a Wiener process.

1. Introduction.

The systematic construction and investigation of filters for Markov chains goes back to Wonham [11], Zakai [12] and Fujisaki, Kallianpur and Kunita [2]. Later the question of finding discrete time approximations for the optimal filter was considered by Clark and Cameron [1] and Newton [7], [8].

At first we introduce in the following filters for continuous time finite state Markov chains. Let $(\Omega, \mathcal{A}, P)$ be the underlying probability space and suppose that the state process $\xi = \{\xi_t, t \in [0, T]\}$ is a continuous time homogeneous Markov chain on the finite state space $\mathcal{S} = \{a_1, a_2, \ldots, a_d\}$. Its d-dimensional probability vector $p(t)$, with components

$$(1) \qquad p_i(t) = P\left(\xi_t = a_i\right)$$

for each $a_i \in \mathcal{S}$, then satisfies the vector ordinary differential equation

$$(2) \qquad \frac{dp}{dt} = A\,p$$

where A is the intensity matrix. In addition, suppose that the m-dimensional observation process $W = \{W_t, t \in [0, T]\}$ is the solution of the stochastic equation

$$(3) \qquad W_t = \int_0^t h\left(\xi_s\right)\,ds + W_t^*,$$

where $W^* = \{W_t^*, t \in [0, T]\}$ with $W_0^* = 0$ is an m-dimensional standard Wiener process with respect to the probability measure P, which is independent of the process ξ. Finally, let $\mathcal{Y}_t$ denote the σ-algebra generated by the observations W_s for $0 \leq s \leq t$. In what follows we shall use superscripts to label the components of vector-valued stochastic processes.

Our task is to filter as much information about the state process ξ as we can from the observation process W. With this aim we shall evaluate the conditional expectation

$$E\left(g\left(\xi_T\right) \mid \mathcal{Y}_T\right)$$

with respect to P for a given function $g : \mathcal{S} \to \Re$.

By application of the Girsanov transformation we obtain a probability measure $\dot{P}$ where

(4) $d\dot{P} = L_T^{-1}\, dP$

with

(5) $L_T = \exp\left(-\frac{1}{2} \int_0^T |h\left(\xi_s\right)|^2 ds + \int_0^T h\left(\xi_s\right)^\top dW_s\right)$

such that W is a Wiener process with respect to $\dot{P}$.

Let us introduce the un-normalized conditional probability X_t^i for the state $a_i \in \mathcal{S}$ at time t as the conditional expectation

(6) $X_t^i = \dot{E}\left(I_{\{a_i\}}\left(\xi_t\right) L_t \mid \mathcal{Y}_t\right),$

$i \in \{1, \ldots, d\}$, $t \in [0, T]$, with respect to the probability measure $\dot{P}$, where $I_{\{a_i\}}(x)$ is the indicator function taking the value 1 when $x = a$ and the value 0 otherwise. It follows from a basic assertion in Fujisaki, Kallianpur and Kunita [2], also known as Kallianpur-Striebel formula, that the conditional probabilities of ξ_t given $\mathcal{Y}_t$ are

(7) $P\left(\xi_t = a_i \mid \mathcal{Y}_t\right) = E\left(I_{\{a_i\}}\left(\xi_t\right) \mid \mathcal{Y}_t\right) = X_t^i \bigg/ \sum_{k=1}^d X_t^k$

for $a_i \in \mathcal{S}$ and $t \in [0, T]$, where the d-dimensional process $X_t = \{X_t^1, \ldots, X_t^d\}$ of the un-normalized conditional probabilities satifies the Zakai equation.

(8) $X_t = p(0) + \int_0^t A\, X_s\, ds + \sum_{j=1}^m \int_0^t H_j\, X_s\, dW_s^j$

for $t \in [0, T]$, which is a homogeneous linear Ito equation. H_j is the $d \times d$ diagonal matrix with iith component $h_j(a_i)$ for $i = 1, \ldots, d$ and $j = 1, \ldots, m$.

The optimal least squares estimate for $g(\xi_t)$ with respect to the observations W_s for $0 \le s \le t$, that is with respect to the σ-algebra $\mathcal{Y}_t$, is given by the conditional expectation

(9) $\Pi_t(g) \;=\; E\left(g\left(\xi_t\right) \mid \mathcal{Y}_t\right)$

$$= \sum_{k=1}^d g\left(a_k\right) X_t^k \bigg/ \sum_{k=1}^d X_t^k,$$

which we call the optimal filter or Markov chain filter.

2. Approximate Filters.

To compute the optimal filter (9) we have to solve the Ito equation (8). In practice, however, it is impossible to detect W completely on $[0, T]$. Electronic devices are often used to obtain increments of integral observations over small time intervals, which in the simplest case are the increments of W in integral form

$$\int_{t_0}^{\tau_1} dW_s^j, \ldots, \int_{t_n}^{\tau_{n+1}} dW_s^j, \ldots,$$

for each $j = 1, \ldots, m$, $\tau_n = n\delta$ for $n = 0, 1, 2, \ldots$. We shall see in the next section that with such integral observations it is possible to construct strong discrete time approximations Y^δ with time step δ of the solution X of the Zakai equation (8). Then for the given function g we can evaluate the expression

$$(10) \qquad \Pi_t^\delta(g) = \sum_{k=1}^{d} g(a_k) Y_t^{\delta,k} \bigg/ \sum_{k=1}^{d} Y_t^{\delta,k}$$

for $t \in [0, T]$, which we shall define to be the corresponding approximate Markov chain filter.

We shall say that a discrete time approximation Y^δ with step size δ converges on the time interval $[0, T]$ with order $\gamma > 0$ to the corresponding solution X of the stochastic differential equation if there exists a finite constant K, not depending on δ, and a $\delta_0 \in (0, 1)$ such that

$$(11) \qquad E\left(\left|X_{\tau_n} - Y_{\tau_n}^\delta\right|\right) \leq K \delta^\gamma$$

for all $\delta \in (0, \delta_0)$ and $\tau_n \in [0, T]$. We note that the expectation in (11) is with respect to the probability measure P under which W is a Wiener process. Analogously we say that an approximate Markov chain filter $\Pi^\delta(g)$ with step size δ converges on the time interval $[0, T]$ with order $\gamma > 0$ to the optimal filter $\Pi(g)$ for a given function g if there exists a finite constant K, not depending on δ, and a $\delta_0 \in (0, 1)$ such that

$$(12) \qquad E\left(\left|\Pi_{\tau_n}(g) - \Pi_{\tau_n}^\delta(g)\right|\right) \leq K \delta^\gamma$$

for all $\delta \in (0, \delta_0)$ and $\tau_n \in [0, T]$. In contrast with (11) we take the expectation in (12) with respect to the original probability measure P.

PROPOSITION. *An approximate Markov chain filter $\Pi^\delta(g)$ with step size δ converges on the time interval $[0, T]$ with order $\gamma > 0$ to the optimal filter*

$\Pi(g)$ *for a given bounded function g if the discrete time approximation* Y^δ
used in it converges on $[0, T]$ *to the solution* X *of the Zakai equation (8)*
with the same order γ.

PROOF.   In view of (12) we need to estimate the error

$$(13) \qquad F_{\tau_n}^\delta (g) \;=\; E\left(\left|\Pi_{\tau_n}(g) - \Pi_{\tau_n}^\delta(g)\right|\right)$$

$$=\; \dot{E}\left(L_{\tau_n}\left|\Pi_{\tau_n}(g) - \Pi_{\tau_n}^\delta(g)\right|\right)$$

for all $\tau_n \in [0, T]$. We shall write

$$(14) \qquad\qquad G_{\tau_n}(f) = \sum_{k=1}^d f(a_k)\, X_{\tau_n}^k$$

and

$$(15) \qquad\qquad G_{\tau_n}^\delta(f) = \sum_{k=1}^d f(a_k)\, Y_{\tau_n}^{\delta,k}$$

for any bounded function $f : \mathcal{S} \to \Re$, $\delta \in (0, \delta_0)$ and $\tau_n \in [0, T]$. Then
similarly to Picard [9] we can use (6), (9) and (10) to rewrite the error (13)
in the form

$$F_{\tau_n}^\delta (g) \;=\; \dot{E}\left(G_{\tau_n}(1)\left|\Pi_{\tau_n}(g) - \Pi_{\tau_n}^\delta(g)\right|\right)$$

$$(16) \qquad =\; \dot{E}\left(G_{\tau_n}(1)\left|\frac{1}{G_{\tau_n}(1)}\left(G_{\tau_n}(g) - G_{\tau_n}^\delta(g)\right.\right.\right.$$

$$\left.\left.\left. + \Pi_{\tau_n}^\delta(g)\left(G_{\tau_n}^\delta(1) - G_{\tau_n}(1)\right)\right)\right|\right)$$

$$\le\; \dot{E}\left(\left|G_{\tau_n}(g) - G_{\tau_n}^\delta(g)\right|\right) + \dot{E}\left(\left|\Pi_{\tau_n}^\delta(g)\right|\,\left|G_{\tau_n}^\delta(1) - G_{\tau_n}(1)\right|\right)$$

$$\le\; K_1 \sum_{k=1}^d \dot{E}\left(\left|Y_{\tau_n}^{\delta,k} - X_{\tau_n}^k\right|\right).$$

Finally, using (11) in (16) gives the estimate $F_{\tau_n}^\delta(g) \le K_2\, \delta^\gamma$ and hence the
desired convergence rate. $\square$

3. Explicit Filters.

It remains to describe discrete time approximations converging with a given
order $\gamma > 0$ to the solution of the Zakai equation (8) which can be used in a
corresponding approximate filter. A systematic presentation of such discrete
time approximations can be found in Kloeden and Platen [3]. Given an

equidistant time discretization of the interval $[0, T]$ with step size $\delta = \Delta = T/N$ for some $N = 1, 2, \ldots$, we define the partition σ-algebra $\mathcal{P}_N^1$ as the σ-algebra generated by the increments

$$(17) \qquad \Delta W_0 = \int_0^\Delta dW_s^j, \quad \ldots, \quad \Delta W_{N-1} = \int_{(N-1)\Delta}^{N\Delta} dW_s^j$$

for all $j = 1, \ldots, m$. Thus $\mathcal{P}_N^1$ contains the information about the increments of W for this time discretization. The simplest discrete time approximation obtained from the Euler scheme (see Maruyama [5]) has for the Zakai equation (8) the form

$$(18) \qquad Y_{\tau_{n+1}}^\delta = \left[I + A\,\Delta + G_n \right] Y_{\tau_n}^\delta$$

with

$$(19) \qquad G_n = \sum_{j=1}^m H_j\,\Delta W_n^j$$

and initial value $Y_0 = X_0$, where I is the $d \times d$ unit matrix. The scheme (18) converges under the given assumptions with order $\gamma = 0.5$. For a general stochastic differential equation this is the maximum order of convergence that can be achieved under the partition σ-algebra $\mathcal{P}_N^1$, as was shown by Clark and Cameron [1]. However, the special multiplicative noise structure of the Zakai equation (8) allows the order $\gamma = 1.0$ to be attained with the information contained in $\mathcal{P}_N^1$. Milstein [6] proposed a scheme of order $\gamma = 1.0$, which for equation (8) has the form

$$(20) \qquad Y_{\tau_{n+1}}^\delta = \left[I + \underline{A}\,\Delta + G_n \left(I + \frac{1}{2} G_n \right) \right] Y_{\tau_n}^\delta$$

where

$$(21) \qquad \underline{A} = A - \frac{1}{2} \sum_{j=1}^m H_j^2.$$

Newton [7] searched for a scheme which is asymptotically the "best" in the class of order 1.0 schemes in the sense that it has the smallest leading error coefficient in an error estimate similar to (11). He obtained the scheme

$$Y_{\tau_{n+1}}^\delta = \left[I + \underline{A}\,\Delta + G_n + \frac{\Delta^2}{2} A^2 + \frac{\Delta}{2} A\,G_n - \frac{\Delta}{2} G_n\,A + G_n \underline{A}\,\Delta + \frac{1}{2} G_n^2 + \frac{1}{6} G_n^3 \right] Y_{\cdot}$$

(22)

which is called asymptotically efficient under $\mathcal{P}_N^1$.

We can obtain higher order convergence by exploiting additional information about the observation process such as contained in the integral observations

$$(23) \quad \Delta Z_0^j = \int_0^\Delta \int_0^s dW_r^j \, ds, \quad \ldots, \quad \Delta Z_{N-1}^j = \int_{(N-1)\Delta}^{N\Delta} \int_{(N-1)\Delta}^s dW_r^j \, ds$$

for all $j = 1, \ldots, m$, easily measured in practice by digital devices. We shall define as the partition σ-algebra $\mathcal{P}_N^{1.5}$ the σ-algebra generated by $\mathcal{P}_N^1$ together with the multiple integrals $\Delta Z_0^j, \ldots, \Delta Z_{N-1}^j$ for all $j = 1, \ldots, m$. The order 1.5 strong Taylor scheme described in Platen [10] and Kloeden and Platen [3] uses for the Zakai equation (8) only the information contained in $\mathcal{P}_N^{1.5}$. It takes the form

$$Y_{\tau_{n+1}}^\delta = \left[I + \underline{A} \, \Delta + G_n + \frac{\Delta^2}{2} A^2 + A \, M_n - M_n \, A + G_n \, \underline{A} \, \Delta + \frac{1}{2} \, G_n^2 + \frac{1}{6} G_n^3 \right] Y_{\tau_n}^\delta$$

(24)

where

$$(25) \qquad\qquad M_n = \sum_{j=1}^m H_j \, \Delta Z_n^j.$$

We note that we obtain the order 1.0 scheme (22) from (24) if we replace the ΔZ_n^j by their conditional expectations under $\mathcal{P}_N^1$ with respect to the probability measure $\dot{P}$, that is we substitute $\frac{1}{2} G_n \, \Delta$ for M_n in (24).

In order to form a scheme of order $\gamma = 2.0$ we need the information from the observation process expressed in the partition σ-algebra $\mathcal{P}_N^2$ which is generated by $\mathcal{P}_N^{1.5}$ together with the multiple Stratonovich integrals

$$J_{(j_1, j_2, 0), n} = \int_{\tau_n}^{\tau_{n+1}} \int_{\tau_n}^{s_3} \int_{\tau_n}^{s_2} \circ dW_{s_1}^{j_1} \circ dW_{s_2}^{j_2} \, ds_3,$$

$$(26) \qquad J_{(j_1, 0, j_2), n} = \int_{\tau_n}^{\tau_{n+1}} \int_{\tau_n}^{s_3} \int_{\tau_n}^{s_2} \circ dW_{s_1}^{j_1} \, ds_2 \circ dW_{s_3}^{j_2},$$

for all $n = 0, 1, \ldots, N - 1$ and $j_1, j_2 = 1, \ldots, m$. Here the symbol "$\circ$" denotes the Stratonovich integration. Electronic devices can extract these Stratonovich integrals from the observation measurements in practical filtering situations. Using this information we can apply the order 2.0 strong Taylor scheme in Kloeden and Platen [3] to the Zakai equation (8) to obtain the approximation

$$Y_{\tau_{n+1}}^\delta = \left[I + \underline{A} \Delta \left(I + \frac{1}{2} \underline{A} \Delta \right) - M_n \underline{A} + \underline{A} M_n \right.$$

$$(27) \qquad +G_n \left(I + \underline{A} \Delta + \frac{1}{2} G_n \left(I + \frac{1}{3} G_n \left(I + \frac{1}{4} G_n \right) \right) \right)$$

$$+ \sum_{j_1, j_2 = 1}^{m} \left(\underline{A} H_{j_2} H_{j_1} J_{(j_1, j_2, 0), n} + H_{j_2} \underline{A} H_{j_1} J_{(j_1, 0, j_2), n} \right.$$

$$\left. + H_{j_2} H_{j_1} \underline{A} (\Delta J_{(j_1, j_2)} - J_{(j_1, j_2, 0)} - J_{(j_1, 0, j_2)}) \right) \Bigg] Y_{\tau_n}^{\delta}.$$

We remark that the corresponding orders of strong convergence of the schemes described above follow from a convergence theorems in Platen [10] or Kloeden, Platen [3].

4. Implicit Filters.

Explicit discrete time approximations can sometimes behave numerically unstable. In such a situation control is lost over the propagation of errors and the approximation is rendered useless. We can then use an implicit discrete time scheme to obtain a numerically stable approximation. Here we state some of the implicit discrete time schemes from Kloeden and Platen [3], [4] applied to the Zakai equation (8). These express an iterate in terms of itself and its predecessor, but since the Zakai equation is linear they can all be rearranged algebraically to express the next iterate just in terms of its predecessor. After rearranging we have from the family of implicit Euler schemes

$$(28) \qquad Y_{\tau_{n+1}}^{\delta} = (I - \alpha A \Delta)^{-1} \left[I + (1 - \alpha) A \Delta + G_n \right] Y_{\tau_n}^{\delta}$$

where $\alpha \in [0, 1]$ denotes the degree of implicitness. The scheme (28) converges with order $\gamma = 0.5$. The family of implicit Milstein schemes, all of which converge with order $\gamma = 1.0$, gives us

$$(29) \quad Y_{\tau_{n+1}}^{\delta} = (I - \alpha \underline{A} \Delta)^{-1} \left[I + (1 - \alpha) \underline{A} \Delta + G_n \left(I + \frac{1}{2} G_n \right) \right] Y_{\tau_n}^{\delta}.$$

In principle to each explicit scheme there corresponds a family of implicit schemes by making implicit the terms involving the nonrandom multiple stochastic integrals such as Δ or $\frac{1}{2} \Delta^2$. As a final example we mention the order 1.5 implicit Taylor scheme yielding

$$Y_{\tau_{n+1}}^{\delta} = \left(I - \frac{1}{2} \underline{A} \Delta \right)^{-1} \left[I + \frac{1}{2} \underline{A} \Delta + G_n \underline{A} \Delta - M_n \underline{A} + \underline{A} M_n \right.$$

$$(30) \qquad \qquad \left. + G_n \left(I + \frac{1}{2} G_n \left(I + \frac{1}{3} G_n \right) \right) \right] Y_{\tau_n}^{\delta}.$$

5. A Numerical Example.

We consider the random telegraphic noise process, that is the two state continuous time Markov chain ξ on the state space $\mathcal{S} = \{-1, +1\}$ with intensity matrix

$$A = \begin{bmatrix} -50.0 & 50.0 \\ 50.0 & -50.0 \end{bmatrix}$$

and initial probability vector $p(0) = (0.9, 0.1)$. Further, we suppose that the observation process W satisfies the stochastic equation (3) with $h(1) = 5$ and $h(-1) = 0$.

Our task is to determine the actual state of the chain on the basis of these observations. We could say that ξ_t has most likely the value $+1$ if $P\left(\xi_t = +1 \,\middle|\, \mathcal{Y}_t\right) \geq 0.5$. We evaluate the conditional probability

$$p_1(t) = P\left(\xi_t = +1 \,\middle|\, \mathcal{Y}_t\right) = E\left(I_{\{+1\}}\left(\xi_t\right) \,\middle|\, \mathcal{Y}_t\right) = \Pi_t\left(I_{\{+1\}}\right),$$

which is the optimal filter here. To obtain an approximation of $\Pi_t\left(I_{\{+1\}}\right)$ we can use a filter $\Pi_t^\delta\left(I_{\{+1\}}\right)$ based on a discrete time approximation. For a comparison of approximate filters we shall suppose that we have here a scenario of a realization of the Markov chain on the interval $[0, 4]$ with $\xi_t = 1$ for $0 \leq t < 0.5$ and $\xi_t = -1$ for $0.5 \leq t \leq 4.0$. Using this realization of the Markov chain we computed the approximate filters $\Pi_t^\delta\left(I_{\{+1\}}\right)$ for the same realization of the Wiener process W^* using the above mentioned schemes with equidistant step size $\delta = \Delta = 2^{-7}$.

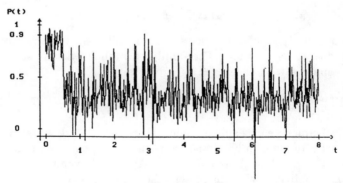

Figure 1. $p_1(t)$ for the explicit order 1.5 strong Taylor filter.

Two calculated $p_1(t)$ paths are plotted in Figures 1 and 2 respectively. The result for the order 1.5 Taylor filter which is an explicit one is plotted in Figure 1. Considerably more sensitive detections of the jump of the Markov chain from state 1 to state -1 at $t = 0.5$ were obtained by implicit schemes.

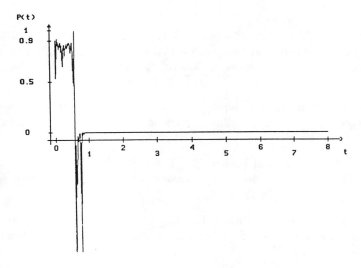

Figure 2. $p_1(t)$ for the implicit order 1.5 Taylor filter.

Figure 2 shows the result with the implicit order 1.5 Taylor scheme. The above numerical example underlines the importance of implicit stochastic numerical schemes. Also the additional information given by multiple stochastic integrals turns out to be substantial for a sensitive detection of a signal.

Authors' addresses

P.E. Kloeden, Deakin University, Geelong, Victoria, 3217, Australia

E. Platen, Australian National University, Canberra, ACT, 2601, Australia and Institute Appl. Anal. Stochastics, Mohrenstr. 39, Berlin, 1086

H. Schurz, Institute Appl. Anal. Stochastics, Mohrenstr. 39, Berlin, 1086

References.

[1] Clark, J.M.C., and Cameron, R.J.   The maximum rate of convergence of discrete approximations for stochastic differential equations. Springer Lecture Notes in Control and Inform. Sc. Vol. 25, (1980), pp. 162-171.

[2] Fujisaki, M., Kallianpur, G., and Kunita, H.   Stochastic differential equations for the nonlinear filtering problem. Osaka J. Math. 9, (1972), 19-40.

[3] Kloeden, P.E., and Platen, E.   The Numerical Solution of Stochastic Differential Equations. Applications of Mathematics Series, Nr. 23. Springer, Heidelberg, (1992).

[4] Kloeden, P.E., and Platen, E.   Higher-order implicit strong numerical schemes for stochastic differential equations. J. Statist. Physics, Vol. 66, No. 1/2 (1992), 283-314.

[5] Maruyama, G.   Continuous Markov processes and stochastic equations. Rend. Circolo Math. Palermo 4, (1955), 48-90.

[6] Milstein, G.N.   Approximate integration of stochastic differential equations. Theor. Prob. Appl. 19, (1974), 557-562.

[7] Newton, N.J.   An asymptotically efficient difference formula for solving stochastic differential equations. Stochastics 19, (1986), 175-206.

[8] Newton, N.J.   Asymptotically efficient Runge-Kutta methods for a class of Ito and Stratonovich equations. SIAM J. Appl. Math. 51, (1991), 542-567.

[9] Picard, J.   Approximation of nonlinear filtering problems and order of convergence. Springer Lecture Notes in Control and Inform. Sc. Vol. 61, (1984), pp. 219-236.

[10] Platen, E.   An approximation method for a class of Ito processes. Lietuvos Matem. Rink. 21, (1981), 121-133.

[11] Wonham, W.M.   Some applications of stochastic differential equations to optimal nonlinear filtering. SIAM J. Control 2, (1965), 347-369.

[12] Zakai, M.   On the optimal filtering of diffusion processes. Z. Wahrsch. verw. Gebiete 11, (1969), 230-343.

Fourier Transform and Cylindrical Hida Distributions

Izumi Kubo and Hui-Hsiung Kuo*

Abstract. The Fourier transform of a cylindrical Hida distribution is expressed in terms of the finite dimensional Fourier transform. Finite dimensional approximations of some Hida distributions are studied.

§1. A heuristic observation

The Fourier transform of the Dirac delta function $\delta(u)$ at 0 on $\mathbb{R}$ is the constant function 1. Thus informally $\delta(u)$ is the inverse Fourier transform of 1, i.e.

$$\delta(u) = \frac{1}{2\pi} \int_{-\infty}^{\infty} e^{iux}\, dx. \tag{1.1}$$

It has been shown in [15] that the Dirac delta function $\delta(u)$ indeed has the following representation in the sense of distribution:

$$\delta(u) = \frac{1}{\sqrt{2\pi}} \sum_{k=0}^{\infty} (-1)^k \frac{1}{2^{2k}k!} H_{2k}\left(\frac{u}{\sqrt{2}}\right), \tag{1.2}$$

where $H_n(u)$ is the Hermite polynomial of degree n. The series representation in (1.2) was used to define the Donsker delta function as a generalized Brownian functional in [8, 15].

Suppose $\Phi(x, y) = f(x)$ is a function on $\mathbb{R}^2$. Then its Fourier transform is given by $\widehat{\Phi}(u, v) = \widehat{f}(u)(2\pi)^{-1/2} \int_{\mathbb{R}} e^{-ivy}\, dy$. By using the representation in (1.1), we can rewrite $\widehat{\Phi}$ as follows:

$$\widehat{\Phi}(u, v) = \widehat{f}(u)\sqrt{2\pi}\, \delta(-v). \tag{1.3}$$

Obviously, (1.3) can be extended to a function Φ defined on $\mathbb{R}^n$ depending only on k variables, $k \leq n$. This leads to a relation between the infinite dimensional Fourier transform of Hida distributions and the finite dimensional Fourier transform in Theorem 5.5.1 [17].

On the other hand, we can rewrite the Fourier transform $\widehat{\Phi}$ in (1.3) in a different form. Let $\langle\!\langle \cdot, \cdot \rangle\!\rangle$ be the pairing of distributions and test functions with respect to the standard Gaussian measure μ_2 on $\mathbb{R}^2$. Suppose φ is a test function. Then from (1.3) we can derive

$$\langle\!\langle \widehat{\Phi}, \varphi \rangle\!\rangle = \int_{\mathbb{R}} \langle\!\langle \widetilde{\delta}_{(u,0)}, \varphi \rangle\!\rangle \widehat{f}(u)\, d\mu_1(u), \tag{1.4}$$

---

* Research supported by NSF Grant DMS-9001859 and LEQSF Grant RD-A-08.

where μ_1 is the standard Gaussian measure on $\mathbb{R}$ and $\widetilde{\delta}_y$ is defined by $\langle\!\langle \widetilde{\delta}_y, \varphi \rangle\!\rangle = \varphi(y)$. Note that this $\widetilde{\delta}_y$ differs from the ordinary δ_y by a constant, i.e. $\delta_y = (2\pi)^{-1} e^{-|y|^2/2} \widetilde{\delta}_y$, $y \in \mathbb{R}^2$. It follows from (1.4) that the Fourier transform $\widehat{\Phi}$ can be rewritten as

$$\widehat{\Phi} = \int_{\mathbb{R}} \widetilde{\delta}_{(u,0)} \widehat{f}(u) \, d\mu_1(u). \tag{1.5}$$

Again, (1.5) can be extended to a function Φ defined on $\mathbb{R}^n$ depending only on k variables, $k \leq n$. In this paper we will generalize (1.5) to the space of Hida distributions. The infinite dimensional analogue of $\widetilde{\delta}_x$ is the delta function introduced by Kubo and Yokoi [12]. We will also study the finite dimensional approximation of Hida distributions.

§2. Fourier transform of Hida distributions

For the discussion of generalized white noise functionals, see [1, 5, 6, 10, 11, 19, 20]. Let $\mathcal{S}(\mathbb{R})$ be the Schwartz space of rapidly decreasing real-valued functions on $\mathbb{R}$. Let μ be the standard Gaussian measure on the dual space $\mathcal{S}'(\mathbb{R})$ of $\mathcal{S}(\mathbb{R})$. By the Wiener-Itô decomposition theorem, we have $(L^2) \equiv L^2(\mu) = \bigoplus_{n=0}^{\infty} K_n$, where K_n is the space of n-fold Wiener integrals $I_n(f)$, $f \in \widehat{L}^2(\mathbb{R}^n)$, the symmetric $L^2(\mathbb{R}^n)$ space. Each $\varphi \in (L^2)$ can be represented uniquely by $\varphi = \sum_{n=0}^{\infty} I_n(f_n)$, $f_n \in \widehat{L}^2(\mathbb{R}^n)$. Moreover, the (L^2)-norm $\|\varphi\|_2$ of φ is given by $\|\varphi\|_2 = (\sum_{n=0}^{\infty} n! |f_n|^2_{L^2(\mathbb{R}^n)})^{1/2}$.

Let $A = -D_u^2 + u^2 + 1$. The second quantization $\Gamma(A)$ of A is densely defined on (L^2) as follows. For $\varphi = \sum_{n=0}^{\infty} I_n(f_n)$, define $(\Gamma(A)\varphi) = \sum_{n=0}^{\infty} I_n(A^{\otimes n} f_n)$. For $p \in \mathbb{R}$, define the norm $\|\varphi\|_{2,p} \equiv \|\Gamma(A)^p \varphi\|_2$. For $p \geq 0$, let $(\mathcal{S})_p \equiv \{\varphi; \|\varphi\|_{2,p} < \infty\}$. For $p < 0$, let $(\mathcal{S})_p$ be the completion of (L^2) with respect to $\|\cdot\|_{2,p}$. The dual space $(\mathcal{S})_p^*$ of $(\mathcal{S})_p$ is $(\mathcal{S})_{-p}$. Let $(\mathcal{S})$ be the projective limit of $\{(\mathcal{S})_p; p \geq 0\}$. Then the dual space $(\mathcal{S})^*$ of $(\mathcal{S})$ is the union of $\{(\mathcal{S})_p^*; p \geq 0\}$ and we have the following continuous inclusions: $(\mathcal{S}) \subset (\mathcal{S})_p \subset (L^2) \equiv (L^2)^* \subset (\mathcal{S})_p^* \subset (\mathcal{S})^*$, $p \geq 0$. We call $(\mathcal{S})$ and $(\mathcal{S})^*$ the spaces of *test functionals* and *Hida distributions*, respectively. $\langle\!\langle \cdot, \cdot \rangle\!\rangle$ will denote the pairing of $(\mathcal{S})^*$ and $(\mathcal{S})$.

For $\Phi \in (\mathcal{S})^*$, its *U-functional* is defined to be $U[\Phi](\xi) = e^{-|\xi|_2^2/2} \langle\!\langle \Phi, e^{(\cdot, \xi)} \rangle\!\rangle$, $\xi \in \mathcal{S}(\mathbb{R})$. A Hida distribution is uniquely determined by its U-functional and, for each $\xi \in \mathcal{S}(\mathbb{R})$, the function $U[\Phi](\lambda\xi)$, $\lambda \in \mathbb{R}$, has an entire extension $U[\Phi](z\xi)$, $z \in \mathbb{C}$ [19].

Now, we turn to the Fourier transform. By using the renormalization $: e^{-i\langle x,y\rangle} :_y = e^{-i\langle x,y\rangle + |x|^2/2}$, the finite dimensional Fourier transform can be written as

$$\widehat{f}(y) = \int_{\mathbb{R}^k} : e^{-i\langle x,y\rangle} :_y f(x) \, d\mu_k(x),$$

where μ_k is the standard Gaussian measure on $\mathbb{R}^k$. Thus it seems to be reasonable to define the Fourier transform on $(\mathcal{S})^*$ as in [14] by:

$$\widehat{\Phi}(y) = \int_{\mathcal{S}'(\mathbb{R})} \, : e^{-i\langle x,y \rangle} :_y \, \Phi(x) \, d\mu(x). \tag{2.1}$$

However, this is not well defined [16, 17]. By taking the U-functional under the integral sign, we get $U[\widehat{\Phi}](\xi) = \langle\!\langle \Phi, e^{-i\langle \cdot, \xi \rangle} \rangle\!\rangle, \xi \in \mathcal{S}(\mathbb{R})$, or equivalently,

$$U[\widehat{\Phi}](\xi) = U[\Phi](-i\xi) \, e^{-\frac{1}{2}|\xi|_2^2}, \quad \xi \in \mathcal{S}(\mathbb{R}). \tag{2.2}$$

Now, suppose $\Phi \in (\mathcal{S})^*$ is given. By the Potthoff-Streit characterization theorem [19] there exists a unique Hida distribution, denoted by $\widehat{\Phi}$, such that (2.2) holds. We call $\widehat{\Phi}$ the *Fourier transform* of Φ.

For examples and properties of Fourier transform, see [7, 16, 17, 18]. Here we only mention the delta function $\widetilde{\delta}_x$ at $x \in \mathcal{S}'(\mathbb{R})$ [12] defined by $\langle\!\langle \widetilde{\delta}_x, \varphi \rangle\!\rangle = \widetilde{\varphi}(x), \varphi \in (\mathcal{S})$, where $\widetilde{\varphi}$ is the unique continuous version of φ. The U-functional of $\widetilde{\delta}_x$ is given by $U[\widetilde{\delta}_x](\xi) = \exp\left[\langle x, \xi \rangle - 2^{-1}|\xi|_2^2\right], \xi \in \mathcal{S}(\mathbb{R})$. For $x \in \mathcal{S}'(\mathbb{R})$, we define the renormalization $: e^{i\langle \cdot, x \rangle} :$ of the meaningless expression $e^{i\langle \cdot, x \rangle}$ to be the Hida distribution with the U-functional given by $U[: e^{i\langle \cdot, x \rangle} :](\xi) = e^{i\langle x, \xi \rangle}, \xi \in \mathcal{S}(\mathbb{R})$. It was shown in [16,17] that for fixed $x \in \mathcal{S}'(\mathbb{R}), (: e^{i\langle \cdot, x \rangle} :)^\frown = \widetilde{\delta}_x$. In particular, when $x = 0$, we have $\widehat{1} = \widetilde{\delta}_0$. In view of (2.1) this gives the following symbolic expression:

$$\widetilde{\delta}_0(y) = \int_{\mathcal{S}'(\mathbb{R})} \, : e^{-i\langle x,y \rangle} :_y \, d\mu(x).$$

This is the counterpart of (1.1) in white noise calculus.

§3. Fourier transform of cylindrical Hida distributions

In view of (2.1) the Fourier transform of Hida distributions is an infinite dimensional generalization of the finite dimensional Fourier transform. Thus it is natural to ask the question how they are related to each other. A relation similar to (1.3) has been obtained in [17]. Here we give another relation which is similar to (1.5).

Suppose Φ is a cylindrical Hida distribution, i.e. it is of the following form: $\Phi = f \circ (\langle \cdot, \zeta_1 \rangle, \ldots, \langle \cdot, \zeta_k \rangle)$, where $f \in \mathcal{S}'(\mathbb{R}^k)$ and $\zeta_1, \ldots, \zeta_k \in L^2(\mathbb{R})$ are linearly independent. The question is to find a relation between $\widehat{\Phi}$ and $\widehat{f}$. But first we need to make sure that Φ is actually a Hida distribution. This has been shown in [9]. It follows also from Watanabe's theorem [21] and the fact that the space $\mathcal{D}^*$ in [21] is contained in $(\mathcal{S})^*$ [20]. However, we give another direct simple proof below.

Lemma 3.1. *Let $f \in \mathcal{S}'(\mathbb{R}^k)$ and let $\zeta_1, \ldots, \zeta_k \in L^2(\mathbb{R})$ be linearly independent. Then $\Phi = f \circ (\langle \cdot, \zeta_1 \rangle, \ldots, \langle \cdot, \zeta_k \rangle)$ is a Hida distribution. Moreover, if $\zeta_1, \ldots, \zeta_k \in L^2(\mathbb{R})$ are orthonormal, then the U-functional of Φ is given by $U[\Phi](\xi) = \langle \hat{f}, \theta_\xi \rangle$, $\xi \in \mathcal{S}(\mathbb{R})$, where $\langle \cdot, \cdot \rangle$ is the pairing of $\mathcal{S}'(\mathbb{R}^k)$ and $\mathcal{S}(\mathbb{R}^k)$ with repect to the Lebesgue measure and*

$$\theta_\xi(y_1, \ldots, y_k) = (2\pi)^{-k/2} e^{i(\langle \zeta_1, \xi \rangle y_1 + \cdots + \langle \zeta_k, \xi \rangle y_k) - \frac{1}{2}|y|^2}. \tag{3.1}$$

Proof. Without loss of generality, we may assume that $\zeta_1, \ldots, \zeta_k$ are orthonormal in $L^2(\mathbb{R})$. Informally, we have

$$\Phi(x) = (2\pi)^{-k/2} \int_{\mathbb{R}^k} e^{i(\langle x, \zeta_1 \rangle y_1 + \cdots + \langle x, \zeta_k \rangle y_k)} \widehat{f}(y) \, dy, \tag{3.2}$$

where $y = (y_1, \ldots, y_k)$. Thus informally the U-functional of Φ is given by

$$U[\Phi](\xi) = (2\pi)^{-k/2} \int_{\mathbb{R}^k} e^{i(\langle \zeta_1, \xi \rangle y_1 + \cdots + \langle \zeta_k, \xi \rangle y_k) - \frac{1}{2}|y|^2} \widehat{f}(y) \, dy, \quad \xi \in \mathcal{S}(\mathbb{R}).$$

Now, we use the following formula: for any $\alpha \in \mathbb{C}$,

$$e^{\alpha u - u^2/2} = e^{\alpha^2/4} \sum_{n=0}^{\infty} \pi^{1/4} 2^{n/2} (n!)^{-1/2} \alpha^n e_n(u),$$

where $e_n(u)$ is the Hermite function of order n. Therefore,

$$\left| A^p \left(e^{\alpha u - u^2/2} \right) \right|_2^2 = e^{|\alpha|^2/2} \sum_{n=0}^{\infty} \sqrt{\pi} (n!)^{-1} 2^n |\alpha|^{2n} (2n+2)^{2p},$$

where A is the operator $A = -D_u^2 + u^2 + 1$ in §2. It is easy to see that there exists a constant C_p depending only on p such that

$$\left| A^p \left(e^{\alpha u - u^2/2} \right) \right|_2 \leq C_p e^{9|\alpha|^2/4}.$$

Finally, since $\widehat{f} \in \mathcal{S}'(\mathbb{R}^k)$, there exists some positive number p such that $|A^{-p} \widehat{f}|_2 < \infty$. Therefore, for any $z \in \mathbb{C}$ and $\xi \in \mathcal{S}(\mathbb{R})$, we have

$$|U[\Phi](z\xi)| \leq (2\pi)^{-k/2} C_p^k |A^{-p} \widehat{f}|_2 e^{9|z|^2 |\xi|_2^2/4}.$$

Thus by the characterization theorem in [19] Φ is a Hida distribution. ∎

Lemma 3.2. *Let* $\zeta_1, \ldots, \zeta_k \in \mathcal{S}(\mathbb{R})$ *be linearly independent. Then for any* $\varphi \in (\mathcal{S})$, *the function* Θ_φ *defined by*

$$\Theta_\varphi(y_1, \ldots, y_k) = \langle\!\langle \widetilde{\delta}_{y_1\zeta_1 + \cdots + y_k\zeta_k}, \varphi \rangle\!\rangle (2\pi)^{-k/2} e^{-\frac{1}{2}|y|^2}$$

is in $\mathcal{S}(\mathbb{R}^k)$. *Here* $y = (y_1, \ldots, y_k) \in \mathbb{R}^k$.

Proof. It was shown in [12, Theorem 4.1.(c)] that for $x \in \mathcal{S}'_p(\mathbb{R}), p \geq 1$,

$$\|\widetilde{\delta}_x\|_{2,-p} \leq \exp[(b_p + |x|^2_{2,-p})/2], \tag{3.3}$$

where $b_p = 2^{-2p}(1 - 2^{-4p})^{-1} \sum_{j=1}^{\infty} (2j)^{-2p}$. We remark that the inequality in [12, Theorem 4.1.(c)] should be

$$\|\delta_y\|_{\mathcal{H}^{(-p)}} \leq \exp\left[\frac{1}{2}\left(\frac{\rho^{2p}}{1-\rho^{4p}}\|\iota_{0,p}\|^2_{H.S.} + |y|^2_{-p}\right)\right].$$

Now, since

$$|y_1\zeta_1 + \cdots + y_k\zeta_k|_{2,-p} \leq 2^{-p}|y|\left(|\zeta_1|^2_{2,0} + \cdots + |\zeta_k|^2_{2,0}\right)^{1/2},$$

it follows from (3.3) immediately that for any positive integer n,

$$\lim_{|y|\to\infty} |y|^n |\Theta_\varphi(y_1, \ldots, y_k)| = 0. \tag{3.4}$$

Moreover, it is easy to check that the function Θ_φ is differentiable with partial derivatives given by

$$\frac{\partial}{\partial y_j}\Theta_\varphi = (2\pi)^{-k/2}e^{-\frac{1}{2}|y|^2}[\langle\!\langle \widetilde{\delta}_{y_1\zeta_1 + \cdots + y_k\zeta_k}, D_{\zeta_j}\varphi \rangle\!\rangle - y_j\langle\!\langle \widetilde{\delta}_{y_1\zeta_1 + \cdots + y_k\zeta_k}, \varphi \rangle\!\rangle].$$

Note that D_{ζ_j} is continuous from $(\mathcal{S})$ into itself. Therefore, by (3.3), the partial derivatives of Θ_φ satisfy the growth condition (3.4). We can show inductively that all the higher derivatives of Θ_φ satisfy the growth condition (3.4). Hence Θ_φ is in $\mathcal{S}(\mathbb{R}^k)$. ∎

Theorem 3.3. *Suppose* $f \in \mathcal{S}'(\mathbb{R}^k)$ *and* $\zeta_1, \ldots, \zeta_k \in L^2(\mathbb{R})$ *are orthonormal. Let* $\Phi = f \circ (\langle \cdot, \zeta_1 \rangle, \ldots, \langle \cdot, \zeta_k \rangle)$. *Then the Fourier transform* $\widehat{\Phi}$ *of* Φ *is given by*

$$\langle\!\langle \widehat{\Phi}, \varphi \rangle\!\rangle = \langle \widehat{f}, \Theta_\varphi \rangle, \quad \varphi \in (\mathcal{S}), \tag{3.5}$$

where $\langle \cdot, \cdot \rangle$ *denotes the pairing of* $\mathcal{S}'(\mathbb{R}^k)$ *and* $\mathcal{S}(\mathbb{R}^k)$ *with respect to the Lebesgue measure and* Θ_φ *is defined by*

$$\Theta_\varphi(y_1, \ldots, y_k) = \langle\!\langle \widetilde{\delta}_{y_1\zeta_1 + \cdots + y_k\zeta_k}, \varphi \rangle\!\rangle (2\pi)^{-k/2} e^{-\frac{1}{2}|y|^2}.$$

Remark. From (3.5) we can write $\widehat{\Phi}$ symbolically as:

$$\widehat{\Phi} = \int_{\mathbb{R}^k} \widetilde{\delta}_{y_1\zeta_1+\cdots+y_k\zeta_k} \widehat{f}(y)\, d\mu_k(y),$$

where μ_k is the standard Gaussian measure on $\mathbb{R}^k$. This is the white noise analogue of (1.5).

Proof. Note that by Lemma 3.1 the U-functional of Φ is given by

$$U[\Phi](\xi) = \langle \widehat{f}, \theta_\xi \rangle, \quad \xi \in \mathcal{S}(\mathbb{R}). \tag{3.6}$$

Then from (2.2) and (3.6), we get

$$\langle\!\langle \widehat{\Phi}, : e^{\langle \cdot, \xi \rangle} : \rangle\!\rangle = \langle \widehat{f}, \theta_{-i\xi} \rangle e^{-\frac{1}{2}|\xi|_2^2}. \tag{3.7}$$

On the other hand, note that

$$U[\widetilde{\delta}_{y_1\zeta_1+\cdots+y_k\zeta_k}](\xi) = \exp(\langle \zeta_1, \xi \rangle y_1 + \cdots + \langle \zeta_k, \xi \rangle y_k - 2^{-1}|\xi|_2^2).$$

Therefore, for $\varphi =: e^{\langle \cdot, \xi \rangle}:$, we have $\Theta_\varphi = \theta_{-i\xi} e^{-\frac{1}{2}|\xi|_2^2}$. This yields that

$$\langle \widehat{f}, \Theta_\varphi \rangle = \langle \widehat{f}, \theta_{-i\xi} \rangle e^{-\frac{1}{2}|\xi|_2^2}. \tag{3.8}$$

It follows from (3.7) and (3.8) that (3.5) holds for any $\varphi =: e^{\langle \cdot, \xi \rangle}:$, $\xi \in \mathcal{S}(\mathbb{R})$. But the linear span of all such φ's is dense in $(\mathcal{S})$. Moreover, observe that both sides of (3.5) are continuous linear functionals on $(\mathcal{S})$. Therefore, the equality in (3.5) holds for all $\varphi \in (\mathcal{S})$. ∎

§4. Finite dimensional approximation of Hida distributions

First we give a simple example to approximate the delta function $\widetilde{\delta}_x$ by the finite dimensional delta functions.

Theorem 4.1. Let $\{\xi_k; k \geq 1\} \subset \mathcal{S}(\mathbb{R})$ be an orthonormal basis for $L^2(\mathbb{R})$. Then for any fixed x in $\mathcal{S}'(\mathbb{R})$

$$\lim_{k \to \infty} (\sqrt{2\pi})^k e^{\frac{1}{2}|\vartheta_k(x)|^2} \delta_{\vartheta_k(x)} \circ (\langle \cdot, \xi_1 \rangle, \ldots, \langle \cdot, \xi_k \rangle) = \widetilde{\delta}_x \quad in \ (\mathcal{S})^*,$$

where $\vartheta_k(x) = (\langle x, \xi_1 \rangle, \ldots, \langle x, \xi_k \rangle)$.

Proof. Let $F_k = (\sqrt{2\pi})^k e^{|\vartheta_k(x)|^2/2} \delta_{\vartheta_k(x)}$ and $\Phi_k = F_k \circ (\langle \cdot, \xi_1 \rangle, \ldots, \langle \cdot, \xi_k \rangle)$. Then the Fourier transform of F_k is given by

$$\widehat{F}_k(y_1, \ldots, y_k) = \exp(-i(\langle x, \xi_1 \rangle y_1 + \cdots + \langle x, \xi_k \rangle y_k) + 2^{-1}|\vartheta_k(x)|^2).$$

Therefore, by Lemma 3.1,

$$U[\Phi_k](\xi) = \exp\left(-\frac{1}{2}\sum_{n=1}^{k}\langle\xi,\xi_n\rangle^2 + \sum_{n=1}^{k}\langle\xi,\xi_n\rangle\langle x,\xi_n\rangle\right).$$

Hence $\lim_{k\to\infty} U[\Phi_k](\xi) = \exp(-2^{-1}|\xi|_2^2 + \langle x,\xi\rangle) = U[\widetilde{\delta}_x](\xi)$ for any $\xi \in \mathcal{S}(\mathbb{R})$. Thus Φ_k converges to $\widetilde{\delta}_x$ in $(\mathcal{S})^*$ as $k \to \infty$. ∎

Now, we consider the approximation of other Hida distributions. Recall that the Hermite function e_k is an eigenfunction of $A = -D_u^2 + u^2 + 1$ with eigenvalue $2k+2$, $k \geq 0$. Let π_k be the orthogonal projection onto the linear span of $e_0, \ldots, e_{k-1}$. For $p \geq 0$, let $\mathcal{S}_p(\mathbb{R})$ be the domain of A^p with the norm $|f|_{2,p} = |A^p f|_{L^2(\mathbb{R})}$.

Lemma 4.2. *For any $q > 0$, $\{e^{-tA}; t \geq 0\}$ is a contraction semigroup on $\mathcal{S}_q'(\mathbb{R})$. Moreover, $|e^{-tA}\zeta|_{2,p} \leq \left(t^{-1}(p+q)\right)^{p+q} e^{-p-q}|\zeta|_{2,-q}$ for any $p, q, t > 0$ and $\zeta \in \mathcal{S}_q'(\mathbb{R})$. Hence $e^{tA}\zeta \in \mathcal{S}(\mathbb{R})$.*

Proof. It is obvious that $\{e^{-tA}; t \geq 0\}$ is a contraction semigroup on $\mathcal{S}_q'(\mathbb{R})$ for any $q > 0$. The inequality can be checked from the fact that

$$\max_{\lambda > 0} \lambda^{p+q} e^{-t\lambda} = \left(t^{-1}(p+q)\right)^{p+q} e^{-p-q}. \quad \blacksquare$$

Let $\mathcal{T}_m$ denote the *trace operator of order m*, i.e.

$$\mathcal{T}_m f = \int_{\mathbb{R}^m} f(t_1, t_1, \ldots, t_m, t_m, \cdot)\, dt_1 \cdots dt_m, \quad f \in \mathcal{S}_p(\mathbb{R})^{\widehat{\otimes}n}, \; p > 0.$$

It has been proved in [4] that $\mathcal{T}_m f \in \mathcal{S}_p(\mathbb{R})^{\widehat{\otimes}(n-2m)}$ for any $f \in \mathcal{S}_p(\mathbb{R})^{\widehat{\otimes}n}$, $p > \frac{1}{2}$ and

$$|\mathcal{T}_m f|_{2,p} \leq a_p^m |f|_{2,p}, \tag{4.1}$$

where $a_p = 2^{-2p}6^{-1/2}\pi$. In the following a_p will denote this number.

Lemma 4.3. *Let $g \in \mathcal{S}_p(\mathbb{R})^{\otimes m}$, $p > \frac{1}{2}$, and $\psi_k(x) = \langle(\pi_k x)^{\otimes m}, g\rangle$. Then for any $k \geq 1$,*

$$\|\psi_k\|_{2,p} \leq (1-a_p)^{-1}\sqrt{2^m m!}\,|g|_{2,p}.$$

Proof. It is easy to check that

$$\psi_k = \sum_{j=0}^{[\frac{m}{2}]} \frac{m!}{2^j(m-2j)!j!} I_{m-2j}(\mathcal{T}_j(\pi_k^{\otimes m}g)).$$

By using (4.1) and the fact that $|\pi_k^{\otimes m}g|_{2,p} \leq |g|_{2,p}$, we get

$$\|\psi_k\|_{2,p} \leq \sum_{j=0}^{[\frac{m}{2}]} \frac{m!}{2^j\sqrt{(m-2j)!j!}} a_p^j |g|_{2,p}. \tag{4.2}$$

It follows from the inequalities $\binom{m}{2j} \le 2^m$ and $\sqrt{(2j)!} \le 2^j j!$ that

$$\frac{1}{\sqrt{(m-2j)!}} \le \frac{\sqrt{2^m}}{\sqrt{m!}} 2^j j!. \tag{4.3}$$

We can derive the inequality in the lemma from (4.2) and (4.3). ∎

For simplicity, let Q_t denote the second quantization $\Gamma(e^{-tA})$.

Lemma 4.4. *Let* $\Phi = I_n(F)$, $F \in \mathcal{S}'_q(\mathbb{R})^{\widehat{\otimes}n}$, $q > 0$. *Then for any* $p > \frac{1}{2}, k \ge 1$,

$$\|(Q_t\Phi)\circ\pi_k\|_{2,p} \le (1-a_p)^{-2} 2^n \|Q_t\Phi\|_{2,p}.$$

Proof. Let $\varphi = Q_t\Phi$ and $f = (e^{-tA})^{\otimes n}F$. Then $\varphi = I_n(f)$ and it can be rewritten as follows:

$$\varphi(x) = \sum_{m=0}^{[\frac{n}{2}]} (-1)^m \frac{n!}{2^m(n-2m)!m!} \langle x^{\otimes(n-2m)}, \mathcal{T}_m f\rangle.$$

Hence we have

$$\varphi(\pi_k x) = \sum_{m=0}^{[\frac{n}{2}]} (-1)^m \frac{n!}{2^m(n-2m)!m!} \langle (\pi_k x)^{\otimes(n-2m)}, \mathcal{T}_m f\rangle.$$

Therefore, by using (4.1) and Lemma 4.3, we obtain

$$\|\varphi\circ\pi_k\|_{2,p} \le (1-a_p)^{-1} \sum_{m=0}^{[\frac{n}{2}]} \frac{n!\sqrt{2^n}}{2^{2m}\sqrt{(n-2m)!}m!} a_p^m |f|_{2,p}. \tag{4.4}$$

Apply the inequality (4.3) to get

$$\|\varphi\circ\pi_k\|_{2,p} \le (1-a_p)^{-1} 2^n \sqrt{n!}|f|_{2,p} \sum_{m=0}^{[\frac{n}{2}]} a_p^m$$

$$\le (1-a_p)^{-2} 2^n |\varphi|_{2,p}. ∎$$

Theorem 4.5. *Let* $\Phi \in (\mathcal{S})^*$. *Then* $(Q_t\Phi)\circ\pi_k \in (\mathcal{S})^*$ *and for any* t

$$\lim_{k\to\infty} (Q_t\Phi)\circ\pi_k = Q_t\Phi \quad in \ (\mathcal{S})^*.$$

Proof. Since $(\mathcal{S})^*$ is the union of $(\mathcal{S})^*_q$, $q > 0$, there exists some $q > 0$ such that $\Phi \in (\mathcal{S})^*_q$. Hence $\Phi = \sum_{n=0}^\infty \Phi_n = \sum_{n=0}^\infty \tilde{I}_n(F_n)$, $F_n \in \mathcal{S}'_q(\mathbb{R})^{\widehat{\otimes}n}$. Here

$\widetilde{I}_n$ is the generalized multiple Wiener integral [2] of order n. By Lemma 4.4, for any $p > \frac{1}{2}, r > 0$, we have

$$\|(Q_t\Phi)\circ\pi_k\|_{2,-q-r} \leq \sum_{n=0}^{\infty} 2^{-(p+q+r)n}\|(Q_t\Phi_n)\circ\pi_k\|_{2,p} \qquad (4.5)$$

$$\leq (1-a_p)^{-2}\sum_{n=0}^{\infty} 2^{-(p+q+r-1)n}\|Q_t\Phi_n\|_{2,p}. \quad (4.6)$$

On the other hand, we can check easily from Lemma 4.2 that

$$\|Q_t\Phi_n\|_{2,p} \leq (t^{-1}(p+q))^{n(p+q)}e^{-n(p+q)}\|\Phi_n\|_{2,-q}. \qquad (4.7)$$

Put (4.7) in (4.6) and apply the Schwarz inequality to get

$$\|(Q_t\Phi)\circ\pi_k\|_{2,-q-r} \leq (1-a_p)^{-2}\|\Phi\|_{2,-q}\left(\sum_{n=0}^{\infty} c_{p,q,r,t}^n\right)^{\frac{1}{2}},$$

where the constant $c_{p,q,r,t}$ is given by

$$c_{p,q,r,t} = 2^{-2(p+q+r-1)}(t^{-1}(p+q))^{2(p+q)}e^{-2(p+q)}.$$

We can choose p, e.g. $p = 1$ and then choose r large enough so that $c_{p,q,r,t} < 1$. Hence $\|(Q_t\Phi)\circ\pi_k\|_{2,-q-r} < \infty$ and so $(Q_t\Phi)\circ\pi_k \in (\mathcal{S})^*$.

To show that $(Q_t\Phi)\circ\pi_k$ converges to $Q_t\Phi$ in $(\mathcal{S})^*$, let n be fixed and $\varphi_n = Q_t\Phi_n$. Then similar to (4.4), we have

$$\|\varphi_n\circ\pi_k - \varphi_n\|_{2,p} \leq (1-a_p)^{-1}\sum_{m=0}^{[\frac{n}{2}]} \frac{n!\sqrt{2^n}}{2^{2m}\sqrt{(n-2m)!m!}}a_p^m|\pi_k^{\otimes(n-m)}f - f|_{2,p},$$

where $\pi_k^{\otimes(n-m)}f$ means that π_k is applied to $n-m$ variables of f. Hence for fixed n, $\lim_{k\to\infty}\|(Q_t\Phi_n)\circ\pi_k - Q_t\Phi_n\|_{2,p} = 0$. On the other hand, similar to (4.5), we have

$$\|(Q_t\Phi)\circ\pi_k - Q_t\Phi\|_{2,-q-r} \leq \sum_{n=0}^{\infty} 2^{-(p+q+r)n}\|(Q_t\Phi_n)\circ\pi_k - Q_t\Phi_n\|_{2,p}. \quad (4.8)$$

From (4.6) and (4.8) we conclude that $\lim_{k\to\infty}\|(Q_t\Phi)\circ\pi_k - Q_t\Phi\|_{2,-q-r} = 0$. Hence $(Q_t\Phi)\circ\pi_k$ converges to $Q_t\Phi$ in $(\mathcal{S})^*$. ∎

Acknowledgement. The authors are very grateful to Professor J. Potthoff for his comments which greatly improve this paper.

References

[1] Hida, T.: *Analysis of Brownian Functionals*. Carleton Mathematical Lecture Notes, no. **13** (1975)

[2] Hida, T.: Generalized multiple Wiener integrals;*Proc. Japan Acad.* **54A** (1978) 55-58

[3] Hida, T.: *Brownian Motion*. Berlin, Heidelberg, New York. Springer-Verlag (1980)

[4] Hida, T., Kuo, H.-H. and Obata, N.: Transformations for white noise functionals;*Preprint* (1991), to appear in *J. Functional Anal.*

[5] Hida, T., Kuo, H.-H., Potthoff, J. and Streit, L.: *White Noise: An Infinite Dimensional Calculus.* Monograph in preparation

[6] Hida, T. and Potthoff, J.: White noise analysis–an overview; in: *White Noise Analysis–Math. and Appls.*, T. Hida, H.-H. Kuo, J. Potthoff and L. Streit (eds.), (1990) 140-165, World Scientific

[7] Ito, Y., Kubo, I. and Takenaka, S.: Calculus on Gaussian white noise and Kuo's Fourier transformation; in: *White Noise Analysis–Math. and Appls.*, T. Hida, H.-H. Kuo, J. Potthoff and L. Streit (eds.), (1990) 180-207, World Scientific

[8] Kallianpur, G. and Kuo, H.-H.: Regularity property of Donsker's delta function; *Appl. Math. Optim.* **12** (1984) 89-95

[9] Kubo, I.: Itô formula for generalized Brownian functionals; *Lecture Notes in Control and Information Sci.* **49** (1983) 156-166, Springer-Verlag

[10] Kubo, I. and Takenaka, S.: Calculus on Gaussian white noise I; *Proc. Japan Acad.* **56A** (1980) 376-380

[11] Kubo, I. and Takenaka, S.: Calculus on Gaussian white noise II; *Proc. Japan Acad.* **56A** (1980) 411-416

[12] Kubo, I. and Yokoi, Y.: A remark on the space of testing random variables in the white noise calculus; *Nagoya Math. J.* **115** (1989) 139-149

[13] Kuo, H. -H.: *Gaussian Measures in Banach Spaces*. Lecture Notes in Math. **463** (1975), Springer-Verlag

[14] Kuo, H.-H.: On Fourier transform of generalized Brownian functionals; *J. Multivariate Anal.* **12** (1982) 415-431

[15] Kuo, H.-H.: Donsker's delta function as a generalized Brownian functional and its application; *Lecture Notes in Control and Information Sci.* **49** (1983) 167-178, Springer-Verlag

[16] Kuo, H.-H.: The Fourier transform in white noise calculus; *J. Multivariate Anal.* **31** (1989) 311-327

[17] Kuo, H.-H.: Lectures on white noise analysis;*Preprint* (1990), to appear in *Soochow J. Math.*

[18] Lee, Y.-J.: Analytic version of test functionals, Fourier transform and a characterization of measures in white noise calculus; *Preprint* (1989), to appear in *J. Functional Anal.*

[19] Potthoff, J. and Streit, L.: A characterization of Hida distributions; *Preprint* (1989), to appear in *J. Functional Anal.*

[20] Potthoff, J. and Yan J.-A.: Some results about test and generalized functionals of white noise; to appear in: *Proc. Singapore Probab. Conf.* (1989), L.Y. Chen (ed.)

[21] Watanabe, S.: Malliavin's calculus in terms of generalized Wiener functionals; *Lecture Notes in Control and Information Sci.* **49** (1983) 284-290, Springer-Verlag

Izumi Kubo
Division of Math. and Inf. Sciences
Faculty of Integrated Arts and Sciences
Hiroshima University, Hiroshima 730
JAPAN

Hui-Hsiung Kuo
Department of Mathematics
Louisiana State University
Baton Rouge, LA 70803
USA

Representation and stability of nonlinear filters associated with Gaussian noises

HIROSHI KUNITA
Department of Applied Science
Kyushu University, Fukuoka 812, Japan

Abstract. This article discusses the nonlinear filtering problem in the case where the noise processes are not Brownian motions (white noises) but are Gaussian processes (colored noises). A likelihood ratio type formula is obtained. Then the formula is applied to proving the stability of the filters.

1. Introduction

Let $(x_t, t \in [0, t])$ be a continuous stochastic process with values in a complete metric space S, called a *system process*. Suppose we wish to observe the process (x_t), but what we can actually observe is a stochastic process perturbed by the noise process $(N_t, t \in [0, t])$, i.e.,

$$(1.1) \qquad Y_t = \int_0^t h(x_s)ds + N_t,$$

where $h(x)$ is a continuous map from S into $\mathbf{R}^d$ and (N_t) is a continuous stochastic process with values in $\mathbf{R}^d$ independent of (x_t). The process (Y_t) is called the *observation process* and (N_t), the *noise process*. The *nonlinear filter* π_t of x_t based on $(Y_s; s \leq t)$ is defined by the conditional distribution:

$$(1.2) \qquad \pi_t(dx) = P(x_t \in dx | \sigma(Y_s; s \leq t)).$$

If the noise process (N_t) is a Brownian motion, several ways of computing the filter (π_t) are known. One is a likelihood ratio type formula often called a *Kallianpur-Striebel formula* ([1]). However, in the real physical system, the noise process (N_t) is not exactly a Brownian motion. It has a continuous derivative $n_t = dN_t/dt$ and the observation process (y_t) is defined by

$$(1.3) \qquad y_t = h(x_t) + n_t.$$

In this paper we consider the case where the noise process is not a Brownian motion, but is a *Gaussian process*. In section 2, we obtain a likelihood ratio type formula for nonlinear filter, assuming that the noise process (N_t)

admits a causal and causal invertible representation by a certain Brownian motion called an *innovation process*. Then it is applied to the filtering problem in the frame work of (1.3), where (n_t) is a simple or multiple Markov Gaussian process. In Section 3, we discuss the stability and the unstability of the nonlinear filter in the case where the law of the noise process is close to that of a Wiener process.

2. Representation of filters

We will first introduce three conditions for a given Gaussian noise process $(N_t) = (N_t^1, \ldots, N_t^d)$.

Condition 1. There exists an r-dimensional standard Wiener process $(W_t) = ((W_t^1, \ldots, W_t^r), t \in [0, t])$ and a kernel $\Psi(t, s) = (\Psi_{ij}(t, s)), i = 1, \ldots, d, j = 1, \ldots, r$ which is continuously differentiable in (t, s) and satisfies

$$(2.1) \qquad N_t^i = M_t^i + \sum_{j=1}^{r} \int_0^t \Psi_{ij}(t, s) dW_s^j, \quad i = 1, \ldots, d,$$

where
$$(2.2) \qquad M_t^i = E[N_t^i | \sigma(N_0)], \quad i = 1, \ldots, d.$$

With the vector notation we denote the above formula by

$$(2.3) \qquad N_t = M_t + \int_0^t \Psi(t, s) dW_s.$$

Let $\mathbf{C}_k^m = \mathbf{C}^m([0, T]; \mathbf{R}^k)$ be the set of all m-times continuously differentiable maps from $[0, t]$ into $\mathbf{R}^k$ such that $\phi_0 = 0$. If $m = 0$, the above space is denoted simply by $\mathbf{C}_k$. We define a continuous linear transformation $\Psi : \mathbf{C}_r \to \mathbf{C}_d$ such that

$$(2.4) \qquad (\Psi\phi)_t = \int_0^t \Psi(t, s)\phi'(s) ds$$

holds if $\phi \in \mathbf{C}_d^1$. For general $\phi \in \mathbf{C}_r$ it is extended by integration by parts as

$$(2.5) \qquad (\Psi\phi)_t = \Psi(t, t)\phi_t - \int_0^t \phi_s \frac{\partial}{\partial s} \Psi(t, s) ds.$$

Note that $(\Psi\phi)_u = (\Psi\psi)_u$ holds for all $u \le t$ if $\phi_s = \psi_s$ holds for all $s \le t$. The transformation with the above property is called a *causal transformation*. Now let K be an another causal transformation from $\mathcal{R}(\Psi)$ into $\mathbf{C}_r$, where $\mathcal{R}(\Psi) = \{\Psi\phi; \phi \in \mathbf{C}_r\}$. It is called an *inverse transformation* of Ψ if $K\Psi\phi = \phi$ holds for all $\phi \in \mathbf{C}_r$.

Condition 2. The transformation Ψ has a causal inverse transformation K. Further, $(K\phi)_t$ is differentiable if $\phi \in \mathbf{C}_d^1 \cap \mathcal{R}(\Psi)$.

Condition 3. (M_t) is continuously differentiable in t. Further, it belongs to $\mathcal{R}(\Psi)$.

We set

$$(2.6) \qquad m_t = \frac{dM_t}{dt},$$

$$(2.7) \qquad (L\phi)_t = \frac{d}{dt}(K\psi)_t, \quad \text{if} \quad \psi_t = \int_0^t \phi_s ds.$$

For almost all ω, the processes $(N_t(\omega)), (M_t(\omega))$ etc. may be regarded as elements of $\mathbf{C}_d$. We often denote them using bold faces as $\mathbf{N}(\omega), \mathbf{M}(\omega)$ etc. or simply $\mathbf{N}, \mathbf{M}$. Then the filter is a functional of $\mathbf{Y}$.

Equation (2.3) is written by $N_t - M_t = (\Psi \mathbf{W})_t$. This means that $N_t - M_t$ is a linear functional of $(W_u; u \leq t)$. Therefore $\sigma(N_s - M_s; s \leq t) \subset \sigma(W_u; u \leq t)$ is satisfied. On the other hand we have $K(\mathbf{N} - \mathbf{M})_t = (K\Psi\mathbf{W})_t = W_t$. This shows $\sigma(N_s - M_s; s \leq t) \supset \sigma(W_s; s \leq t)$, so that the equality $\sigma(N_s - M_s; s \leq t) = \sigma(W_s; s \leq t)$ holds for all t. The Wiener process (W_t) with the above property is called the *innovations process* of the noise process $(N_t - M_t)$.

Now let $\mathbf{C} = \mathbf{C}([0,t]; S)$ be the set of all continuous maps from $[0,t]$ into S. Elements of $\mathbf{C}$ are denoted by $\mathbf{x} = (x(t); t \in [0,t])$. The law of the system process (x_t) can be defined on the space $\mathbf{C}$, which is denoted by $P_{\mathbf{X}}$. Now if $\mathbf{x} \in \mathbf{C}, (h(x(t)); t \in [0,t])$ can be regarded as an element of $\mathbf{C}_d$, which we denote by $h(\mathbf{x})$.

We introduce a statistical hypothesis for the noise process.

Hypothesis $(\mathbf{H_0})$. The noise process (N_t) satisfies Conditions 1-3.

Theorem 2.1. *Assume that the noise process (N_t) satisfies Hypothesis (H_0). Assume further that $(\int_0^t h(x_s)ds)$ belongs to $\mathcal{R}(\Psi)$ a.s. Then the filter (π_t) is represented by*

$$(2.8) \qquad \pi_t(g)(\mathbf{Y}) = \frac{\displaystyle\int \alpha_t(\mathbf{x}, \mathbf{Y})g(x(t))dP_{\mathbf{X}}}{\displaystyle\int \alpha_t(\mathbf{x}, \mathbf{Y})dP_{\mathbf{X}}},$$

where

$$\alpha_t(\mathbf{x}, \mathbf{Y}) = \exp\left\{\int_0^t (L(h(\mathbf{x}) + \mathbf{m})_s, d\widehat{Y}_s) - \frac{1}{2}\int_0^t |L(h(\mathbf{x}) + \mathbf{m})_s|^2 ds\right\},$$

(2.9)

and

$$(2.10) \qquad \widehat{Y}_t = \int_0^t Lh(\mathbf{x})_s ds + \int_0^t (L\mathbf{m})_s ds + W_t.$$

Proof. Define $\widehat{Y}_t = (KY)_t$. Applying the operator K to each term of (1.1), we find that $\widehat{Y}_t$ satisfies (2.10). Now we have $\Psi K \Psi \phi = \Psi \phi$ since K is the inverse of Ψ. Therefore $\Psi K \psi = \psi$ holds for all $\psi \in \mathcal{R}(\Psi)$. Since $Y \in \mathcal{R}(\Psi)$ holds, we have $\Psi K Y = Y$ or $\Psi \widehat{Y} = Y$ a.s., which proves $\sigma(Y_s; s \leq t) = \sigma(\widehat{Y}_s; s \leq t)$ as before. Therefore the filter based on $(Y_s; s \leq t)$ coincides with the filter based on $(\widehat{Y}_s; s \leq t)$. Since (W_t) is a standard Brownian motion, the Kallianpur-Stribel formula [1] shows that the filter of (x_t) based on $(\widehat{Y}_s; s \leq t)$ is represented by (2.8) and (2.9). The proof is complete.

As an application we will consider the case where (N_t) has the continuous derivative (n_t) and the observation process y_t is defined by (1.3). We introduce:

Hypothesis (H_1). The Gaussian noise process (n_t) satisfies a linear stochastic differential equation

$$(2.11) \qquad dn_t = F_t n_t dt + G_t dW_t.$$

Here F_t is a $d \times d$ matrix and G_t is a $d \times r$ matrix with rank r, both of which are differentiable in t. (W_t) is an r-dimensional standard Wiener process where r is less than or equal to d.

Before we state a theorem, some notation is necessary. Given a $d \times r$ matrix G with rank r, $G'G$ is a linear map from $\mathbf{R}^r$ onto itself, so that it has the inverse map $(G'G)^{-1}$. Define the $r \times d$ matrix G^- by $G^- = (G'G)^{-1}G'$. Then G^-G is the identity map. For convenience we call G^- the left inverse of G. Of course if G is a $d \times d$ nonsingular matrix, G^- coincides with the inverse matrix G^{-1}.

Theorem 2.2. *Assume that $h_t = h_t(\mathbf{x}) \equiv h(x_t)$ is differentiable and $h'_t - F_t h_t$ belongs to $\mathcal{R}(G_t)$ for all t. Then under Hypothesis (H_1), the filter (π_t) is represented by (2.8), where $\alpha_t(\mathbf{x}, \mathbf{Y})$ is replaced by*

$$(2.12) \qquad \alpha_t(\mathbf{x}, \mathbf{y}) = \exp \left\{ \int_0^t (G_s^-(h'_s - F_s h_s), G_s^-(dy_s - F_s y_s ds)) \right.$$
$$\left. - \frac{1}{2} \int_0^t |G_s^-(h'_s - F_s h_s)|^2 ds \right\}.$$

Proof. We will show that Hypothesis (H_1) implies Hypothesis (H_0). Let $\Phi(t, s)$ be the fundamental solution of the linear differential equation $dx/dt = F_t x$. Then the solution of equation (2.11) is represented by

$$(2.13) \qquad n_t = \Phi(t, 0)n_0 + \int_0^t \Phi(t, s)G_s dW_s.$$

Furthermore, $N_t = \int_0^t n_s\,ds$ is represented by (2.3), setting $M_t = (\int_0^t \Phi(u,0)du)n_0$ and $\Psi(t,s) = (\int_s^t \Phi(u,s)du)G_s$. Now define the continuous linear transformation $K : \mathbf{C}_d^1 \to \mathbf{C}_r$ such that

(2.14)
$$(K\phi)_t = \int_0^t G_s^-(d\phi_s' - F_s\phi_s'\,ds)$$

holds if $\phi \in \mathbf{C}_d^2$. This K is causal and is an inverse of Ψ. Indeed, if $\phi \in \mathbf{C}_r^1$, we have

(2.15)
$$K\left(\int_0 du \int_0^u \Phi(u,s)G_s\phi_s'\,ds\right)_t = \phi_t.$$

Therefore, Hypothesis (H_0) is satisfied. Further, by our assumption, h_t is differentiable and $h_t' - F_t h_t \in \mathcal{R}(G_t)$. This implies $(\int_0^t h(x_s)ds) \in \mathcal{R}(\Psi)$.

The operator L is given by

(2.16)
$$(L\phi)_t = G_t^-(\phi_t' - F_t\phi_t).$$

Further we have $(K\mathbf{M})_t = 0$. Therefore, the equality (2.12) follows from (2.9) immediately. The proof is complete.

We shall next consider the case where the noise process (n_t) is a one dimensional k-ple Markov process, (k is a positive integer greater than 1), such that it satisfies an k-th order stochastic differential equation

(2.17)
$$L_t n_t = \dot{W}_t,$$

where

(2.18) $\quad L_t = a_k(t)(\frac{d}{dt})^k + a_{k-1}(t)(\frac{d}{dt})^{k-1} + \cdots + a_1(t)(\frac{d}{dt}) + a_0(t),$

$a_k(t)$ is strictly positive and $\dot{W}_t = (\frac{d}{dt})W_t$. More precisely, we introduce the following:

Hypothesis $(\mathbf{H_k})$. The one dimensional Gaussian noise process (n_t) is $(k-1)$-times differentiable and satisfies

(2.19) $\quad a_k(t)dn_t^{(k-1)} + \left(a_{k-1}(t)n_t^{(k-1)} + \cdots + a_1(t)n_t^{(1)} + a_0(t)n_t\right)dt$
$$= dW_t,$$

where $(n_t^{(j)}), j = 1,\ldots,k-1$ are j-th derivatives of (n_t).

Set $\mathbf{n}_t = (n_t, n_t^{(1)}, \ldots, n_t^{(k-1)})$ and

(2.20)
$$F_t = \begin{pmatrix} 0 & 1 & 0 & \cdots & 0 \\ 0 & 0 & 1 & \cdots & 0 \\ \multicolumn{5}{c}{\cdots\cdots\cdots\cdots\cdots\cdots} \\ \multicolumn{3}{c}{\cdots\cdots\cdots} & 0 & 1 \\ b_0 & b_1 & \cdots\cdots & & b_{k-1} \end{pmatrix}, \qquad G_t = \begin{pmatrix} 0 \\ 0 \\ \vdots \\ \vdots \\ b_k \end{pmatrix},$$

where $b_i = -a_i/a_k$ for $i = 0, \ldots, k-1$ and $b_k = 1/a_k$. Then $\mathbf{n}_t$ satisties

$$(2.21) \qquad\qquad d\mathbf{n}_t = F_t\mathbf{n}_t dt + G_t dW_t.$$

Therefore $(\mathbf{n}_t)$ is a simple Markov process. It is easily verified that it satisfies Hypothesis (H_1).

Now we assume that h_t is (k-1)-times differentiable . Set

$$(2.22) \qquad \mathbf{h}_t = (h_t, \ldots, h_t^{(k-1)}), \quad \mathbf{y}_t = (y_t, \ldots, y_t^{(k-1)}).$$

Then we have $\mathbf{y}_t = \mathbf{h}_t + \mathbf{n}_t$. Therefore, we can regard $(\mathbf{n}_t)$ as a noise process and $(\mathbf{y}_t)$ as an observation process. Further, the first (k-1)-th components of $\mathbf{h}'_t - F_t\mathbf{h}_t$ are 0. Therefore $\mathbf{h}'_t - F_t\mathbf{h}_t$ belongs to $\mathcal{R}(G_t)$. The corresponding transformations K and L defined by (2.14) and (2.16) satisfy

$$(2.23) \qquad (L\mathbf{h})_t = L_t(h(x_t)), \qquad (K\mathbf{y})_t = \int_0^t a_k(s)dy_s^{(k-1)}.$$

Therefore, under Hypothesis (H_k), Theorem 2.2 is valid if we replace $\alpha_t(\mathbf{x}, \mathbf{y})$ by

$$(2.24) \qquad \exp\left\{\int_0^t L_s(h(x_s))a_k(s)dy_s^{(k-1)} - \frac{1}{2}\int_0^t |L_s(h(x_s))|^2 ds\right\}.$$

3. Stability of filters

We shall introduce a family of hypotheses with parameter $\epsilon > 0$ for the noise process.

Hypothesis $(\mathbf{H_0})_\epsilon$. The noise process (N_t) satisfies Conditions 1-3, where we have the representation

$$(3.1) \qquad\qquad N_t = \int_0^t m_s^\epsilon ds + \int_0^t \Psi^\epsilon(t, s)dW_s$$

instead of (2.3).

We denote by K^ϵ and L^ϵ the corresponding transformations. The filter of x_t under Hypothesis $(H_0)_\epsilon$ is denoted by π_t^ϵ. Let (π_t^0) be a stochastic process with values in the space of probability measures on $\mathbf{R}^d$. The family of processes $(\pi_t^\epsilon), \epsilon > 0$ is said to converge a.s. to (π_t^0) as $\epsilon \to 0$ if $\pi_t^\epsilon(g)$ converges to $\pi_t^0(g)$ for any t and bounded continuous function g.

Now suppose that the family of laws of the noise process (N_t) converges weakly to that of a certain Wiener process, (Wiener noise process), as $\epsilon \to 0$. If the corresponding family of filters (π_t^ϵ) converges a.s. to the filter of (x_t) associated with the Wiener noise process, the filtering problem is called *stable*. Generally speaking, it is not always stable. It may occur that $(\pi_t^\epsilon), \epsilon > 0$ converge to (π_t^0) but (π_t^0) is a filter of some other filtering problem. (See the Example after Corollary 3.3.)

In this section, we discuss three problems. The first is the weak convergence of the laws of the noise process. The second is the convergence of the filters. The third is the stability of the filtering problem. It turns out that conditions needed for the first problem and the second are different to each other. Thus, it may occur that the family of laws of noise processes does not converge weakly but the family of filters converges a.s. to a filter of another filtering problem.

We introduce three conditions.

Condition 4. As $\epsilon \to 0$ $\Psi^\epsilon(t, s) \to \Psi^0(s)$ for all $t > s$ and $M_t^\epsilon \to M_t^0$ for all t.

Condition 5. As $\epsilon \to 0$ $L^\epsilon h(\mathbf{x})_t \to \widehat{h}_t^0$ and $(L^\epsilon \mathbf{m}^\epsilon)_t \to \widehat{m}_t^0$ for all t.

Condition 6. The rank of $\Psi^0(s)$ is r for all s. Further $h_s = \Psi^0(s)\widehat{h}_s$ and $M_t^0 = \int_0^t \Psi^0(s)\widehat{m}_s^0 ds$ hold for all t and s.

Theorem 3.1. *(1) Assume Condition 4. Then the laws of (N_t) converge weakly to that of the Wiener process given by*

$$(3.2) \qquad N_t^0 = M_t^0 + \int_0^t \Psi^0(s)dW_s.$$

(2) Assume Condition 5. Then $(\pi_t^\epsilon), \epsilon > 0$ converges a.s. to the filter of (x_t) based on

$$(3.3) \qquad \widehat{Y}_t^0 = \int_0^t (\widehat{h}_s^0 + \widehat{m}_s^0)ds + W_t.$$

(3) Assume Conditions 4-6. Then $(\pi_t^\epsilon), \epsilon > 0$ converges a.s. to the filter of (x_t) based on

$$(3.4) \qquad Y_t^0 = \int_0^t h(x_s)ds + N_t^0.$$

Proof. The first assertion (1) is obvious. For the proof of (2) observe that $\pi_t^\epsilon(g)$ is represented by (2.8) replacing $\alpha_t(\mathbf{x}, \mathbf{Y})$ by the following $\alpha^\epsilon(\mathbf{x}, \mathbf{Y})$:

$$(3.5) \quad \exp\left\{\int_0^t (L^\epsilon(h(\mathbf{x}) + \mathbf{m}^\epsilon)_s, d(K^\epsilon \mathbf{Y})_s) - \frac{1}{2}\int_0^t |L^\epsilon(h(\mathbf{x}) + \mathbf{m}^\epsilon)_s|^2 ds\right\}.$$

Set $\widehat{Y}_t^\epsilon = (K^\epsilon \mathbf{Y})_t$. Since $\widehat{Y}_t^\epsilon = \int_0^t L^\epsilon(h(\mathbf{x}) + \mathbf{m}^\epsilon)_s ds + W_t$, $(\widehat{Y}_t^\epsilon)$ converges to $(\widehat{Y}_t^0)$, which satisfies (3.3). Therefore, $\lim_{\epsilon \to 0} \alpha_t^\epsilon(\mathbf{x}, \mathbf{Y}) = \widehat{\alpha}_t^0(\mathbf{x}, \widehat{Y})$ exists and it is represented by

$$(3.6) \quad \widehat{\alpha}_t^0(\mathbf{x}, \widehat{Y}) = \exp\left\{\int_0^t (\widehat{h}_s^0 + \widehat{m}_s^0, d\widehat{Y}_s^0) - \frac{1}{2}\int_0^t |\widehat{h}_s^0 + \widehat{m}_s^0|^2 ds\right\}.$$

This proves that $\pi_t^\epsilon(g)$ converge to $\widehat{\pi}_t^0(g)$ a.s. and it is represented by (2.8), replacing $\alpha_t(\mathbf{x}, \mathbf{Y})$ by $\widehat{\alpha}_t^0(\mathbf{x}, \widehat{\mathbf{Y}})$. Then the Kallianpur-Striebel formula [1] shows that $(\widehat{\pi}_t^0)$ is the filter of (x_t) based on the observation process $(\widehat{Y}_t^0)$.

Finally we show (3). By Condition 6, we have clearly two relations $Y_t^0 = \int_0^t \Psi^0(s) d\widehat{Y}_s^0$ and $\widehat{Y}_t^0 = \int_0^t \Psi^0(s)^{-1} dY_s^0$. Therefore, the filter based on $(\widehat{Y}_s^0; s \le t)$ coincides with the filter based on $(Y_s^0; s \le t)$. The proof is complete.

We shall apply the above theorem to the case where $n_t = dN_t/dt$ is a Markov Gaussian process. We introduce:

Hypothesis $(\mathbf{H_1})_\epsilon$. The Gaussian noise process (n_t) satisfies a linear stochastic differential equation

$$(3.7) \qquad dn_t = F_t^\epsilon n_t dt + G_t^\epsilon dW_t,$$

on $\mathbf{R}^d$, where the F_t^ϵ are $d \times d$ matrices with ranks d and the G_t^ϵ are $d \times r$ matirices with ranks r.

Theorem 3.2. *Let (π_t^ϵ) be the filter of (x_t) under Hypothesis $(H_1)_\epsilon$.*

(1) Assume that the family of the pairs $(F_t^\epsilon, G_t^\epsilon), \epsilon > 0$ satisfies the following (a) and (b).

(a) As $\epsilon \to 0, -(F_s^\epsilon)^{-1} G_t^\epsilon \to \Psi^0(t)$ holds for all s, t.

(b) Let $\Phi^\epsilon(t, s)$ be the fundamental solution of equation $\frac{dx}{dt} = F_t^\epsilon x$. Then $\lim_{\epsilon \to 0} \Phi^\epsilon(t, s) = 0$ holds for any $t > s$.

Then, the laws of the noise process (N_t) converge weakly to the law of a Wiener process of (3.2) where $M_t^0 = 0$.

(2) Assume

(c) $(G_s^\epsilon)^- \to 0$ and $-(G_s^\epsilon)^- F_s^\epsilon \to H^0(s)$ hold for all s as $\epsilon \to 0$.

Then the filters (π_t^ϵ) converge a.s. to the filter based on (3.3) where $\widehat{h}_s = H^0(s)h(x(s))$ and $\widehat{m}_s^0 = 0$.

(3) Assume further Conditions (a)-(b) and one of the following:

(d) The rank of $\Psi^0(t)$ is r for all t and $\Psi^0(t)H^0(t)h_t = h_t$ for all t.

(d') The ranks of $\Psi^0(t)$ and $H^0(t)$ are d for all t.

Then (π_t^ϵ) converges a.s. to the filter based on (3.4).

Proof. Note that $\Phi^\epsilon(t, s) = I + \int_s^t \Phi^\epsilon(t, u) F_u^\epsilon du$. Then we have from (a),

$$(3.8) \qquad \Phi^\epsilon(t, s)(F_s^\epsilon)^{-1} G_s^\epsilon = (F_s^\epsilon)^{-1} G_s^\epsilon + \Psi^\epsilon(t, s) + o(1).$$

The left hand side converges to 0 by (b). Therefore, $\lim_{\epsilon \to 0} \Psi^\epsilon(t, s) = \Psi^0(s)$. Further $\lim_{\epsilon \to 0} M_t^\epsilon = 0$ by (b). Then the first assertion follows from Theorem 3.1 (1).

Assume next (c). The operator L^ϵ is given by $(L^\epsilon \phi)_t = (G_t^\epsilon)^-(\phi_t' - F_t^\epsilon \phi_t)$. Then $(L^\epsilon h)_t$ converges to $\widehat{h}_t = H^0(t)h_t$, so that Condition 5 is satisfied

($\widehat{m}_t^0 = 0$ in this case). Therefore, (π_t^ϵ) converges a.s. to the filter based on $(\widehat{Y}_t^0)$ of (3.3) by Theorem 3.1 (2).

For the proof of (3), we shall check Condition 6. Since $H^0(s)\Psi^0(s) = I_r$, both (d) and (d') imply $\Psi^0(s)\widehat{h}_s = \Psi^0(s)H^0(s)h_s = h_s$. The proof is complete.

Remark. In the second assertion of Theorem 3.2, we do not have to assume that the family of laws of the noise process converges. In particular, Condition (b) is not necessary. Indeed suppose that $F^\epsilon = \epsilon^{-1}F$. If the real part of an eigenvalue of F is positive, then the fundamental solution of the matrix $\epsilon^{-1}F$ diverges so that the family of laws of the noise process can not converge. Even so, under condition (c), the family of filters converges.

Corollary 3.3. *Suppose that F_t^ϵ and G_t^ϵ are given by $F_t^\epsilon = \epsilon^{-1}F, G_t^\epsilon = \epsilon^{-1}G$, where F is a stable matrix, that is, the real parts of all eigenvalues are negative. Set $\Psi^0 = -F^{-1}G$ and $H^0 = -G^-F$. Then the conclusions of Theorem 3.2 (1), (2) are valid. In particular, if the ranks of F and G are d then the filtering problem is stable.*

Example. We give an unstable example. Set $F = \begin{pmatrix} -1 & 0 \\ \lambda & -1 \end{pmatrix}$, $G = \begin{pmatrix} 1 \\ 0 \end{pmatrix}$ and $F^\epsilon = \epsilon^{-1}F, G^\epsilon = \epsilon^{-1}G$. Then we have $F^{-1} = \begin{pmatrix} -1 & 0 \\ -\lambda & -1 \end{pmatrix}$ and $G^- = (1, 0)$. Therefore we have $\Psi^0 = \begin{pmatrix} 1 \\ \lambda \end{pmatrix}$ and $H^0 = (1, 0)$. Consequently,

$$(3.9) \quad \widehat{Y}_t^0 = \int_0^t h_1(s)ds + W_t, \quad Y_t^0 = \int \begin{pmatrix} h_1(s) \\ h_2(s) \end{pmatrix} ds + \begin{pmatrix} 1 \\ \lambda \end{pmatrix} W_t.$$

Then the filters based on $(\widehat{Y}_t^0)$ and (Y_t^0) are clearly different to each other.

We next consider the case where (n_t) is a k-ple Markov process. We introduce:

Hypothesis $(\mathbf{H_k})_\epsilon$. The one dimensional Gaussian noise process (n_t) is (k-1)-times differentiable and satisfies

$$(3.10) \quad a_k^\epsilon dn_t^{(k-1)} + \left(a_{k-1}^\epsilon n_t^{(k-1)} + \cdots + a_1^\epsilon n_t^{(1)} + a_0^\epsilon n_t \right) dt = dW_t.$$

Theorem 3.4. *Let (π_t^ϵ) be the filter of (x_t) under Hypothesis $(H_k)_\epsilon$. Assume:*

(e) $a_j^\epsilon \to a_j^0$ for $j = 0, \ldots, k$, where $a_0^0 \neq 0$ and $a_k^0 = 0$.

(f) Let λ_j^ϵ be solutions of the characteristic equation $\sum_{j=0}^k a_j^\epsilon \lambda^j = 0$. Then $\sup_j \mathcal{R}e(\lambda_j^\epsilon) \to -\infty$ as $\epsilon \to 0$.

Then we have the following two assertions.

(1) The family of laws of (N_t) converges weakly to that of the Wiener process

$(N_t^0) = ((a_0^0)^{-1} W_t).$

(2) The family of filters (π_t^ϵ) converges a.s. to the filter of (x_t) based on the observation process

$$(3.11) \qquad \widehat{Y}_t^0 = \sum_{j=0}^{k-1} a_j^0 \int_0^t h_s^{(j)} ds + W_t.$$

In particular if $a_j^0 = 0$ for any $j = 1, \ldots, k-1$, then the family of filters (π_t^ϵ) converges to the filter of (x_t) based on the observation process (3.4).

    Proof. Consider the vector process $\mathbf{n}_t = (n_t, n_t^{(1)}, \ldots, n_t^{(k-1)})$ as in the previous section. Associated with (3.10), define F^ϵ and G^ϵ by (2.20). Then we have $-(F^\epsilon)^{-1} G^\epsilon = (1/a_0^\epsilon, 0, \ldots, 0)'$ which converges to $\Psi^0 = (1/a_0^0, 0, \ldots, 0)'$. Hence Condition (a) of Theorem 3.2 is satisfied. Further Condition (f) implies Condition (b). Consequently, the family of laws of the vector noise process $(\mathbf{N}_t) = (\int_0^t \mathbf{n}_s ds)$ converges weakly to that of the Wiener process $((1/a_0^0, \ldots, 0)' W_t)$ by Theorem 3.2 (1). On the other hand, $(G^\epsilon)^-$ converges to 0 and $-(G^\epsilon)^- F^\epsilon$ converges to $H^0 = (a_0^0, a_1^0, \ldots, a_{k-1}^0)$. Then the process (3.3) is represented by (3.11). Therefore, by Theorem 3.2 (2), (π_t^ϵ) converges a.s. and the limit coincides with the filter based on $(\widehat{Y}_t^0)$ of (3.11). The last assertion is obvious.

References

[1] M. Fujisaki, G. Kallianpur and H. Kunita, Stochastic differential equations for the nonlinear filtering problem, Osaka J. Math., 9(1972), 19-42.

[2] G. Kallianpur and C. Striebel, Estimations of stochastic processes, Arbitrary system process with additive white noise observation errors, Ann. Math. Statist., 39(1968), 785-801.

[3] H. Kunita, *Stochastic flows and stochastic differential equations*, Cambridge Univ. Press, 1990.

[4] H. Kunita, The stability and approximation problems in nonlinear filtering theory. Stochastic Analysis, Liber Amicorum for Moshe Zakai, ed. by E. Mayer-Wolf et al., Academic press, 1991, 311-330.

On Central Limit Theory for Families of Strongly Mixing Additive Random Functions *

M.R. Leadbetter and Holger Rootzén

Abstract

The paper considers distributional limits for families of additive random interval functions $\zeta_T = \zeta_T(I)$, under an array form of strong mixing. This provides a natural general setting for discussing the central limit problem in a variety of situations, including sums of strongly mixing arrays, and certain random measures of interest in continuous parameter extremal theory. In particular previous results on array sums are extended, providing also insights into the role of various mixing conditions used in earlier works.

1 Introduction and notation

In one sense this paper is a revisitation of dependent central limit theory - a field which saw activity in the 1960's and early 1970's with the appearance of the books [4] [8] and a variety of papers (e.g. [6], [13], [14], [1], [2], [3], and more recently [10], [17]), and which has found widespread application (e.g. in [5], [15], to mention just two out of very many applications). In particular we present some results which complement and extend existing central limit theory for array sums, exhibiting strong mixing as a central basic condition for these problems. This fact is not too obvious from much of the literature since various stronger mixing conditions (especially "ϕ-mixing") are often assumed, but used more to facilitate convenient (though sometimes overly restrictive) formulations for application rather than the central results obtained.

For the "bounded variance cases" it is shown in [13] that a variety of mixing conditions (including a form of strong mixing) give the same results for the central limit problem as apply to iid arrays (cf. [12]). In that work natural variance conditions are used, requiring some detailed assumptions and significant effort in calculation. Here we use essentially the same simple characteristic function arguments to obtain results for the general central limit problem under strong mixing, but do not undertake the work needed to reformulate the conditions comparably to those in [13]. A more specific

*Research partially supported by the AFOSR Contract No. F49620 85C - 0144.

comparison with [13] in the bounded variance case is contained in Section 3. However our wider aim is to provide a simple unifying framework for a variety of such central limit problems in discrete and continuous time including the general problem for array sums and, for example, random measures such as occupation times in rare sets. For such applications it is of interest to see how the degree of "rareness" of the set alters the type of limit law obtained.

Specifically we shall throughout be concerned with additive random functions $\zeta_T(I)$ (or $\zeta_n(I)$) defined on semiclosed intervals $I = (a, b] \subset (0, 1]$, for each $T > 0$ (for $n = 1, 2, \ldots$). For example, $\zeta_n(a, b] = \sum_{k_n a < i \le k_n b} X_i^{(n)}$ gives array sums $\sum_1^{k_n} X_i^{(n)}$ for $a = 0$, $b = 1$ and $\zeta_T(I) = \int_{T \cdot I} 1_{(\xi_t > u_T)} dt$ determines the exceedance random measure (cf. [11]) for a process ξ_t above a family of levels u_T. Limits in distribution for additive random functions ζ_T will be characterized under a natural array form of strong mixing, defined below. The framework of additive random functions was used in related contexts in [1] and [6]. In the latter paper infinite divisibility of distributional limits was obtained (using an entirely different approach from that here) under an assumption of "uniform ergodicity", subsequently shown ([16]) to be equivalent to strong mixing in important cases. Such a result is also included below, using present methods, under the array strong mixing condition (Theorem 3.1).

Section 2 of the paper contains the specific form of strong mixing and the basic factorization lemma for the characteristic functions within that framework. In Section 3 the reductions of theorems under dependence to those under independence are thereby explicitly obtained. Section 4 concerns the more definitive results involving stationarity and Section 5 contains comments concerning applications and functional limit theorems.

2 Array strong mixing and basic results

Let $\{\zeta_T, T > 0\}$ be a family of additive random functions as above, defined on semiclosed intervals $I = (a, b] \subset (0, 1]$, i.e. $\zeta(a, b] + \zeta(b, c] = \zeta(a, c]$ when $0 \le a \le b \le c \le 1$. Extend the definition of ζ_T to finite unions of intervals by additivity. Strong mixing for these families will be defined as follows:

Write $\mathcal{B}_{s,t}^T = \sigma\{\zeta_T(u, v] : s \le u < v \le t\}$, for $0 \le s < t \le 1$ (where σ denotes the generated σ-field) and, for $0 < \ell < 1$,

$$\alpha_{T,\ell} = \sup\{|P(A \cap B) - P(A)P(B)| : A \in \mathcal{B}_{0,s}^T, \ B \in \mathcal{B}_{s+\ell,1}^T, \ 0 \le s \le 1-\ell\}.$$

Then $\{\zeta_T\}$ is called *strongly mixing* (s.m.) if $\alpha_{T,\ell_T} \to 0$ for some $\ell_T \to 0$ as $T \to \infty$ (or equivalently if $\alpha_{T,\epsilon} \to 0$ for each fixed $\epsilon > 0$).

Note that this definition is potentially slightly more general than in some earlier works where the mixing coefficient depends only on the separation

ℓ. Note also that ([18]),

$$\beta_{T,\ell} = \sup\{|EXY - EXEY|: X,\, Y \text{ resp. } \mathcal{B}_{0,s} \text{ and } \mathcal{B}_{s+\ell,1} - \text{measurable}$$

$$\text{complex r.v.'s, } |X|,\, |Y| \le 1\}$$

satisfies $\alpha_{T,\ell} \le \beta_{T,\ell} \le 16\alpha_{T,\ell}$ so that $\{\zeta_T\}$ is s.m. if and only if $\beta_{T,\ell_T}, \to 0$, for some $\ell_T \to 0$.

The above strong mixing assumption and notation will be employed throughout. Also, by an "interval" we shall mean a semiclosed subinterval $(a, b]$ of $(0, 1]$. It will be further assumed throughout that

(2.1) $\sup\{P\{|\zeta_T(I)| > \epsilon\}: I \text{ interval}, m(I) \le \ell_T\} \to 0 \quad \text{as} \quad T \to \infty$,

for each $\epsilon > 0$, m denoting Lebesgue measure. Note that (2.1) is equivalent to the condition

(2.2) $\gamma_T = \sup\{1 - \mathcal{E}\exp(-|\zeta_T(I)|): I \text{ interval}, m(I) \le \ell_T\} \to 0$,

as $T \to \infty$, which follows since e.g.,

$$(1 - e^{-\epsilon}) \sup_{m(I) \le \ell_T} P\{|\zeta_T(I)| \ge \epsilon\} \le \gamma_T \le 1 - e^{-\epsilon} + \sup_{m(I) \le \ell_T} P\{|\zeta_T(I)| \ge \epsilon\}.$$

Now let $\{k_T\}$ be a family of integers such that

(2.3) $$k_T(\alpha_{T,\ell_T} + \gamma_T) \to 0.$$

Since the terms in parentheses tend to zero, (2.3) holds for bounded k_T, but clearly $k_T \to \infty$ may be chosen so that it also holds.

The following key basic asymptotic independence result leads to the main theorems and is proved along now classical lines.

Lemma 2.1 Let $\{k_T\}$ satisfy (2.3) and I_j, $1 \le j \le k_T$ be disjoint intervals (which may change with T) such that $m(I_j) > 2\ell_T$. Then

(2.4) $$\mathcal{E}\exp\{i\sum_{j=1}^{k_T} t_j\zeta_T(I_j)\} - \prod_{j=1}^{k_T} \mathcal{E}\exp\{it_j\zeta_T(I_j)\} \to 0 \quad \text{as} \quad T \to \infty,$$

for any t_j which may depend on T, but are uniformly bounded in j, T.

Proof: If $I_j = (a_j, b_j]$ write $I_j' = (a_j, b_j - \ell_T], I_j^* = (b_j - \ell_T, b_j]$. First note that

(2.5) $$\sum |\zeta_T(I_j^*)| \xrightarrow{P} 0 \quad \text{as} \quad T \to \infty,$$

where an undesignated range in a sum, product, union etc. is to be taken from 1 to k_T. For by an obvious induction on the (alternative form of the) mixing condition,

(2.6) $$|\mathcal{E}\exp\{it\sum|\zeta_T(I_j^*)|\} - \prod\mathcal{E}\exp\{it|\zeta_T(I_j^*)|\}| \le 16k_T\alpha_{T,\ell_T} \to 0$$

since the I_j^* are separated by at least ℓ_T. Hence if (2.5) is shown assuming independence of $\zeta_T(I_j^*)$, it holds in general. But under independence, using the inequality

$$(2.7) \qquad |\prod_1^n x_i - \prod_1^n y_i| \leq \sum_1^n |x_i - y_i|, \quad 0 \leq x_i, y_i \leq 1,$$

it follows that

$$\begin{aligned} 0 \ &\leq \ 1 - \mathcal{E}\exp\{-\sum |\zeta_T(I_j^*)|\} = 1 - \prod \mathcal{E}\exp\{-|\zeta_T(I_j^*)|\} \\ &\leq \ \sum_j (1 - \mathcal{E}\exp\{-|\zeta_T(I_j^*)|\}) \ \leq \ k_T \gamma_T \to 0 \end{aligned}$$

from which (2.5) follows at once. (Actually, the mixing assumption is not needed to prove (2.5). However, the argument used is needed later.) Now,

$$\begin{aligned} (2.8) \qquad |\mathcal{E}\exp\{i\sum t_j \zeta_T(I_j')\} \ &- \ \mathcal{E}\exp\{i\sum t_j \zeta_T(I_j)\}| \\ &\leq \ \mathcal{E}|1 - \exp\{i\sum t_j \zeta_T(I_j^*)\}| \to 0 \end{aligned}$$

by dominated convergence since $\sum t_j \zeta_T(I_j^*) \xrightarrow{P} 0$ by (2.5) and the uniform boundedness of the t_j. Further

$$(2.9) \quad |\mathcal{E}\exp\{i\sum t_j \zeta_T(I_j')\} - \prod \mathcal{E}\exp\{it_j \zeta_T(I_j')\}| \leq 16 k_T \alpha_{T,\ell_T} \to 0,$$

the intervals I_j being separated by ℓ_T.

Finally let (X_j, X_j'), $1 \leq j \leq k_T$, be independent pairs with $(X_j, X_j') \stackrel{d}{=} (\zeta_T(I_j), \zeta_T(I_j'))$ for each j. Then

$$\begin{aligned} (2.10) \quad |\prod \mathcal{E}\exp\{it_j \zeta_T(I_j')\} &- \prod \mathcal{E}\exp\{it_j \zeta_T(I_j)\}| \\ &= \ |\mathcal{E}\exp\{i\sum t_j X_j'\} - \mathcal{E}\exp\{i\sum t_j X_j\}| \\ &\leq \ \mathcal{E}|1 - \exp\{i\sum t_j (X_j - X_j')\}| \end{aligned}$$

which tends to zero since $X_j - X_j' \stackrel{d}{=} \zeta_T(I_j^*)$ and (X_j, X_j') are independent for $1 \leq j \leq k_T$, and the argument giving (2.5) shows that $\sum |X_j - X_j'| \xrightarrow{P} 0$ and hence $\sum t_j(X_j - X_j') \xrightarrow{P} 0$. The result (2.4) now follows from (2.8)-(2.10). $\qquad \square$

The main condition, (2.3), of the lemma is primarily chosen for simplicity and tractability. For cases when ζ_T converges to a random measure it also seems close to being optimal. However, e.g. for normal limits, still weaker (but more complicated) conditions may be useful.

Specifically, the requirement $k_T \gamma_T \to 0$ may be weakened as follows. Let $I_j = I_{T,j}$, $1 \le j \le k_T$, be specified disjoint intervals, and as in the proof of Lemma 2.1 set $I_j^* = (b_j - \ell_T, b_j]$ if $I_j = (a_j, b_j]$. Inspection of the proof of Lemma 2.1 shows that $k_T \gamma_T \to 0$ is used only to prove that

$$(2.11) \qquad \sum t_j \zeta_T(I_j^*) \xrightarrow{P} 0,$$

for uniformly bounded t_j's and with the $\zeta_T(I_j^*)$, $1 \le j \le k_T$, assumed independent. From Loève's degenerate convergence criterion [12, p. 329] it can be seen that this holds if and only if there exists a $\tau > 0$ such that

$$(2.12) \qquad \sum |E\{\zeta_T(I_j^*) 1_{\{|\zeta_T(I_j^*)| < \tau\}}\}| \quad \to \quad 0,$$

$$(2.13) \qquad \sum \mathrm{var}\{\zeta_T(I_j^*) 1_{\{|\zeta_T(I_j^*)| < \tau\}}\} \quad \to \quad 0,$$

$$(2.14) \qquad \sum P(|\zeta_T(I_j^*)| \ge \epsilon) \quad \to \quad 0, \quad \text{each } \epsilon > 0.$$

For example, (2.12) and (2.14) may be used to show that (2.12) holds with $t_j \zeta_T(I_j^*)$ replacing $\zeta_T(I_j^*)$, and similar arguments show that $\zeta_T(I_j^*)$ may be replaced by $t_j \zeta_T(I_j^*)$ also in (2.13) and (2.14).

We thus have the following result.

Corollary 2.2 *Assume $k_T \alpha_{T,\ell_T} \to 0$, and let $I_j, 1 \le j \le k_T$, be disjoint intervals with $m(I_j) > 2\ell_T$, as before. If furthermore (2.12)-(2.14) hold for some $\tau > 0$, then (2.4) follows.*

Finally, in applications one is often not interested in establishing (2.4) for completely arbitrary (bounded) choices of t_j's. This may permit further slight weakenings of conditions. For example, in the obviously interesting case when all t_j's are equal, it is sufficient to require

$$\sum \mathcal{E}\{\zeta_T(I_j^*) 1_{\{|\zeta_T(I_j^*)| < \tau\}}\} \to 0$$

instead of (2.12).

3   Associated independent arrays, and the central limit problem

As noted, central limit theory is typically obtained under mixing assumptions by reducing the problem to independence. Here we make this reduction totally explicit, by associating independent arrays with the strongly mixing family.

Specifically let I be an interval (which may depend on T) and $\{I_j\}$ a partition of I into k_T (satisfying (2.3)) disjoint subintervals $I_j (= I_{T,j})$ with

$\max m(I_j) \to 0$ and (for convenience) $m(I_j) > 2\ell_T$. We shall refer to the family $\{I_{T,j}\}$ as a k_T-partition of I, it being understood that (2.3) holds. For each T let $\zeta_{T,j}$, $1 \leq j \leq k_T$, be independent r.v.'s with $\zeta_{T,j} \stackrel{d}{=} \zeta_T(I_j)$. $\{\zeta_{T,j}\}$ will be termed an *independent array* (of size k_T) associated with $\zeta_T(I)$. Finally a k_T-partition $\{I_j\}$ will be termed uniformly asymptotically negligible (u.a.n.) for $\zeta_T(I)$ if $\max P\{|\zeta_T(I_j)| \geq \epsilon\} \to 0$ as $T \to \infty$, each $\epsilon > 0$. This of course means that the associated array $\{\zeta_{T,j}\}$ is u.a.n. in the usual sense (cf. also the "null array" of [9]).

Such independent arrays associated with $\{\zeta_T\}$ are of course not uniquely defined. However as can be seen, the main results are independent of the choice of array. Lemma 2.1 immediately gives the following result.

Theorem 3.1 *Let $\{\zeta_T\}$ be strongly mixing and satisfy (2.1). Let I be an interval and $\{\zeta_{T,i}\}$ an independent array for $\{\zeta_T(I)\}$ based on the k_T - partition $\{I_j\}$. Then $\zeta_T(I)$ has the same limit in distribution (if any) as $\sum \zeta_{T,j}$. In particular if the partition $\{I_j\}$ is u.a.n., any limit is infinitely divisible.*

Note that this result does not depend on the particular choice of partition as long as its size satisfies the growth rate restriction (2.3). In particular if any corresponding independent sum has a limit in distribution, all such array sums have the same limit.

It may be useful as a simple application to digress at this point to compare the special case of this result for array sums under "bounded variance assumptions" with a corresponding result of [13]. Specifically [13] considers uniformly bounded r.v.'s $X_{Nj}, 1 \leq j \leq k_N$, (with zero means and finite variances σ_{Nj}^2) satisfying a strong mixing condition (potentially slightly more restrictive than the "array form" used here), with a certain rate of decay of the mixing function. The classical bounded variance conditions

$$(3.1) \qquad\qquad \sigma_N^2 = \max_j \sigma_{Nj}^2 \to 0,$$

$$(3.2) \qquad\qquad \Sigma_N^2 = \text{var}\left(\sum_j X_{Nj}\right) \leq c < \infty$$

are assumed and it is shown that

(a) the class of distributional limits for $\sum X_{Nj}$ is that with characteristic functions (c.f.'s) $e^{\psi(t)}$ where

$$\psi(t) = \int [(e^{itx} - 1 - itx)/x^2] dK(x)$$

for some (multiple of a d.f.) K and

(b) $\sum X_{Nj} \stackrel{d}{\to} X$ (with c.f. e^ψ as above) if $K_N \stackrel{v}{\to} K$ ($\stackrel{v}{\to}$ denoting convergence at continuity points of K) where $K_N(x) = \sum_j \int_\infty^x y^2 dF_{N,j}$, and $F_{N,j}$ denotes the d.f.'s of certain partial sums Y_{nj} of $\sum X_{Nj}$ formed from (separated) groups of consecutive terms.

The same conclusions follow from Theorem 3.1 (without the uniform boundedness and mixing rate assumptions) when (2.1) holds (which can be guaranteed by an obvious simple variance condition) and (3.2) is replaced by the assumption $\sum_j \mathrm{var} Y_{Nj} \leq c < \infty$. This assumption in terms of variances of the partial sums Y_{Nj} (corresponding to the partitions I_j in our framework) seems slightly less natural than (3.2) (used in [13]) involving the variance of the entire array sum. However the use of (3.2) is perhaps less compelling in view of (a) the further assumptions and calculation required and (b) the fact that the second part of the result is in any case necessarily couched in terms of the (d.f.'s of the) partial sums Y_{nj}. Note also that the results both in [13] and here do not depend on the partial sum groupings within a wide range of choices (referred to as "admissible" in [13], and corresponding to the allowable partition choices in the present work).

Many of the classical results for the general central limit problem (cf. [12, Chapter 22]) can be immediately restated for mixing arrays by means of Theorem 3.1. For example the "Central convergence criterion" of [12] applies as follows, using the notation of [12], writing e^ψ for the canonical representation of an infinitely divisible characteristic function, with $\psi = (\alpha, \Psi)$ denoting

$$\psi(t) = it\alpha + \int (e^{itx} - 1 - \frac{itx}{1+x^2})\frac{1+x^2}{x^2} d\Psi(x)$$

where α is real and Ψ is a multiple of a distribution function. For a k_T-partition $\{I_j\}$ of an interval I we also write $F_{T,j}$ for the d.f. of $\zeta_T(I_j)$ and, for $\tau > 0$,

$$(3.3) \qquad \begin{aligned} a_{T,j}(\tau) &= \mathcal{E}\{\zeta_T(I_j)1_{\{|\zeta_T(I_j)|<\tau\}}\} = \int_{|x|<\tau} x dF_{T,j} \\ \sigma^2_{T,j}(\tau) &= \mathcal{E}\{\zeta^2_T(I_j)1_{\{|\zeta_T(I_j)|<\tau\}}\} - a^2_{T,j} \\ &= \int_{|x|<\tau} x^2 dF_{T,j} - a^2_{T,j}(\tau). \end{aligned}$$

Theorem 3.2 *Let $\{\zeta_T\}$ be strongly mixing and satisfy (2.1). Let I be an interval and $\{I_j\} = \{I_{T,j}\}$ a u.a.n. k_T-partition for $\zeta_T(I)$. Then $\zeta_T(I) \xrightarrow{d} \eta$ with c.f. $e^\psi, \psi = (\alpha, \Psi)$ if and only if*

(i) $$\sum F_{T,j}(x) \to \int_{-\infty}^x [(1+y^2)/y^2] d\Psi, \quad x < 0,$$

$$\sum (1 - F_{T,j}(x)) \to \int_x^\infty [(1+y^2)/y^2] d\Psi, \quad x > 0,$$

(ii) $\sum \sigma^2_{T,j}(\tau) \to \Psi(0) - \Psi(0-)$ *as $T \to \infty$ and then $\tau \to 0$,*

(iii) for a fixed τ such that $\pm\tau$ are continuity points of Ψ,

$$\sum a_{T,j}(\tau) \to \alpha + \int_{|x|<\tau} x\,d\Psi - \int_{|x|\geq\tau} \frac{1}{x}\,d\Psi.$$

Criteria for normal limits may be stated in the present context as follows (using the above notation).

Theorem 3.3 *Let $\{\zeta_T\}$ be strongly mixing and satisfy (2.1). Let I be an interval, $\{I_j\}$ $(= \{I_{T,j}\})$ any k_T -partition of I, and suppose $\zeta_T(I) \xrightarrow{d} \eta$, a r.v., as $T \to \infty$. Then η is normal and $\{I_j\}$ is u.a.n. if and only if*

(3.4) $\sum P\{|\zeta_T(I_j)| \geq \epsilon\} \to 0$ *as $T \to \infty$,   each $\epsilon > 0$.*

In that case η is $N(\alpha, \sigma^2)$ with

(3.5) $\alpha = \lim_{T\to\infty} \sum a_{T,j}(\tau),\quad \sigma^2 = \lim_{T\to\infty} \sum \sigma^2_{T,j}(\tau),\quad \tau > 0,$

where $a_{T,j}(\tau)$ and $\sigma_{T,j}(\tau)$ are given by (3.3)

Proof: This follows easily using Theorem 3.1, from the "Normal convergence criterion" and its corollary of [12, sec. 22.4].   □

It is evident from Lemma 2.1 that distributional limits for $\zeta_T(I)$ and $\zeta_T(J)$ must be independent when I and J are disjoint intervals. This suggests that if e.g. $\zeta_T(I)$ is normal and $J \subset I$, $\zeta_T(J)$ must also be normal (e.g. in view of the necessary normality of independent components of a normal r.v.). This is explicitly shown as follows.

Corollary 3.4 *Let $\{\zeta_T\}$ be strongly mixing and satisfy (2.1), and suppose that for some interval I, $\zeta_T(I) \xrightarrow{d} \eta_I \sim N(\alpha, \sigma^2)$ as $T \to \infty$. Then if $\zeta_T(J) \xrightarrow{d} \eta_J$ for an interval $J \subset I$, the limit η_J is also normal.*

Proof: By the theorem (3.4) holds and an independent array for $\zeta_T(I)$ is u.a.n. Hence the same is true for the independent array "induced" in the obvious way for $\zeta_T(J)$ so that η_J is normal by the theorem.   □

The existence of a limit for $\zeta_T(I)$ is assumed in Theorem 3.3 and then necessary and sufficient conditions stated for normality of that limit. Sufficient conditions may also be given for the existence of a limit, as in the following result.

Theorem 3.5 *Let $\{\zeta_T\}$ be strongly mixing and let $\{I_j\}$ be a u.a.n. k_T -partition of an interval I. Suppose that*

(3.6) $\sum a_{T,j}(\tau) \to \alpha,\quad \sum \sigma^2_{T,j}(\tau) \to \sigma^2,\quad as\ T \to \infty,$

for some τ, $\sigma > 0$, α, where $a_{T,j}$ and $\sigma_{T,j}$ are defined by (3.3). Then $\zeta_T(I) \xrightarrow{d} \eta$, where η is $N(\alpha, \sigma^2)$. For a centered $(a_{T,j}(\infty) = 0$, for all $j)$ and normalized $(\sum \sigma_{T,j}^2(\infty) \to \sigma^2)$ array, (3.6) may be replaced by the Lindeberg condition $\sum E\{\zeta_{T,j}^2 1_{\{|\zeta_{T,j}| > \epsilon\}}\} \to 0$, for all $\epsilon > 0$, as $T \to \infty$.

Proof: This is immediate from Theorem 3.1 and the normal convergence criterion of [12, Section 22.4]. □

4 Stationarity

It is of interest to note the yet more definitive results obtainable under the weak stationarity assumption

(4.1) $\zeta_T(h + I) \stackrel{d}{=} \zeta_T(I)$, each h and interval I with I, $h + I \subset (0, 1]$.

Theorem 4.1 *Let $\{\zeta_T\}$ be strongly mixing and stationary in the sense of (4.1) and suppose that (2.1) holds. If $\zeta_T(I) \xrightarrow{d} \eta_I$, a r.v., for some (non-degenerate) interval I, then such convergence occurs for all intervals I and η_I is infinitely divisible with characteristic function $\mathcal{E} \exp(it\eta_I) = \varphi(t)^{m(I)}$ where φ is the characteristic function of $\eta_{(0,1]}$.*

Proof: We assume for simplicity that convergence $\zeta_T(I) \xrightarrow{d} \eta_I$ occurs when $I = (0, 1]$ and prove it for general I. Let $k_T \to \infty$ satisfy (2.3). Then by Lemma 2.1 for intervals $I_j = ((j-1)/k_T, j/k_T]$ we have $\mathcal{E}^{k_T} \exp\{it\zeta_T(I_1)\} \to \psi(t)$. If $I = (0, a]$ for some a, $0 < a < 1$, write $n_T = [k_T a]$. Then clearly $\mathcal{E}^{n_T} \exp\{it\zeta_T(I_1)\} \to \psi(t)^a$ and Lemma 2.1 implies that $\mathcal{E}\exp\{it\zeta_T(0, a_T]\} \to \psi(t)^a$, where $a_T = n_T/k_T \to a$. Since $0 \leq a - a_T < k_T^{-1} < \ell_T$ (for large T), (2.1) implies that $\zeta_T(a_T, a] \xrightarrow{P} 0$, and it follows that $\mathcal{E} \exp\{it\zeta_T(I)\} \to \psi(t)^a$. Thus the result follows for intervals $I = (0, a]$ and by stationarity for intervals $(a, b] \subset (0, 1]$. □

Much of the work in central limit theory for mixing processes has been concerned with normalized sums of a single sequence. As an example, we give a short proof of results of [7] and [19]. Let $X_1, X_2, \ldots$ be strictly stationary with $EX_1 = 0$, $EX_1^2 < \infty$. Set $S_n = \sum_{i=1}^n X_i$, $s_n^2 = \text{Var}(S_n)$ and

$$\zeta_n(a, b] = s_n^{-1} \sum_{na < i \leq nb} X_i, \quad 0 \leq a < b \leq 1.$$

In this context, $\{X_i\}$ is strongly mixing if

$$\alpha_\ell' = \sup\{|P(AB) - P(A)P(B)|: A \in \mathcal{B}_n, B \in \mathcal{B}^{n+\ell}\} \to 0,$$

as $\ell \to \infty$, where $\mathcal{B}_n = \sigma(\ldots, X_{n-1}, X_n)$, $\mathcal{B}^{n+\ell} = \sigma(X_{n+\ell}, X_{n+\ell+1}, \ldots)$ and the supremum is over A, B, and n. With the previous notation, clearly

$\alpha_{n,\ell} \leq \alpha'_{[n\ell]}$ so that in particular $\alpha_{n,\ell_n} \to 0$ for any sequence ℓ_n with $n\ell_n \to \infty$.

Theorem 4.2 *Suppose $\{X_i\}$ is strongly mixing and $s_n^2 \to \infty$, $n \to \infty$. Then the following are equivalent,*

$$(4.2) \qquad\qquad S_n \xrightarrow{d} N(0,1), \quad n \to \infty,$$

and

$$(4.3) \qquad\qquad \lim_{a \to \infty} \sup_n E\{s_n^{-2} S_n^2 1_{\{|s_n^{-1} S_n| > a\}}\} = 0, \quad and$$

s_n^2 *is regularly varying with index* 1.

Proof: If $m(I) < \ell - n^{-1}$ then, for $\epsilon > 0$,

$$P(|\zeta_n(I)| > \epsilon) \leq \frac{\mathcal{E}\zeta_n^2(0,\ell]}{\epsilon^2} \leq \frac{(n\ell)^2 \mathcal{E} X_1^2}{s_n^2 \epsilon^2},$$

and since $s_n \to \infty$, there exist ℓ_n with $n\ell_n \to \infty$ such that $\gamma_n \to 0$, with γ_n given by (2.2). Hence, by Theorem 4.1, if (4.2) holds then $S_{[n\beta]}/s_n = \zeta_n(0,\beta] \xrightarrow{d} N(0,\beta)$, for $\beta \in (0,1)$. Since also $S_{[n\beta]}/s_{[n,\beta]} \xrightarrow{d} N(0,1)$, it follows that

$$(4.4) \qquad\qquad s_{[n\beta]}^2/s_n^2 \to \beta, \quad n \to \infty,$$

proving that s_n^2 is regularly varying with index one. Further, the remaining part of (4.3) (i.e. uniform integrability of S_n^2/s_n^2) follows at once from

$$S_n^2/s_n^2 \xrightarrow{d} Z^2$$

where Z is $N(0,1)$ and $\mathcal{E} S_n^2/s_n^2 = 1 = \mathcal{E} Z^2$.

Conversely, suppose (4.3), and then also (4.4), is satisfied. Then, for $r \in (0,1), \epsilon > 0$,

$$\limsup_n r^{-1} \mathcal{E}\zeta_n^2(0,r] 1_{\{|\zeta_n(0,r]| > \epsilon\}} \leq \sup_n \mathcal{E}\{s_{[nr]}^{-2} S_{[nr]}^2 1_{\{|s_{[nr]}^{-1} S_{[nr]}| > 2\epsilon/\sqrt{r}\}}\}.$$

Hence, by (4.3), for any $r_n \to 0$ slowly enough,

$$r_n^{-1} \mathcal{E}\zeta_n^2(0,r_n] 1_{\{|\zeta_n(0,r_n]| > \epsilon\}} \to 0,$$

as $n \to \infty$. Thus, if $\{I_j\}$ is as in the proof of Theorem 4.2, k_n, ℓ_n may be chosen such that the conditions of the last part of Theorem 3.5 are satisfied, with $\sigma^2 = 1$, and hence (4.2) holds.   □

In the proof that (4.3) implies (4.2), the condition of regular variation is not needed, see [7], [19]. This may be seen by a straightforward argument, similar to the proof of Lemma 2.1.

5 Multivariate and functional limits

Let ζ_T be strongly mixing and satisfy (2.1) and let $\zeta_T(I_j) \xrightarrow{d} \eta_{I_j}$, $j = 1, \ldots, p$ for each of j disjoint intervals $I_1, \ldots, I_p$ and r.v.'s $\eta_{I_1} \ldots \eta_{I_p}$. Then it follows from Lemma 2.1 that

$$(5.1) \qquad (\zeta_T(I_1), \ldots, \zeta_T(I_p)) \xrightarrow{d} (\eta_1, \ldots, \eta_p)$$

where $\eta_1, \ldots, \eta_p$ are independent and $\eta_j \overset{d}{=} \eta_{I_j}$.

Suppose now that $\zeta_T(I) \xrightarrow{d} \eta_I$, a r.v., for each interval I in some semiring $\mathcal{C}$, and that $I_j \in \mathcal{C}$, $j = 1, \ldots, p$ but I_j are not necessarily disjoint. Since intersections of sets in $\mathcal{C}$ are in $\mathcal{C}$ and differences of sets of $\mathcal{C}$ are finite disjoint unions of sets in $\mathcal{C}$ it follows by obvious arguments using additivity of $\zeta_T(I)$ that (5.1) again holds though $\eta_1, \ldots, \eta_p$ are not independent in general.

As examples we briefly consider two cases - first when limits are normal and second where ζ_T is positive and countably additive i.e. a random measure.

(a) *Normal convergence.* Here we take $\mathcal{C}$ to be all intervals $I = (a, b] \subset (0, 1]$ and assume that $\zeta_T(I) \xrightarrow{d} \eta_I$ for each such I where $\eta_{(0,1]}$ is normal. Then η_I is normal for each interval I (by Corollary 3.4) with mean α_I and variance σ_I^2 given from Theorem 3.3 by

$$\alpha_I = \lim_{T \to \infty} \sum_{j:I_{T,j} \subset I} a_{T,j}(\tau), \quad \sigma_I^2 = \lim_{T \to \infty} \sum_{j:I_{T,j} \subset I} \sigma_{T,j}^2(\tau),$$

where $\{I_{Tj}\}$ is any k_T-partition of $(0, 1]$, and $a_{T,j}$ and $\sigma_{T,j}$ are defined by (3.3). Thus (5.1) holds for any fixed intervals $I_1, \ldots, I_p$ and the components $\eta_1, \ldots, \eta_p$ of the limit are independent if $I_1, \ldots, I_p$ are disjoint. In particular the finite dimensional distributions of the processes $\xi_T(t) = \zeta_T(0, t]$, $0 \le t \le 1$, converge to those of a normal process $\xi(t)$ on $(0, 1]$ with independent increments, mean function $\mathcal{E}\xi(t) = m_t = \alpha_{(0,t]}$ and variance $\sigma_t^2 = \sigma_{(0,t]}^2$.

If in addition $\zeta_T(0, t]$ is right continuous and satisfies a tightness condition (e.g. an increment condition as in [4, Theorem 15.6]) then of course the processes $\zeta_T(t) = \zeta_T(0, t]$ converge in distribution to the normal process $\xi(t)$ in $D(0, 1]$.

In the stationary case convergence of $\zeta_T(I)$ follows for all I from that for any one I, e.g. $(0, 1]$, as in Section 4. For example if $\zeta_T(0, 1] \xrightarrow{d} \eta$, $N(\alpha, \sigma^2)$ then by Theorem 4.1, $\zeta_T(0, t]$ has c.f. $\phi_t(u) = e^{t(i\alpha u - \sigma^2 u^2/2)}$ so that the limit $\xi(t)$ is a Wiener process with an added linear term αt.

(b) *Random measures.* Suppose now that ζ_T is a random measure on the Borel subsets of $(0, 1]$ for each T (cf. [9]) and that $\zeta_T(I) \xrightarrow{d} \eta_I$ for each interval, where η_I is a r.v. such that $\eta_{I_n} \overset{d}{\to} 0$ whenever I_n are intervals such that $I_n \downarrow \{c\}$, some $0 < c \leq 1$. Again it follows that (5.1) holds if $\{\zeta_T\}$ is strongly mixing and satisfies (2.1), the components of the limits being independent for disjoint $I_1, \ldots, I_p$. It then follows from [9, Lemma 5.1] that there exists a random measure ζ such that $\zeta_T \overset{d}{\to} \zeta$ and $\zeta(I) \overset{\alpha}{=} \eta_I$ for each interval I. Clearly ζ has independent increments (and no fixed atoms).

If in addition ζ_T satisfies the weak stationarity condition (4.1) it follows from Theorem 4.1 that the assumption $\zeta_T(I) \overset{d}{\to} \eta_I$ needs be made only for one I, e.g. $I = (0,1]$, and the condition $\eta_{I_n} \overset{d}{\to} 0$ when $I_n \downarrow \{c\}$ automatically holds since by Theorem 4.1 , η_{I_n} has c.f. $= (\varphi(t))^{m(I_n)} \to 1$ in a neighborhood of $t = 0$. Thus under stationarity the convergence of $\zeta_T(I)$ in distribution for some I guarantees full convergence of ζ_T to some random measure ζ having independent increments. A detailed account of the possible (Compound Poisson type) forms for ζ may be found in [11].

Acknowledgements. We are very grateful to Professors Richard Bradley, Ildar Ibragimov, Walter Philipp and Murray Rosenblatt for very helpful conversations concerning existing works in this area.

References

[1] Bergström, H., A comparisons method for distribution functions of sums of independent and dependent random variables, *Theor. Probab. Appl.,* **15** (1970), 430-457.

[2] Bergström, H., On the convergence of sums of random variables in distribution under mixing conditons, *Period. Math. Hung.,* **2** (1972), 173-190.

[3] Bergström, H., Reduction of the limit problem for sums of random variables under a mixing condition, *Proc. 4th Conf. on Probab. Theory,* Brasov (1973), 107-120.

[4] Billingsley, P., *Convergence of Probability Measures,* Wiley, New York (1968).

[5] Bradley, R. C., Asymptotic normality of some kernel type estimates of probability density, *Statist. Probab. Letters,* **1** (1983), 295-300.

[6] Cogburn, R., Conditional probability operators, *Ann. Math. Statist.,* **33** (1962), 634-658.

[7] Denker, M., Uniform integrability and the central limit theorem for strongly mixing processes. In: *Dependence in Probability and Statistics*, Eds. Eberlein, E. and Taqqu, M.S., Birkhäuser, Boston (1986), 269-274.

[8] Ibragimov, I. and Linnik, Yu.V., *Independent and Stationary Sequences of Random Variables*, Walter Noordhoff, Groningen (1971).

[9] Kallenberg, O., *Random Measures*, Academic Press, New York (1983).

[10] Krieger, H.A., A new look at Bergström's Theorem on convergence in distribution for sums of dependent random variables, *Israel J. Math.*, 47 (1984), 32-64.

[11] Leadbetter, M.R. and Hsing, T., Limit theorems for strongly mixing stationary random measures, *Stoch. Proc. Appl.* 36 (1990), 231-243.

[12] Loève, M. *Probability Theory*, 4th ed., Springer, New York (1977).

[13] Philipp, W., The central limit problem for mixing sequences of random variables, *Z. Wahrsch. verw. Geb.*, 12 (1969), 155-171.

[14] Rosén, B., On the central limit problem for sums of dependent random variables, *Z. Wahrsch. verw. Geb.*, 7 (1967), 48-82.

[15] Rosenblatt, M., Some comments on narrow band-pass filters, *Quarterly Appl. Math.*, 18 (1961), 387-393.

[16] Rosenblatt, M., Uniform ergodicity and strong mixing, *Z. Wahrsch. verw. Geb.*, 24 (1972), 79-84.

[17] Samur, J., Convergence of sums of mixing triangular arrays of random vectors with stationary rows, *Ann. Probab.*, 12 (1984), 390-426.

[18] Volkonski, V.A. and Rozanov, Yu.A., Some limit theorems for random functions I, *Theor. Probab. Appl.*, 4 (1959), 178-197.

[19] Yori, K. and Yoshihara, K.-I., A note on the central limit theorem for stationary strong-mixing sequences, *Yokohama Math. J.*, 34 (1986), 143-146.

Positive Generalized Functions on Infinite Dimensional Spaces

Yuh-Jia Lee*

Abstract

In this paper, a new class of generalized functions $\mathcal{E}^*$ on the dual of a certain nuclear space is introduced. Then it is shown that positive members in $\mathcal{E}^*$ are represented by measures satisfying certain growth conditions. The results also generalize Kontrat'ev-Yokoi Theorem.

1 Introduction

Recently the representations of positive generalized functions in terms of measures on infinite dimensional spaces, such as abstract Wiener spaces and the space $\mathcal{S}'$ of tempered distributions or, more generally, the dual of a nuclear space in a Gel'fand triple, have been studied by many authors [4, 7, 8, 15, 18, 20] and proved to be useful in applications to quantum field theory [1, 2, 3, 6, 17]. It can also be shown that such measures satisfy a certain "growth" condition or moment condition depending on the choice of test functions. In this note, we shall introduce a new class of positive generalized functions on infinite dimensional spaces which are also characterized by measures.

To start with, let us introduce the space of test functions as follows.

First, let us describe the underlying infinite dimensional space. Let H be a real separable Hilbert space with norm $|\cdot|$ and inner product $\langle\cdot,\cdot\rangle$ and A a densely defined positive self-adjoint operator on H with Hilbert-Schmidt type inverse. We shall also assume that the Hilbert-Schmidt norm $\|A^{-1}\|_{HS}$ of A^{-1} is less than 1.

For any $p \in \mathbf{N}$, the positive integers, let $E_p = \mathcal{D}(A^p)$, the domain of A^p, and E_{-p} the completion of H with respect to the norm $|x|_{-p} = |A^{-p}x|$. Then E_p is a real separable Hilbert space with dual E_{-p} and

---

*Research support by the National Science Council of R.O.C.

Mailing address: Department of Mathematics, National Cheng-Kung University, Tainan, Taiwan, R.O.C.

$E_p \subset E_q$ for $p \geq q$. Set $E = \bigcap_{p \in \mathbf{N}} E_p$ and endow E with the projective limit topology. Then E is a nuclear space [5] with dual space given by $E^* = \bigcup_{p \in \mathbf{N}} E_{-p}$. It is well-known that E^* carries a centered Gaussian measure μ with characteristic function $C(\xi) = \exp(-\frac{1}{2}|\xi|^2)$, where $\xi \in E$, and, for each $p \in \mathbf{N}$, (H, E_{-p}) forms an abstract Wiener space with Wiener measure μ.

For a real linear space V, we shall denote by $\mathcal{C}V$ the complexification of V. If V is a normed linear space, the norm of $\mathcal{C}V$ is taken to be the Euclidean norm induced by the norm $\|\cdot\|_V$ of V. For notational convenience, we shall also use $\|\cdot\|_V$ to deonte the norm of $\mathcal{C}V$.

For each $p \in \mathbf{N}$, let $\mathcal{E}_p$ denote the space of complex- valued functions f defined on E_{-p} satisfying the following properties:

(1.1) There exists an entire (analytic) function $\tilde{f}$ defined on $\mathcal{C}E_{-p}$ such that $\tilde{f}(x) = f(x)$ for all $x \in E_{-p}$. (We shall call $\tilde{f}$ the analytic extension of f.)

(1.2) There exist constants c, c' depending only f such that

$$|\tilde{f}(z)| \leq c \exp(c'|z|_{-p}),$$

for all $z \in E_{-p}$.

For $p, m \in \mathbf{N}$, define a norm $\|\cdot\|_{p,m}$ in $\mathcal{E}_p$ by

$$\|f\|_{p,m} = \sup_{z \in \mathcal{C}E_{-p}} [|\tilde{f}(z)| \exp(-m|z|_{-p})]$$

and denote by $\mathcal{E}_{p,m}$ the class of functions f in $\mathcal{E}_p$ such that $\|f\|_{p,m} < \infty$. Then $(\mathcal{E}_{p,m}, \|\cdot\|_{p,m})$ forms a Banach space and $\mathcal{E}_{p,m} \subset \mathcal{E}_{p,n}$ for $n \geq m$. Moreover, $\mathcal{E}_p = \bigcup_{m \in \mathbf{N}} \mathcal{E}_{p,m}$. Endow $\mathcal{E}_p$ with the inductive limit topology of the family of Banach spaces $\{\mathcal{E}_{p,m} : m \in \mathbf{N}\}$, $\mathcal{E}_p$ becomes a locally convex topological algebra (see [11]). It is easy to see that a linear functional on $\mathcal{E}_p$ is continuous if and only if it is bounded on every space $\mathcal{E}_{p,m}$.

Now set $\mathcal{E} = \bigcap_{p \in \mathbf{N}} \mathcal{E}_p$, and endow it with the projective limit topology induced by the family $\{\mathcal{E}_p : p \in \mathbf{N}\}$. Then $\mathcal{E}$ becomes a topological algebra. $\mathcal{E}$ will serve as the space of test functions. Denote the dual spaces of $\mathcal{E}$ and $\mathcal{E}_p$ by $\mathcal{E}^*$ and $\mathcal{E}_p^*$, respectively. Denote the pairing of $\mathcal{E}^*$ and $\mathcal{E}$ by $\ll \cdot, \cdot \gg$ and the pairing of $\mathcal{E}_p^*$ and $\mathcal{E}_p$ by $\ll \cdot, \cdot \gg_p$. Members of $\mathcal{E}^*$ are referred as the generalized functions. The spaces $\mathcal{E}^*$ and $\mathcal{E}_p^*$ are topologized respectively by the weak-* topology. This completes the construction of the test and generalized functions.

Calculus of generalized functions can be studied in a similar way as demonstrated in [11]. Among the examples of generalized functions the most important class are those represented by measures of exponential growth. More precisely, suppose η is a Borel measure on $(E^*, \mathcal{B})$ satisfying the following properties:

(1.3) The measurable support of η is contained in some E_{-p} $(p \in \mathbf{N})$.

(1.4) $\int_{E_{-p}} \exp(m|x|_{-p})\eta(dx) < \infty$ for all $m \in \mathbf{N}$.

Then η generates a generalized function, denoted also by η, such that

$$\ll \eta, \varphi \gg = \int_{E_{-p}} \varphi(x)\eta(dx) \tag{1}$$

Clearly, η is positive.

In this paper, we shall show that the converse of the above result holds and that every positive member in $\mathcal{E}^*$ can be represented in the form (1) by a measure η satisfying conditions (1.3) and (1.4). It is also shown that the results imply Kontrat'ev-Yokoi Theorem (see section 4).

2   Basic properties of test functions

In this section, some intrinsic properties of functions in $\mathcal{E}$ are studied. For related results we refer the reader to our previous paper [11]. To fix the notations, let X be a complex normed linear space, we denote the space of all n-linear operators on X by $\mathcal{L}^n[X]$. For simplicity, we deonte $T(y_1, \ldots, y_n)$ by $Ty_1, \ldots, y_n$ for $T \in \mathcal{L}^n[X]$. If $y_1 = y_2 = \cdots = x$, we write $T(x, \ldots, x) = Tx^n$.

PROPOSITION 2.1 *Let* $f \in \mathcal{E}_{p,m}$. *Then we have*

(a) $\tilde{f}$, *the analytic extenstion of* f, *is inifinitely Fréchet differentiavle in* CE_{-p} *and, for each integer n and* $y_1, \ldots, y_n \in E_{-p}$, $z \in CE_{-p}$, *we have*

$$|D^n \tilde{f}(z)y_1 \cdots y_n| \leq \|f\|_{p,m} m^n \exp[m|z|_{-p} + \sum_{k=1}^{n} |y_k|_{-p}]. \tag{2}$$

Consequently, $D^n f(\cdot)y_1 \ldots y_n \in \mathcal{E}_{p,m}$ *and*

$$\|D^n \tilde{f}(z)\|_{\mathcal{L}^n[E_{-p}]} \leq \|f\|_{p,m}(em)^n \exp(m|z|_{-p}). \tag{3}$$

(b) *The Taylor series* $\sum_{n=0}^{\infty} \frac{1}{n!} D^n f(0)x^n$ *converges to* f *in* $\mathcal{E}_{p,3m}$.

(c) $\bar{f} \in \mathcal{E}_{p,3m}$.

(d) *The Wiener-Ito decomposition of* f *converges to* f *in* $\mathcal{E}_{p,3m}$.

(e) *Let* $\{e_n : n \in \mathbf{N}\}$ *be the CONS for H consisting of eigenvectors of A with corresponding eigenvalues* $0 < \beta_1 \leq \beta_2 \leq \cdots$. *Define* $P_n x = \sum_{k=1}^{n}(x, e_k)e_k$. *Then* $f(P_n x)$ *converges to* f *in* $\mathcal{E}_p$ *as* $n \to \infty$ *for any* $f \in \mathcal{E}_{p+1}$, *where* $(\cdot, \cdot)$ *denotes the pairing of* E^* *and* E *and the pairing of* E_{-p} *and* E_p.

PROOF: (a) The infinite Fréchet differentiability of $\tilde{f}$ follows from the definition of $\mathcal{E}_p$ given in section 1. To prove the inequality (2), we first apply the Cauchy formula to $g(\lambda_1, \ldots, \lambda_n) = \tilde{f}(z + \lambda_1 y_1 + \cdots + \lambda_n y_n)$ and obtain

$$\mathrm{D}^n \tilde{f}(z) y_1 \cdots y_n$$
$$= \left(\tfrac{1}{2\pi i}\right)^n \int_{|\lambda_1|=r_1} \cdots \int_{|\lambda_n|=r_n} \frac{\tilde{f}(z+\lambda_1 y_1 + \cdots + \lambda_n y_n)}{\lambda_1^2 \cdots \lambda_n^2} d\lambda_1 \cdots d\lambda_n \qquad (4)$$

where $r_1, r_2, \ldots$ are arbitrary positive numbers. Then the inequality (2) follows from (4) by choosing $r_k = m^{-1}$ for $k = 1, \ldots, n$.

As a consequence of (2), we immediately conclude that the function $x \to \mathrm{D}^n f(x) y_1 \cdots y_n$ is in $\mathcal{E}_{p,m}$. The inequality (3) follows from (2) by choosing $y_k \in \mathrm{E}_{-p}$ with $|y_k|_{-p} = 1$.

(b) For $f \in \mathcal{E}_{p,m}$, define, for $x \in \mathrm{E}_{-p}$, $\mathrm{S}_\mathrm{N}(x) = \sum_{n=0}^{\mathrm{N}} \frac{1}{n!} \mathrm{D}^n \tilde{f}(0) x^n$. Then we have $\tilde{\mathrm{S}}_\mathrm{N}(z) = \sum_{n=0}^{\mathrm{N}} \frac{1}{n!} \mathrm{D}^n \tilde{f}(0) z^n$ for $z \in C\mathrm{E}_{-p}$ and $|\tilde{\mathrm{S}}_\mathrm{N}(z)| \leq \|f\|_{p,m} e^{me|z|_{-p}} \leq \|f\|_{p,m} e^{3m|z|_{-p}}$ so that $\mathrm{S}_\mathrm{N} \in \mathcal{E}_{p,3m}$.

Further, by (3), we have $\|\mathrm{S}_\mathrm{N} - f\|_{p,3m} \leq [\sum_{n=\mathrm{N}+1}^{\infty} (\tfrac{e}{3})^n] \|f\|_{p,m} \to 0$.

(c) For $x \in \mathrm{E}_{-p}$, write $f(x) = \sum_{n=0}^{\infty} \frac{1}{n!} \mathrm{D}^n \tilde{f}(0) x^n$. Define $\overline{f}(x) = \sum_{n=0}^{\infty} \frac{1}{n!} \overline{\mathrm{D}^n \tilde{f}(0)} x^n$. Then $\overline{f} \in \mathcal{E}_{p,3m}$ by (b).

(d) Recall that the Wiener-Ito decomposition of $f \in \mathcal{E}_{m,p}$ can be formulated in the following form:

$$f(x) = \sum_{n=0}^{\infty} \frac{1}{n!} \int_{\mathrm{E}_{-p}} \mathrm{D}^n \mu f(0)(x + iy)^n \mu(dy) \qquad (5)$$

where $\mu f = \mu * f$ (see [12]). Using the estimation (3), one sees easily that the sum on the right hand side of (5) converges in $\mathcal{E}_{p,3m}$.

(e) Let $x, y \in \mathrm{E}_{-p}$. By the mean value theorem, there is a number $t' \in [0, 1]$ such that $f(x) - f(y) = (x - y, \mathrm{D}f(t'x + (1 - t')y))$ and, consequently, we have

$$|f(x) - f(y)| \leq \|f\|_{p,m} (me)^{m(|x|_{-p} \vee |y|_{-p})} (|x - y|_{-p}). \qquad (6)$$

Now let $f \in \mathcal{E}_{p+1,m}$ and apply the estimation (6), we obtain

$$\|f \circ \mathrm{P}_n - f\|_{p,m+1} \leq \|f\|_{p+1,m} \left(\sum_{k=n+1}^{\infty} \beta_k^{-2} \right)^{\frac{1}{2}} \to 0.$$

Since $f \circ \mathrm{P}_n \in \mathcal{E}_{p+1,m} \subset \mathcal{E}_{p,m+1}$, we have shown that $f \circ \mathrm{P}_n \to f$ in $\mathcal{E}_{p,m+1}$ for $f \in \mathcal{E}_{p+1,m}$. This completes the proof of (e). $\qquad \square$

Some implications of Proposition 2.1 are summarized in the following

PROPOSITION 2.2 (a) *Let f be a complex-valued function defined on E^*. Then $f \in \mathcal{E}$ if and only if f is infinitely Fréchet differentiable in E^* and,* *for any* $p \in \mathbf{N}$, *there exist constants* c *and* K, *depending only on* f *and* p, *such that* $\|\mathrm{D}^n f(0)\|_{\mathcal{L}^n[\mathrm{E}_{-p}]} \le c\mathrm{K}^n$.

(b) *If* $f \in \mathcal{E}$, *then its Taylor series expansion* $\sum_{n=0}^{\infty} \frac{1}{n!} \mathrm{D}^n f(0) x^n$ *converges to* f *in* $\mathcal{E}$.

(c) *If* $f \in \mathcal{E}$, *so is* $\overline{f}$.

(d) *The Wiener-Ito decomposition of* f *converges to* f *in* $\mathcal{E}$ *if* $f \in \mathcal{E}$.

(e) *If* $f \in \mathcal{E}$, *then* $f \circ \mathrm{P}_n \to f$ *in* $\mathcal{E}$ *as* $n \to \infty$, *where* P_n *is the projection as given in Proposition 2.1.*

To represent a positive generalized function, we also need the following proposition.

PROPOSITION 2.3 *Let* $\mathcal{P}$ *denote the collectin of polynomials of the form* $Q((x, e_1), (x, e_2), \ldots, (x, e_k))$, *where* k *is an arbitrary nonnegative integer and* Q *an arbitrary polynomial of* k *variables. Then* $\mathcal{P}$ *is dense in* $\mathcal{E}$.

PROOF: Let P_n be the projection as given in Proposition 2.1. Then $f \circ \mathrm{P}_n \to f$ in $\mathcal{E}$ by Proposition 2.2(e). For each n and N, define $\mathrm{S}_{\mathrm{N},n}(x) = \sum_{j=0}^{\mathrm{N}} \frac{1}{j!} \mathrm{D}^j f(0)(\mathrm{P}_n x)^j$. Then $\mathrm{S}_{\mathrm{N},n} \in \mathcal{P}$ and $\mathrm{S}_{\mathrm{N},n} \to f \circ \mathrm{P}_n$ in $\mathcal{E}$ as $\mathrm{N} \to \infty$ according to Proposition 2.2(e). Consequently, $\lim_{n \to \infty} \lim_{\mathrm{N} \to \infty} \mathrm{S}_{\mathrm{N},n} = f$ in $\mathcal{E}$. Thus $\mathcal{P}$ is sequentially dense in $\mathcal{E}$. □

COROLLARY 2.4 *Let* $\mathcal{A}$ *denote the algebra spanned by the set of finite linear combinations of* $\{\exp[i(x, \xi)] : \xi \in \mathrm{E}\}$. *Then* $\mathcal{A}$ *is dense in* $\mathcal{E}$.

PROOF: As a consequence of Proposition 2.3, it suffices to prove that each member in $\mathcal{P}$ can be approximated by a sequence in $\mathcal{A}$. Note that the mapping $(\varphi, \psi) \to \varphi \cdot \psi$ from $\mathcal{E} \times \mathcal{E}$ into $\mathcal{E}$ is continuous. We only need to prove that, for a given $\xi \in \mathrm{E}$, there exists a sequence $\{\varphi_n\} \subset \mathcal{A}$ such that $\varphi_n \to \tilde{\xi}$ in $\mathcal{E}$, where $\tilde{\xi}(x) = (x, \xi)$. Observe that $\epsilon^{-1}[e^{i(x, \epsilon\xi)} - 1] \to i(x, \xi)$ in $\mathcal{E}$ as $\epsilon \to 0$. The corollary follows by choosing $\varphi_n(x) = n[e^{i(x, n^{-1}\xi)} - 1]$. □

3   Positive generalized function in $\mathcal{E}^*$

A generalized function $\mathrm{F} \in \mathcal{E}^*$ is called positive if $\ll \mathrm{F}, \varphi \gg \, \ge 0$ for all $\varphi \in \mathcal{E}$ with $\varphi(x) \ge 0$.

THEOREM 3.1 *Suppose that* F *is a positive generalized function in* $\mathcal{E}^*$. *Then there exists a unique finite measure* ν_{F} *on* $(\mathrm{E}^*, \mathcal{B})$ *such that*

$$\ll \mathrm{F}, \varphi \gg \, = \int_{\mathrm{E}^*} \varphi(x) \nu_{\mathrm{F}}(dx)$$

for all $\varphi \in \mathcal{E}$, *where* $\mathcal{B}$ *is the Borel field in* E^*.

PROOF: For $\xi \in E$, let $K(\xi) = \ll F, e^{i\hat{\xi}} \gg$. Then the positivity of F implies that K is positive definite. Further, K is continuous in E since the function $g(\xi) = e^{i\hat{\xi}}$ is an $\mathcal{E}$-valued continuous function (in fact, one can see easily that, for $\xi_1, \xi_2 \in E$, $\|e^{i\hat{\xi}_1} - e^{i\hat{\xi}_2}\|_{p,1} \le |\xi_1 - \xi_2|_p$ so that $g(\xi)$ is uniformly continuous in E).

Thus, by Bochner-Minlos theorem [14], there exists a unique measure ν_F on $(E^*, \mathcal{B})$ such that, for all $\xi \in E$,

$$\ll F, e^{i\hat{\xi}} \gg = \int_{E^*} e^{i(x,\xi)} \nu_F(dx). \tag{7}$$

The rest of the proof is carried out in a well-known pattern (see, for example, [20]). First we extend (7) linearly to $\mathcal{A}$, so that

$$\ll F, \varphi \gg = \int_{E^*} \varphi(x) \nu_F(dx) \tag{8}$$

holds for $\varphi \in \mathcal{A}$.

Next, let $\psi \in \mathcal{E}$ and choose a sequence $\{\varphi_n\} \subset \mathcal{A}$ such that $\varphi_n \to \psi$ in $\mathcal{E}$. It follows that $|\varphi_n - \psi|^2 = (\varphi_n - \psi)(\overline{\varphi}_n - \overline{\psi})$ converges to zero in $\mathcal{E}$. Similarly, one can also prove taht $|\varphi_n - \varphi_m|^2 \to 0$ in $\mathcal{E}$ as $n, m \to \infty$. Consequently, $\int_{E^*} |\varphi_n(x) - \varphi_m(x)|^2 \nu_F(dx) = \ll F, |\varphi_n - \varphi_m|^2 \gg \to 0$ as $n, m \to \infty$. This shows that $\{\varphi_n\}$ is a Cauchy sequence in $L^2(E^*, \nu_F)$. Therefore there exists a function $\psi' \in L^2(E^*, \nu_F)$ so that $\varphi_n \to \psi'$ in $L^2(E^*, \nu_F)$, which in turn implies that there exists a subsequence $\{\varphi_n'\} \subset \{\varphi_n\}$ such that $\varphi_n'(x) \to \psi'(x)$ a.e. (ν_F). Since $\varphi_n(x) \to \psi(x)$ also uniformly on bounded sets in E^*, $\psi'(x) = \psi(x)$ a.e. (ν_F).

Finally, since $\int_{E^*} \varphi_n'(x) \nu_F(dx) \to \int_{E^*} \psi'(x) \nu_F(dx)$, we have $\ll F, \psi \gg = \int_{E^*} \psi'(x) \nu_F(dx) = \int_{E^*} \psi(x) \nu_F(dx)$. $\square$

In the next theorem, we shall characterize the measure ν_F which represents the positive generalized function F. First we need a lemma.

Let $\tilde{\mathcal{E}}_p$ denote the closure of $\mathcal{E}$ in $\mathcal{E}_p$ with respect to the topology of $\mathcal{E}_p$ and set $\tilde{\mathcal{E}}_{p,m} = \mathcal{E}_{p,m} \cap \tilde{\mathcal{E}}_p$. Then $\{(\tilde{\mathcal{E}}_{p,m}, \|\cdot\|_{p,m})\}$ are also Banach spaces and $\tilde{\mathcal{E}}_p = \cup_{m \in \mathbf{N}} \tilde{\mathcal{E}}_{p,m}$. $\tilde{\mathcal{E}}_p$ is the inductive limit $\tilde{\mathcal{E}}_p = \varinjlim \tilde{\mathcal{E}}_{p,m}$ of $\tilde{\mathcal{E}}_{p,m}$ for each p. Moreover, $\mathcal{E} = \cap_{p \in \mathbf{N}} \tilde{\mathcal{E}}_p$ and $\mathcal{E}$ becomes the reduced topological projective limit $\varprojlim \tilde{\mathcal{E}}_p$. Then, according to Theoerm 6 in [9, §22.6, pp.290], $\mathcal{E}^* = \cup_{p \in \mathbf{N}} \tilde{\mathcal{E}}_p^*$ and $\mathcal{E}^*$ is the inductive limit $\mathcal{E}^* = \varinjlim \tilde{\mathcal{E}}_p^*$ of $\tilde{\mathcal{E}}_p^*$. As a result, we obtain the following

LEMMA 3.2 (a) *For each p, $\tilde{\mathcal{E}}_p = \cup_{m \in \mathbf{N}} \tilde{\mathcal{E}}_{p,m}$ and $\tilde{\mathcal{E}}_p = \varinjlim \tilde{\mathcal{E}}_{p,m}$ so that $\tilde{\mathcal{E}}_p^* = \cap_{m \in \mathbf{N}} \tilde{\mathcal{E}}_{p,m}^*$ and $\tilde{\mathcal{E}}_p^* = \varprojlim \tilde{\mathcal{E}}_{p,m}^*$.*

(b) $\mathcal{E} = \cap_{m \in \mathbf{N}} \tilde{\mathcal{E}}_p$ and $\mathcal{E} = \varprojlim \tilde{\mathcal{E}}_p$ is the reduced topological projective limit of $\tilde{\mathcal{E}}_p$. Further, $\mathcal{E}^* = \cup_{p \in \mathbf{N}} \tilde{\mathcal{E}}_p^*$ and $\mathcal{E}^*$ is the topological inductive limit $\mathcal{E}^* = \varinjlim \tilde{\mathcal{E}}_p^*$ of $\tilde{\mathcal{E}}_p^*$.

REMARK 3.3 Lemma 3.2 implies that if $\mathrm{F} \in \mathcal{E}^*$ then there exists $p \in \mathbf{N}$ such that $\mathrm{F} \in \tilde{\mathcal{E}}_{p,m}^*$ for all $m \in \mathbf{N}$. $\qquad\square$

THEOREM 3.4 A measure η on $(\mathrm{E}^*, \mathcal{B})$ is a member in $\mathcal{E}^*$ if and only if there exists $p \in \mathbf{N}$ such that the measurable support of η is contained in E_{-p} and, for all $m \in \mathbf{N}$

$$\int_{\mathrm{E}_{-p}} e^{m|x|_{-p}} \eta(dx) < \infty. \tag{9}$$

PROOF: (Necessity). Suppose that η is a measure in $\mathcal{E}^*$. Then, by Lemma 3.2, there exists an $r \in \mathbf{N}$ such that $\eta \in \tilde{\mathcal{E}}_r^*$. Let $p = r + 1$. We claim that the measurable support of η contains in E_{-p} and that the condition (9) holds. Let $\mathrm{Q}(x) = |x|_{-p}^2$. Then $\mathrm{Q} \in \mathcal{E}_{p,1}$ with its analytic extension given by $\tilde{\mathrm{Q}}(z) = |x|_{-p}^2 - |y|_{-p}^2 + 2i \langle x, y \rangle_{-p}$, where $z = x + iy$ and $\langle \cdot, \cdot \rangle_{-p}$ denotes the inner product of E_{-p}. For $k = 1, 2, 3, \ldots$, let $g_k(x) = \sum_{j=0}^k (2j + 2)^{-2p}(x, e_j)^2$. Then $g_k \in \mathcal{E}$ with analytic extension $\tilde{g}_k(z) = \sum_{j=0}^k (2j + 2)^{2p}(z, e_j)^2$ for $z \in \mathcal{C}\mathrm{E}^*$. Moreover, for $\epsilon > 0$, $\sup_{z \in \mathcal{C}\mathrm{E}_{-r}} |\tilde{g}_k(z) - \tilde{\mathrm{Q}}(z)| e^{-\epsilon|z|_{-r}} \le \left[\sum_{j=k+1}^\infty (2j+2)^{-2} \right] (2\epsilon^{-2}) \to 0$ as $k \to \infty$. Apply the above convergence with $\epsilon = \frac{1}{n}$ and recall that $\mathcal{E}$ and $\mathcal{E}_r$ are topological algebras. One sees easily that g_k^n converges to Q^n with respect to the norm $\| \cdot \|_{r,1}$ for every $n \in \mathbf{N}$. This shows that $\mathrm{Q}^n \in \tilde{\mathcal{E}}_{r,1}$ which, in turn, implies that $\mathrm{Q}^n \in \tilde{\mathcal{E}}_{r,m}$ for all $m \ge 1$ and $n \in \mathbf{N}$.

Now we turn to the estimation of the integral (9). First we observe that, for $|x|_{-p} \ge 1$, $\exp\left[m|x|_{-p} \right] \le 1 + \sum_{j=1}^\infty \left[\frac{m^{2j}}{(2j)!} + \frac{m^{2j-1}}{(2j-1)!} \right] |x|_{-p}^{2j}$, for $m \ge 1$. Then we obtain

$$\int_{\mathrm{E}^*} e^{m|x|_{-p}} \eta(dx)$$

$$= \int_{|x|_{-p} < 1} e^{m|x|_{-p}} \eta(dx) + \int_{|x|_{-p} \le 1} e^{m|x|_{-p}} \eta(dx)$$

$$\le (e^m + 1) \ll \eta, 1 \gg + \sum_{j=1}^\infty \left[\frac{m^{2j}}{(2j)!} + \frac{m^{2j-1}}{(2j-1)!} \right] \ll \eta, \mathrm{Q}^j \gg_{r,m}$$

$$(\ll \cdot, \cdot \gg_{r,m} \text{ denotes the pairing of } \tilde{\mathcal{E}}_{r,m}^* \text{ and } \tilde{\mathcal{E}}_{r,m})$$

$$\le (e^m + 1) \ll \eta, 1 \gg + \|\eta\|_{\tilde{\mathcal{E}}_{r,m}^*} \left[\sum_{j=1}^\infty \left[\frac{m^{2j}}{(2j)!} + \frac{m^{2j-1}}{(2j-1)!} \right] \|\mathrm{Q}^j\|_{r,m} \right]$$

$$\leq \quad (e^m + 1)\ll \eta, 1\gg + \|\eta\|_{\bar{\mathcal{E}}_{r,m}^*} \left[\sum_{j=1}^{\infty} \left(1 + \frac{2j}{m} \right) 4^{-j} \right] < \infty.$$

Since $|x|_{-p} = \infty$ for $x \in E^* \backslash E_{-p}$, the above condition implies that the measurable support of η is contained in E_{-p}.

(Sufficiency). Suppose that the measurable support η is contained in E_{-p}. Then, for all $m \in N$ and $\varphi \in \mathcal{E}_{p,m}$, we have

$$
\begin{aligned}
|\ll \eta, \varphi \gg| &= \left| \int_{E_{-p}} \varphi(x)\eta(dx) \right| \\
&\leq \int_{E_{-p}} |\varphi(x)| e^{-m|x|_{-p}} e^{m|x|_{-p}} \eta(dx) \\
&\leq \|\varphi\|_{p,m} \left[\int_{E_{-p}} e^{m|x|_{-p}} \eta(dx) \right].
\end{aligned}
$$

It follows that $\eta \in \cap_{m \in N} \mathcal{E}_{p,m}^* = \mathcal{E}^*$. $\qquad\qquad\qquad\qquad\qquad \square$

4   Remarks on Kontrat'ev-Yokoi Theorem

For $p \in N$, let (E_p) be the collection of functions f in $(H) := L^2(E^*, \mu)$ such that

$$\|f\|_{2,p} := \left\{ \sum_{n=0}^{\infty} \frac{1}{n!} \|D^n \mu f(0)\|_{\mathcal{L}_2^n[E_{-p}]}^2 \right\}^{\frac{1}{2}} \tag{10}$$

is finite, where $\mathcal{L}_2^n[E_{-p}]$ denotes the space of Hilbert-Schmidt n-linear operators on E_{-p}. Let (E_{-p}) denote the completion of (H) with respect to the norm $\|f\|_{2,p}$ defined by (10). Then (E_p) and (E_{-p}) are Hilbert spaces, and $(E_p)^* = (E_{-p})$. Set $(E) = \cap_{p \in N}(E_p)$ and endow it with the projective limit topology. Then (E) becomes a complete countably normed linear spaces, and it serves as the space of test functions in white noise analysis (see [10, 12, 13, 16, 19]). Let $(E)^*$ be the dual of (E). Then $(E^*) = \cup_{p \in N}(E_p)$ and $(E) \subset (H) \subset (E)^*$ forms a Gel'fand triple. Members of $(E)^*$ are called generalized white noise functionals when $E = \mathcal{S}$.

Kontrat'ev [7] and Yokoi [20] proved independently the following

THEOREM 4.1 If F is a positive member in $(E)^*$, then there exists a unique measure ν_F on $(E^*, \mathcal{B})$ such that $\ll F, \varphi \gg = \int_{E^*} \varphi(x)\nu_F(dx)$.

Furthermore, along the line of the proof of Theorm 5.1 in Lee [12], one can also characterize the measure ν_F without difficulty in the following way.

THEOREM 4.2 *Let η be a Borel measure on E^*. Then $\eta \in (E)^*$ if and only if there exists $p \in N$ such that the measurable support of η is contained in E_{-p} and $\int_{E_{-p}} \exp\left(\frac{1}{2}|x|^2_{-p}\right) \eta(dx) < \infty.$*

The connection between the measures in $(E)^*$ and $\mathcal{E}^*$ can be seen from the lemma below.

LEMMA 4.3 (a) $\mathcal{E} \subset (E)$ *and the embedding is continuous.* (b) $(E)^* \subset \mathcal{E}^*$.

PROOF: (b) is clearly a consequence of (a). It suffices to prove (a).
Let $f \in \mathcal{E}$. Then we have

$$
\begin{aligned}
\|f\|^2_{2,p} &= \sum_{n=0}^{\infty} \frac{1}{n!} \|D^n \mu f(0)\|^2_{\mathcal{L}^n_2[E_{-p}]} \\
&\leq \sum_{n=0}^{\infty} \frac{1}{n!} \|D^n \mu f(0)\|^2_{\mathcal{L}^n[E_{-p-1}]} \|A^{-1}\|^{-2n}_{HS} \\
&\leq K^2_{p,m} \|f\|^2_{p+1,m} \quad \text{(by Proposition 2.1)}, \quad (11)
\end{aligned}
$$

where $K_{p,m} = \left[\int_{E_{-p-1}} e^{m|x|-p-1} \mu(dx)\right] e^{\frac{1}{2}(em)^2}$. (a) then follows from the inequality (11). $\qquad \Box$

At the end of the paper, we show that Lemma 4.3 can be used to prove Kontrat'ev-Yokoi Theorem. Let F be a positive generalizede function in $(E)^*$. Then F is also a positive generalized function in $\mathcal{E}^*$. Hence, by Theorem 3.1, there exists a unique Borel measure ν_F such that

$$
\ll F, \varphi \gg = \int_{E^*} \varphi(x) \nu_F(dx) \quad (12)
$$

for all $\varphi \in \mathcal{E}$. Since $\mathcal{E}$ is dense in (E) (by Proposition 2.3), it follows by the similar arguments as given in the proof of Theorem 3.1 that the identity (12) also holds for $\varphi \in (E)$ by extension. This again proves the Kontrat'ev-Yokoi Theorem.

References

[1] Albeverio, S., Hida, T., Potthoff, J. and Streit, L.: The vacuum of the Hoegh-Krohn model as a generalized white noise functional; Phys. Lett. B 217(1989) 511-514.

[2] Albeverio, S., Hida, T., Potthoff, J., Röckner, M. and Streit, L.: Dirichlet forms in terms of white noise analysis I-Construction and QFT examples; Rev. Math. Phys. 1(1990) 291-312.

[3] Akbeverio, S., Hida, T., Potthoff. J., Röckner, M. and Streit, L.: Dirichlet forms in terms of white noise analysis II-Construction and QFT examples; Rev. Math. Phys. 1 313-323.

[4] Yu.M. Berezansky, Yu.G. Kondrat'ev, "Spectral Methods in Infinite dimensional Analysis", Naukova Duma, Kiev, 1988.

[5] I.M. Gel'fand and N.Ya. Vilenkin, "Generalized Functions", Vol.4, Academic Press, New York.London, 1964.

[6] T. Hida, J. Potthoff and L. Streit, Dirchlet Forms and White Noise Analysis, Commun. Math. Phys. 116, 235-245. (1988).

[7] Ju.G. Kondrat'ev, Nuclear Spaces of Entire Functions in Problems of Infinite-demensional Analysis, Soviet Math. Dokl., Vol.22(1980), No.2

[8] Ju.G. Kondrat'ev and Ju.S. Samoilenko, An Integral Representation for Generalized Positive Definite Kernels in Infinitely Many Variables, Soviet Math. Dokl., Vol.17(1976), No.2.

[9] G. Köthe, Topological Vector Space I, Springer-Verlag, New York, Heidelberg Berlin, 1966.

[10] I. Kubo and Y. Yokoi: A remark on the space of testing random variables in the white noise Calculus, Nagoya Math. J., 116(1989), 139-149.

[11] Y.-J. Lee, Generalized functions on infinite dimensional spaces and its application to white noise calculus, J. Funct. Anal. 82(1989), 429-464.

[12] Y.-J. Lee, Analytic Version of Test Functionals, Fourier Transform and a Characterization of Neasures in White Noise Claculus, J. Funct. Anal., 100(1991), 359-380.

[13] Y.-J. Lee, A characterization of generalized functions on infinite dimensional spaces and Bargman-Segal analytic functions, In "Gaussian Random field", The Third Nagoya Lévy Seminar, Edited by K. Itô and T. Hida, 272-284, World Scientific, 1991.

[14] R.A. Minlos, Generalized Random Processes and Their Extension to a Measure, Selected Transl. in Math. Stat. & Prob., Vol.3(1962), 291-313.

[15] J. Potthoff, On Positive Generalized Functionals, J. Funct. Anal. 74, 1987.

[16] J. Potthoff and L. Streit, A Characterization of Hida Distributions, to appear in J. Funct. Anal.

[17] L. Streit and T. Hida, Generalized Brownian Functionals and Feynman Integral, Stochastic Process. Appl., 16(1983), 55-69.

[18] H. Sugita, Positive Generalized Wiener Functionals and Potential Theory over Abstract Wiener Spaces, Osaka J. Math., 25(1988), 665-696.

[19] J.A. Yan, A Characterization of White Noise Functionals, Preprint, 1990.

[20] Y. Yokoi, Positive Generalized White Noise Functionals, Hiroshima Math. J., Vol.20(1990), No.1, 137-157.

STRONG SOLUTIONS OF STOCHASTIC BILINEAR EQUATIONS WITH ANTICIPATING DRIFT IN THE FIRST WIENER CHAOS[1]

JORGE A. LEON and VICTOR PEREZ-ABREU

Abstract. This article deals with solutions of stochastic bilinear equations having Gaussian anticipating drift living in chaos of order one. The solution is given in terms of explicit expressions for the kernels of the multiple Wiener integrals in the chaos expansion.

Introduction.

Solutions of real-valued anticipating stochastic differential equations have recently been studied by several authors (see Pardoux [8] and references therein). In particular, if the stochastic differential equation is bilinear with both coefficients non-random, it is possible to use the chaos expansion in terms of multiple Wiener integrals to obtain explicit expressions for the kernels of the solution with non-anticipating or anticipating initial condition. This approach has been useful in finite as well as infinite dimensional spaces (Shiota [11], Nualart and Zakai [7], Miyahara [5], Zhang [12] and Pérez-Abreu [9]).

The aim of this paper is to use the chaos expansion approach to obtain an explicit expression for the strong solution of a bilinear real-valued stochastic differential equation with a drift A, which may be anticipating, having a Gaussian distribution in the first Wiener chaos and with an initial condition not necessarily non-anticipating.

---
[1] Research partially supported by CONACYT Grant 20E9105

In Section 1 of this paper we establish notation and main results, differing the proofs to Section 2. In Section 3 we briefly indicate some extensions of our results, which could use recent results by Kallianpur et al. [3]. Applications to the study of weak convergence of solutions and the case of a random diffusion coefficient will be considered elsewhere.

1.- Notation and main result.

Let $W = \{W_t, 0 \leq t \leq 1\}$ be a real-valued Wiener process defined on an arbitrary probability space (Ω, F, P) and let F^W be the σ-field generated by W. For an integer $m \geq 1$ we denote by $I_m(f_m)$ the m-th multiple Wiener integral of the symmetric kernel $f_m \in L^2[0,1]^m$. It is well known (see [6]) that a stochastic process F in $L^2(\Omega \times [0,1], F^W \times B[0,1])$ admits the representation

$$F_t = I(f_0) + \sum_{m=1}^{\infty} I_m(f_m^t) , \qquad (1.1)$$

where $I(f_0) = f_0 = E(F)$. Let $\tilde{f}_m^t(t_1, \ldots, t_m)$ denote the symmetrization of f_m with respect to the $m + 1$ variables $t, t_1, \ldots, t_m$. If the kernels of the process F satisfy the condition

$$\sum_{m=0}^{\infty} (m + 1)! \, \| f_m^{\cdot} \|_{L^2([0,1]^{m+1})}^2 < \infty , \qquad (1.2)$$

then F is Skorohod integrable (see [6] for this an other approaches to anticipating integrals) and its Skorohod integral is given by the expression (equality in $L^2(\Omega)$)

$$\delta(F) = \int_0^1 F_s \, \delta W_s := \sum_{m=0}^{\infty} I_{m+1}(\tilde{f}_m^\circ(\circ)). \qquad (1.3)$$

This paper concerns with the strong solution of the real-valued bilinear stochastic equation

$$dX_t = A(t)X_t \, dt + B(t)X_t \, \delta W_t , \qquad X_0 = \zeta , \qquad (1.4)$$

where for $0 \leq t \leq 1$, $A(t) = I_1(a^t(\circ))$, for a family of kernels $a^\circ(\circ) \in L^2([0,1]^2)$, $B(t)$ is a deterministic process in $L^2[0,1]$ and the initial condition ζ is anticipating. A strong solution of (1.4) is a process X_t satisfying the following two conditions

$$E(\int_0^1 |X_t|^2 \, dt \,) < \infty \,, \tag{1.5}$$

$$X_t = \zeta + \int_0^t A(s)X_s \, ds + \int_0^1 B(s)X_s 1_{[0,t]}(s) \, \delta W_s. \tag{1.6}$$

The drift $A(t)$ is a zero mean gaussian process which is allowed to be anticipating. The main result of this paper is the Theorem 1.1 which gives solution to equation (1.4) in terms of the kernels in the chaos expansion. Using this approach, the case of $A(t)$ non-random has been considered by Shiota [11] in dimension 1 and by Pérez-Abreu [9] in Hilbert spaces.

We introduce the following notation: Given $t_1,\ldots,t_m$ and a symmetric function f_{m-s} of m-s variables ($s < m$), we denote by $f_{m-s}(\hat{t}_{i_1},\ldots,\hat{t}_{i_s})$ the function f_{m-s} evaluated in t's other than $t_{i_1},\ldots,t_{i_s}$. Given a function h, we shall denote by $h^{\otimes m}$ the m-th tensor product of h with itself. Let

$$C_t = \exp \left(\int_0^t \int_0^s a^s(r) \, B(r) \, dsdr \right), \tag{1.7}$$

$$h_t(r) = \int_0^t a^s(r) \, ds, \tag{1.8}$$

and

$$Y_0^t = \exp \left(1/2 \int_0^1 h_t^2(r) \, dr \right). \tag{1.9}$$

We shall assume that the initial condition lives in finite chaos, i.e. there exists a non-negative integer M such that $\zeta = \sum_{m=0}^M I_m(g_m)$, where $g_m \in L^2[0,1]^m$. Writing $g_M = 0$ for $m > M$, the following functions are well defined

$$S_n^t(t_1,\ldots,t_n) = \sum_{k=0}^\infty (k + 1)\ldots(k + n) \{$$

$$<g_{n+k}(t_1,\ldots,t_n,\circ) \,, \, h_t^{\otimes k}(\circ) >_{L^2[0,1]^k}\} \tag{1.10}$$

and $S_0^t = \sum_{k=0}^{\infty} <g_k, \ h_t^{\otimes k}(\circ) >_{L^2[0,1]^k}$.

THEOREM 1.1.   The equation (1.4) has a unique strong solution X in $L^2(\Omega \times [0,1])$.   Moreover X admits the representation

$$X_t = \sum_{m=0}^{\infty} I_m(\phi_m^t)$$

where the kernels ϕ_n's are given by

$$\phi_0^t = C_t Y_0^t S_0^t \tag{1.11}$$

$$n! \phi_n^t(t_1, \ldots, t_n) = \sum_{j=1}^{n} \sum_{1 < i_1 < i_2 \ldots < i_j < n} (-1)^{j-1} B(t_{i_1}) \ldots B(t_{i_j}) \{$$

$$1_{(t_{i_1}, \ldots, t_{i_j} < t)} \ (n-j)! \ \phi_{n-j}^t(\hat{t}_{i_1}, \ldots, \hat{t}_{i_j}) \ \}$$

$$+ \ C_t Y_0^t \{ h_t^{\otimes n}(t_1, \ldots, t_n) S_0^t + \sum_{j=1}^{n} \sum_{1 < i_1 < i_2 \ldots < i_j < n} \{$$

$$h_t^{\otimes(n-j)}(\hat{t}_{i_1}, \ldots, \hat{t}_{i_j}) S_j^t(t_{i_1}, \ldots, t_{i_j}) \} \}. \tag{1.12}$$

REMARKS.   a).- There have been several attempts to study solutions of real-valued anticipating stochastic differential equations (see [1], [2], [8]).   Most of these methods are heavily based on the fact of being in dimension one.   Our approach of chaos expansion could be extended to higher and infinite dimensions as we indicate in Section 3.
b).-   In [8] sufficient conditions are given for the uniqueness of the solution .
c).-   In [4] conditions are given for equivalence of evolution and strong solutions.   It is straightforward to see that even when A(t) lives in the first chaos, the corresponding random evolution system does not live in finite chaos.   It is an open problem to find the distribution properties of such a random evolution system.

2. Proofs.

For the proof of Theorem 1.1 we shall use the following two results.   The first Lemma reduces the stochastic problem to the solution of an infinite system of deterministic integral equations.

LEMMA 2.1. Let $X_t = \sum_{m=0}^{\infty} I_m(f_m^t)$ be such that

$$\sum_{m=0}^{\infty} (m+1)! \; \| f_m \|_{L^2([0,1]^{m+1})}^2 < \infty , \qquad (2.1)$$

Then, X_t is a solution of (1.4) if and only if $\underline{f}^t = (f_0^t, f_1^t, \ldots)$ is the solution of the following infinite system of deterministic integral equations:

$$f_0^t = g_0 + \int_0^t \int_0^1 a^r(\theta) \; f_1^r(\theta) \; d\theta dr \qquad (2.2)$$

$$f_n^t(t_1,\ldots,t_n) = g_n(t_1,\ldots,t_n) + (1/n) \sum_{i=1}^{n} \int_0^t a^r(t_i) \; f_{n-1}^r(\hat{t}_i) dr$$

$$+ (n+1) \int_0^t \int_0^1 a^r(\theta) \; f_{n+1}^r(\theta, t_1,\ldots,t_n) \; d\theta dr$$

$$+ (1/n) \sum_{i=1}^{n} B(t_i) \; 1_{\{t_i < t\}} \; f_{n-1}^{t_i}(\hat{t}_i) , \; n \geq 1. \qquad (2.3)$$

Proof. Let X satisfy (2.1). Then $B(\circ)X_\circ 1_{[0,t]}(\circ)$ is Skorohod integrable and we assume that X_t satisfies (1.6) . From the expression for the Skorohod integral (1.3) and the product formula for multiple Wiener integrals (see Shigekawa [10]) we obtain

$$\sum_{k=0}^{\infty} I_k(f_k^t) = \sum_{k=0}^{M} I_k(g_k^t) + \int_0^t I_1(a^r(\circ)) \; [\sum_{k=0}^{\infty} I_k(f_k^r)] \; dr$$

$$+ \int_0^1 B(r) \sum_{k=0}^{\infty} I_k(f_k^r) \; 1_{[0,t]}(r)\delta W_r = \sum_{k=0}^{M} I_k(g_k^t)$$

$$+ I_1(\int_0^t a^r(\circ) \; f_0^r \; dr) + \sum_{k=0}^{\infty} I_{k+1}(B(\circ)1_{[0,t]}(\circ)f_k^\circ)$$

$$+ \sum_{k=1}^{\infty} \sum_{m=0}^{1} k!/(k-m)! I_{1+k-2m}(\int_0^t a^r(\circ)\otimes_m f_k^r(\circ)dr), \qquad (2.4)$$

where

$$(\int_0^t a^r(\circ)\otimes_0 f_k^r(\circ)dr) \; (s_1,\ldots,s_{k+1}) = \int_0^t a^r(s_1)f_k^r(s_2,\ldots s_{k+1})dr$$

and

$$(\int_0^t a^r(\circ) \otimes_1 f_k^r(\circ)dr) \ (s_1,\dots,s_{k-1})$$

$$= \int_0^t \int_0^1 a^r(\theta) f_k^r(\theta,s_1,\dots s_{k-1})d\theta dr \ .$$

Expressions (2.2) and (2.3) are then obtained identifying the kernels of multiple Wiener integrals of the same order in the first and last side of (2.4). □

LEMMA 2.2.   The unique solution of the deterministic system (2.2)-(2.3) is given by the kernels (1.11)-(1.12).

Proof.   We   shall   use   the   following   straightforward expressions

$$\frac{d}{dt}(C_t Y_0^t) = C_t Y_0^t \left(\int_0^t a^t(r) \ B(r)dr + \int_0^1 a^t(r) \ h_t(r)dr \right) \qquad (2.5)$$

$$\frac{d}{dt} h_t(r) = a^t(r) \qquad (2.6)$$

$$\frac{d}{dt} S_n^t(t_1,\dots,t_n) = \int_0^1 a^t(s) \ S_{n+1}^t(t_1,\dots,t_n,s) \ ds \ . \qquad (2.7)$$

First, using f_0^t and f_1^t given by (1.11) and (1.12) we have that, for $u \in [0,1]$,

$$\int_0^u \int_0^1 a^t(t_1) \ f_1^t(t_1) \ dt_1 dt = \int_0^u C_t Y_0^t S_0^t \left\{ \int_0^t a^t(t_1) \ B(t_1) \ dt_1 + \right.$$

$$\int_0^1 a^t(t_1) \ h_t(t_1) \ dt_1 \right\} \ dt + \int_0^u C_t Y_0^t (\int_0^1 a^t(t_1) \ S_1^t(t_1) \ dt_1)dt$$

$$= \int_0^u S_0^t \frac{d}{dt}(C_t Y_0^t) \ dt + \int_0^u (C_t Y_0^t) \frac{d}{dt}(S_0^t) \ dt \ = f_0^u - g_0^u \ ,$$

that is, f_0^t and f_1^t satisfy (2.2).   Next using the expression for f_2^t given by (1.12) we have that, for $u, s \in [0,1]$,

$$2! \int_0^u \int_0^1 a^t(\theta) \ f_2^t(\theta,s) \ d\theta dt = 2! \int_0^u f_1^t(s) \int_0^1 a^t(\theta) \ B(\theta) \ d\theta dt$$

$$+ 1_{[s<u]} B(s) \int_s^u \int_0^1 a^t(\theta) f_1^t(\theta) d\theta dt$$

$$- \int_0^u \int_0^1 a^t(\theta) 1_{[\theta,t<s]} B(\theta) B(s) f_0^t d\theta dt + \int_0^u \int_0^1 a^t(\theta) C_t Y_0^t \ [S_2^t(\theta,s)$$

$$+ h_t(s)S_1^t(\theta) + h_t(\theta)S_1^t(s) + h_t(\theta)h_t(s)S_0^t]d\theta dt$$

$$+ 1_{[s<u]}B(s)(f_0^u - f_0^s) + C_u Y_0^u S_1^u - g_1(s) + f_0^u h_u(s) - \int_0^u a^t(s)f_0^t dt \ ,$$

from which it follows that f_0^t, f_1^t and f_2^t given by (1.11)-(1.12) satisfy (2.3) for $n = 1$. The proof for a general n follows by induction as above, since the corresponding equations have kernels of order n-1, n and n+1.

Finally, straightforward computations yield that there exists a constant K_M, depending on h, B, C, Y_0 and $g_0, \cdots g_M$, such that for each $n \geq 1$

$$(n+1)! \ \|\phi_n\|_{L^2([0,1]^{n+1})} \leq$$

$$\left\{ \exp\left(1 + \|B\|_{L^2([0,1])} \right) \right\}^{n+1} \left\{ \|\phi_0\|_{L^2([0,1])} + K_M \right\} \ ,$$

and therefore the kernels ϕ_n's satisfy (2.1).

3. Extensions.

There are several possible directions for extensions of our results. For example, we could consider $A(t)$ belonging to the m-th chaos and obtain the corresponding infinite system of deterministic equations satisfied by the kernels of the solution. In this case, and in the one considered in this paper, the chaos expansion approach might be useful to study weak convergence of anticipating solutions in terms of convergence of the kernels. On the other hand, it is possible to go to higher and infinite dimensions where no results are available for solutions of anticipating stochastic differential equations. To work in infinite dimensional Hilbert spaces, we must use the Skorohod integral on Hilbert spaces constructed through multiple Wiener integrals with respect to a cylindrical Brownian motion, as it has recently been defined in [3].

Acknowledgements

The authors would like to thank David Nualart for several helpful discussions. Part of this work was done while the first author was visiting CIMAT and the second author was visiting CINVESTAV. The authors express their thanks for the support of both centers.

References

[1]   Buckdahn, R. (1988). Quasilinear   partial   stochastic differential equations without nonanticipation requirement. Preprint Nr. 176, Humboldt-Universitaet Berlin.

[2]   Buckdahn, R. (1989). Transformations   on   the   Wiener space and Skorohod-type stochastic differential equations. Seminarbericht Nr. 105, Humboldt-Universitaet Berlin.

[3]   Kallianpur, G. and V. Pérez-Abreu (1992). The Skorohod integral and the derivative operator of functionals of a cylindrical Brownian motion. *Appl. Math. Optim.* 25: 11-29.

[4]   León, J. A. (1989). On   equivalence   of   solutions   to stochastic   differential   equations   with   anticipating evolution systems. *Stoch. Anal. and Appl.* 8: 363-387.

[5]   Miyahara, Y. (1981). Infinite   dimensional   Langevin equation and Fokker-Planck equation. *Nagoya Math. J.* 81: 177-223.

[6]   Nualart, D. and E. Pardoux (1988). Stochastic calculus with anticipating integrands. *Probab. Theory Rel. Fields* 78: 535-581.

[7]   Nualart, D. and M. Zakai (1989). Generalized Brownian functionals and the solution to a stochastic partial differential equation. *J. Funct. Anal.* 84: 279-296.

[8]   Pardoux, E. (1990). Applications   of   anticipating stochastic   calculus   to   stochastic   differential   equations. *Lecture Notes in Math.* 1444 :63-105.

[9]   Pérez-Abreu, V. (1992). Anticipating   solutions   of stochastic   bilinear   equations   in   Hilbert   spaces. *Proceedings of the IV Latin American Meeting in Probability and Mathematical Statistics, Bernoulli Society* (to appear).

[10]   Shigekawa, I. (1980).   Derivatives   of   Wiener functionals and absolute continuity of induced measures. *J. Math. Kyoto Univ.* 20: 263-289.

[11]   Shiota, Y. (1986). A   linear   stochastic   integral equation containing the extended Ito integral.   *Math. Rep. Toyama Univ.* 9: 43-65.

[12]   Zhang, W. (1990). Analytical   analysis   of   a   stochastic partial   differential   equation.   *Utilitas   Mathematica*   37: 223-232.

Centro de Investigación y de Estudios Avanzados del IPN
Apdo. Postal 14-740, México 07000 D. F., México.

Centro de Investigación en Matemáticas A. C.
Apdo. Postal 402, Guanajuato Gto., 36000, México.

Structure of Periodically Distributed Stochastic Sequences

ANDRZEJ MAKAGON   and   HABIB SALEHI

Abstract. It is shown that every periodically distributed (PD) stochastic sequence with a period T can be represented as a linear combination of the coordinates of an associated T-variate strictly stationary sequence. This result extends a theorem of Gladyshev for periodically correlated sequences to stochastic sequences with possibly infinite second moment. It is proved that if a PD sequence is SαS then the associated strictly stationary sequence is also SαS, and that it shares the regularity properties of the underlying PD sequence. As a byproduct new proofs of most of Gladyshev's results in [2] are obtained.

The purpose of this paper is to give a structural theorem for periodically distributed stochastic sequences in terms of an associated multivariate stationary random sequence.

Let $\mathbb{Z}$, $\mathcal{N}$ and $\mathbb{C}$ denote the set of all integers, positive integers and complex numbers, respectively. A **stochastic process** is a collection $X=\{X(x): x \in \mathfrak{X}\}$ of complex random variables indexed by a set $\mathfrak{X}$. A stochastic process indexed by $\mathfrak{X}=\{0,...,q-1\} \times \mathbb{Z}$ is called a **q-variate stochastic sequence** (for $\mathfrak{X}=\mathbb{Z}$ it is simply called a **stochastic sequence**). A q-variate stochastic sequence $X=\{X(j,n): j \in \{0,...,q-1\}, n \in \mathbb{Z}\}$ is **strictly stationary** if for any $N \in \mathcal{N}$, any $n_1,...,n_N \in \mathbb{Z}$, $j_1,....j_N \in \{0,...,q-1\}$ and $\lambda_1,...,\lambda_N \in \mathbb{C}$, the random variables $\sum_{k=1}^{N} X(j_k,n_k)\lambda_k$ and $\sum_{k=1}^{N} X(j_k,n_k+1)\lambda_k$ have the same distribution.

In [3] Hurd introduced the notion of a periodically nonstationary process, which is a generalization of strict stationarity. Below this concept is stated for sequences under the phrase 'periodically distributed'.

DEFINITION 1. A stochastic sequence $X=\{X(n): n \in \mathbb{Z}\}$ is called **periodically distributed (PD)** with a period $T \in \mathcal{N}$, if for any $N \in \mathcal{N}$ and any choice of $n_1,...,n_N \in \mathbb{Z}$ the random vectors $<X(n_1),...,X(n_N)>$ and $<X(n_1+T),...,X(n_N+T)>$ are identically distributed.

Note that "PD with every period T" is equivalent to "strictly stationary".

For stochastic sequences with finite second moments the following weaker definitions of periodicity and stationarity are also considered. A stochastic sequence $X=\{X(n): n \in \mathbb{Z}\}$ is **periodically correlated (PC) with a period T** if $\mathbb{E}X(n)=0$, $\mathbb{E}|X(n)|^2<\infty$, and its correlation function $R(n,m)=\mathbb{E}[X(n)\overline{X(m)}]$, satisfies the relation $R(n,m)=R(n+T,m+T)$,

$n,m \in \mathbb{Z}$. A q-variate stochastic sequence $X=\{X(j,n): j \in \{0,...,q-1\}$, $n \in \mathbb{Z}\}$ is called **wide sense stationary** if $EX(j,n)=0$, $E|X(j,n)|^2<\infty$, and the cross-correlation function $R_{j,k}(n,m)=E[X(j,n)\overline{X(k,m)}]$ satisfies the relation $R_{j,k}(n,m)=R_{j,k}(n+1,m+1)$, $k,j \in \{0,...,q-1\}$, $n,m \in \mathbb{Z}$.

A stochastic process $X=\{X(x): x \in \mathfrak{X}\}$ is **symmetric α stable (SαS)**, $0<\alpha \leq 2$, if for eech $N \in \mathcal{N}$, $x=\{x_1,...x_N\} \in \mathfrak{X}^N$ and $\lambda=\{\lambda_1,...,\lambda_N\} \in \mathbb{C}^N$, the random variable $X(\lambda,x)=\sum_{p=1}^{N}\lambda_p X(x_p)$ is SαS, i.e., its characteristic function $\varphi(z)=E\exp[i\mathrm{Re}(\overline{z}X(\lambda,x))]$, $z \in \mathbb{C}$, has the form $\varphi(s+it)=\exp(- \int_0^{2\pi}|s\cos u+t\sin u|^{\alpha}\Gamma_{\lambda,x}(du))$, where $\Gamma_{\lambda,x}$ is a nonnegative symmetric finite Borel measure on $[0,2\pi)$. For $\alpha=2$ they coincide with zero mean Gaussian processes. X is **isotropic SαS** if the characteristic function of $X(\lambda,x)$ is given by $\exp(-c(\lambda,x)|z|^{\alpha})$, $z \in \mathbb{C}$, where $c(\lambda,x)$ is a nonnegative constant. An SαS random variable X is isotropic if and only if for each real θ, X and $e^{i\theta}X$ are identically distributed. If $X=\{X(x): x \in \mathfrak{X}\}$ is an SαS process with $1<\alpha \leq 2$ then in the space $\mathcal{M}_X$ of linear combinations of $X(x)$, $x \in \mathfrak{X}$, the function $\| \sum_{p=1}^{N}\lambda_p X(x_p) \|_{\alpha}$ defined by $\{\Gamma_{\lambda,x}([0,2\pi))\}^{1/\alpha}$ for $1<\alpha<2$, and $\{E|\sum_{p=1}^{N}\lambda_p X(x_p)|^2\}^{1/2}$, for $\alpha=2$, is a norm equivalent to convergence in probability. If $X_1,...,X_n$ are independent random variables in $\mathcal{M}_X$ then (cf. [1], [7])

$$\| \sum_{k=1}^{n}X_k \|_{\alpha}^{\alpha} = \sum_{k=1}^{n} \| X_k \|_{\alpha}^{\alpha} , \tag{1}$$

For X, a PC sequence with period T, Gladyshev [2] has shown the existence of a T-variate wide sense stationary sequence **Y** such that

$$X(n)= \sum_{j=0}^{T-1}Y(j,n)\exp(-2\pi ijn/T), \quad n \in \mathbb{Z}. \tag{2}$$

The main object of this paper is to show that the representation (2) remains valid for an arbitrary PD sequence with **Y** strictly stationary and the equality in law. An explicit construction of the T-variate sequence **Y** is provided. The construction shows that if X is an SαS, so is the sequence **Y**, and that in this case the sequence **Y** shares the regularity properties of the sequence X. Periodically distributed SαS sequences have also been recently studied by Hurd and Mandrekar [4] with emphasis on spectral analysis.

THEOREM 2. Let $X=\{X(n): n \in \mathbb{Z}\}$ be a PD sequence with a period T and let X_k, $k=0,1,...,T-1$, be T independent copies of the sequence X. For each $j=0,...,T-1$ and $n \in \mathbb{Z}$, let

$$Y(j,n)=T^{-1}\sum_{k=0}^{T-1}X_k(n+k)\exp(2\pi ij(n+k)/T). \tag{3}$$

Then:
(i) $Y=\{Y(j,n): j \in \{0,...,T-1\}, n \in \mathbb{Z}\}$ is a T-variate strictly stationary

sequence, and

(ii)   the sequence

$$\sum_{j=0}^{T-1} Y(j,n)\exp(-2\pi ijn/T), \quad n \in \mathcal{Z}, \tag{4}$$

has the same finite dimensional distributions as the sequence X.
Moreover, if X is additionaly SαS (isotropic SαS), then so is **Y**.

PROOF.   Let $N \in \mathcal{N}$, $n_1,...,n_N \in \mathcal{Z}$, $j_1,...j_N \in \{0,...,q\text{-}1\}$ and
$\lambda_1,...\lambda_N \in \mathcal{C}$. Since $X_0=\{X_0(n): n \in \mathcal{Z}\}$ is PD and the sequences
$X_k=\{X_k(n): n \in \mathcal{Z}\}$, k=0,1,...,T-1, are independent and identically
distributed we have

$$T\sum_{p=1}^{N} Y(j_p,n_p)\lambda_p$$
$$= \sum_{k=1}^{T-1} \sum_{p=1}^{N} X_k(n_p+k)\exp(2\pi ij_p(n_p+k)/T)\lambda_p$$
$$\quad + \sum_{p=1}^{N} X_0(n_p)\exp(2\pi ij_p n_p/T)\lambda_p$$
$$\overset{d}{=} \sum_{k=1}^{T-1}\sum_{p=1}^{N} X_k(n_p+k)\exp(2\pi ij_p(n_p+k)/T)\lambda_p$$
$$\quad + \sum_{p=1}^{N} X_0(n_p+T)\exp(2\pi ij_p(n_p+T)/T)\lambda_p$$
$$\overset{d}{=} \sum_{k=1}^{T-1}\sum_{p=1}^{N} X_{k\text{-}1}(n_p+k)\exp(2\pi ij_p(n_p+k)/T)\lambda_p$$
$$\quad + \sum_{p=1}^{N} X_{T\text{-}1}(n_p+T)\exp(2\pi ij_p(n_p+T)/T)\lambda_p$$
$$= T\sum_{p=1}^{N} Y(j_p,n_p+1)\lambda_p$$

where $\overset{d}{=}$ denotes equality in distribution. It follows that **Y** is a T-variate
strictly stationary sequence. (ii) follows from the observation that
$\sum_{j=0}^{T-1} Y(j,n)\exp(-2\pi ijn/T)=X_0(n)$, $n \in \mathcal{Z}$, and the fact that the
processes X_0 and X have the same finite dimensional distributions.

Assume now that the process X is additionaly SαS (isotropic SαS).
Since the sequences X_k, k=0,...,T-1, are independent and SαS (isotropic
SαS), then clearly **Y** is SαS (isotropic SαS). $\square$

REMARK 3. If X=$\{X(n): n \in \mathcal{Z}\}$ is a PD sequence of real-valued
random variables with a period T, then Theorem 2 guarantees the
existence of a 2T-variate real-valued strictly stationary sequence
Y=$\{Y(j,n): j \in \{0,...,2T\text{-}1\}, n \in \mathcal{Z}\}$ such that   X and the sequence
$\sum_{j=0}^{T-1}\{Y(2j,n)\cos(2\pi ijn/T)+Y(2j+1,n)\sin(2\pi ijn/T)\}$, $n \in \mathcal{Z}$,   have the
same finite dimensional distributions. Indeed, the sequence defined by

$$Y(2j,n)=T^{-1}\sum_{k=0}^{T-1} X_k(n+k)\cos(2\pi ij(n+k)/T) , \quad \text{and}$$
$$Y(2j+1,n)=T^{-1}\sum_{k=0}^{T-1} X_k(n+k)\sin(2\pi ij(n+k)/T) ,$$

j=0,...,T-1, $n \in \mathcal{Z}$, can be used for this purpose.

In the next theorem we show that with a slight modification the construction given in Theorem 2 can be applied to PC sequences so that the resulting sequence $\mathbf{Y}$ becomes wide sense stationary. This will provide alternative proofs of Gladyshev's representation (13) in [2] and Theorems 1 and 3 given there.

Let $X=\{X(n): n \in \mathbb{Z}\}$ be a nonzero sequence of zero mean square integrable random variables. X is called **purely nondeterministic** if $\bigcap_k M_X(k)=\{0\}$; following [2] it is called **completely nondeterministic** if it is purely nondeterministic and $X(k) \notin M_X(k-1)$, for every $k \in \mathbb{Z}$. Here $M_X(k)=\overline{sp}\{X(n): n \leq k\}$ where the closure is in $L^2(\Omega, P)$. In the sequel $\mathbf{H}$ will stand for $L^2(\Omega, P)$.

THEOREM 4. (cf. [2]). Let $X=\{X(n): n \in \mathbb{Z}\} \subseteq \mathbf{H}$ be a PC sequence with a period T and let for every $j=0,...,T-1$ and $n \in \mathbb{Z}$, $Y(j,n)$ be an element of the Hilbert space $\mathbf{K}=\mathbf{H}^{\oplus T}$, the orthogonal sum of T copies of $\mathbf{H}$, defined by the formula

$$Y(j,n)=T^{-1} \overset{T-1}{\underset{k=0}{\oplus}} [X(n+k)\exp(2\pi ij(n+k)/T)]. \qquad (5)$$

Then

(i) $\mathbf{Y}=\{Y(j,n): j \in \{0,...,T-1\}, n \in \mathbb{Z}\}$ is a T-variate wide sense stationary sequence in $\mathbf{K}$,

(ii) for every $n \in \mathbb{Z}$,

$$X(n)=\sum_{j=0}^{T-1} Y(j,n)\exp(-2\pi ijn/T) \qquad (6)$$

(here $\mathbf{H}$ is identified with $\mathbf{H} \oplus \{0\} \oplus ... \oplus \{0\}$),

(iii) X is purely nondeterministic if and only if $\mathbf{Y}$ is purely nondeterministic, i.e., $\bigcap_k M_{\mathbf{Y}}(k)=\{0\}$, where $M_{\mathbf{Y}}(k)=\overline{sp}\{Y(j,n): j \in \{0,...,T-1\}, n \leq k\}$, and

(iv) X is completely nondeterministic if and only if $\mathbf{Y}$ is purely nondeterministic of full rank (*full rank* means $\dim(M_{\mathbf{Y}}(k) \ominus M_{\mathbf{Y}}(k-1))=T$ for all $k \in \mathbb{Z}$).

PROOF. (i) Since X is PC with a period T, for every $r \in \mathbb{Z}$ the function $R(r+k,k)=(X(r+k),X(k))_{\mathbf{H}}$ is periodic in the variable k with a period T and hence it admits the representation

$$R(r+k,k)= \sum_{s=0}^{T-1} B_s(r)\exp(2\pi isk/T), \quad k \in \mathbb{Z}, r \in \mathbb{Z}, \qquad (7)$$

with $B_s(r)=T^{-1}\sum_{k=0}^{T-1} R(r+k,k)\exp(-2\pi isk/T)$. Since $(Y(p,n),Y(q,m))_{\mathbf{K}}$ $=T^{-2}\sum_{k=0}^{T-1}(X(n+k),X(m+k))_{\mathbf{H}}\exp[2\pi i(pn+pk-qm-qk)/T]$, substituting k by k-m, and using the periodicity of the expression under the summation sign, we obtain that $R_{p,q}(n,m) = (Y(p,n),Y(q,m))_{\mathbf{K}} = T^{-1}B_{q-p}(n-m)\exp[2\pi ip(n-m)/T]$ for every $p,q \in \{0,...,T-1\}$ and $n,m \in \mathbb{Z}$.

This shows that $\mathbf{Y}$ is a T-variate wide sense stationary sequence.

(ii) follows from

$$\sum_{j=0}^{T-1} Y(j,n)\exp(-2\pi ijn/T) = \sum_{j=0}^{T-1} T^{-1}\bigoplus_{k=0}^{T-1}[X(n+k)\exp(2\pi ij(n+k-n)/T)]$$

$$= X(n)\oplus 0\oplus\ldots\oplus 0.$$

(iii) Property (iii) is an immediate consequence of the obvious inclusions

$$M_X(n)\oplus\{0\}\oplus\ldots\oplus\{0\}\subseteq M_Y(n)$$
$$\subseteq M_X(n)\oplus M_X(n+1)\oplus\ldots\oplus M_X(n+T-1),\qquad n\in\mathbb{Z}.$$

(iv) Let $M_Y(j,n)=\overline{sp}\big\{M_Y(n-1)\cup\{Y(k,n): k=0,\ldots,j\}\big\}$, $j\in\{0,\ldots,T-1\}$, $n\in\mathbb{Z}$, and let $M_Y(-1,n)=M_Y(n-1)$, $n\in\mathbb{Z}$. Then $M_Y(n)\ominus M_Y(n-1)$ $=\bigoplus_{j=0}^{T-1}(M_Y(j,n)\ominus M_Y(j-1,n))$, $n\in\mathbb{Z}$. By (iii), $\mathbf{Y}$ is purely nondeterministic of full rank iff $\mathbf{Y}$ is purely nondeterministic and for every $n\in\mathbb{Z}$ and $j=0,\ldots,T-1$, $Y(j,n)\notin M_Y(j-1,n)$.

First assume that X is completely nondeterministic. Then by (iii), $\mathbf{Y}$ is purely nondeterministic, and therefore it suffices to prove that $Y(j,n)\notin M_Y(j-1,n)$ for all $n\in\mathbb{Z}$ and $j=0,\ldots,T-1$. Suppose to the contrary that

$$Y(j,n)=\lim_{N\to\infty}\Big(\sum_{r=-N}^{n-1}\sum_{k=0}^{T-1}a_{k,r,N}Y(k,r)+\sum_{k=0}^{j-1}b_{k,N}Y(k,n)\Big).$$

Substituting for $Y(\cdot,\cdot)$ from (5) and examining the coordinates of both sides we obtain that for every $p=0,\ldots,T-1$, $X(n+p)\exp(2\pi ij(n+p)/T)$ is the limit, as $N\to\infty$, of

$$\sum_{r=-N}^{n-1}\sum_{k=0}^{T-1}a_{k,r,N}X(r+p)\exp(2\pi ik(r+p)/T)$$
$$+\sum_{k=0}^{j-1}b_{k,N}X(n+p)\exp(2\pi ik(n+p)/T).$$

Since by assumption $X(n+p)\notin M_X(n+p-1)$, we conclude that

$$\exp(2\pi ij(n+p)/T)=\lim_{N\to\infty}\Big(\sum_{k=0}^{j-1}b_{k,N}\exp(2\pi ik(n+p)/T)\Big),$$

$p=0,\ldots,T-1$, which is not possible because the vectors $\mathbf{e}_k=$ $[\exp(2\pi ik(n+0)/T),\ldots,\exp(2\pi ik(n+T-1)/T)]$, $k=0,\ldots,T-1$, are orthogonal elements of $\mathbb{C}^T$.

Conversely assume that $\mathbf{Y}$ is purely nondeterministic of full rank. Then for every $n\in\mathbb{Z}$ and $j=0,\ldots,T-1$, $Y(j,n)\notin M_Y(n,j-1)$. Therefore one has $M_Y(n-1)\cap\overline{sp}\{Y(j,n): j=0,\ldots,T-1\}=\{0\}$, and consequently $X(n)$ is not in $M_X(n-1)$. □

REMARK 5. The representation (6) for a PC sequence was first obtained by Gladyshev (see (13) in [2]). The construction of the stationary sequence **Y** given by Gladyshev is different from our construction, which is reminiscent of the concept of the 'random shift' introduced by Hurd in [3]. In particular in Gladyshev's construction all $Y(j,n)$'s are in the space generated by $\{X(n): n \in \mathbb{Z}\}$, while in our case the space generated by X is essentially smaller that the space generated by the sequence **Y**. However, an advantage of our construction is that the stationary sequence **Y** defined by (5) shares the regularity properties of the sequence X, while this is not so for Gladyshev's construction. Moreover from the proof of Theorem 4 it follows that the correlation matrix of the sequence **Y** is given by

$$R_{p,q}(n)=(Y(p,n),Y(q,0))_K= T^{-1}B_{q-p}(n) \exp(2\pi ipn/T),$$

$p,q \in \{0,...,T-1\}$, $n \in \mathbb{Z}$, where the B_s's are defined in (7). This is exactly the matrix that appears in Gladyshev [2], Theorem 1. Therefore Theorems 1 and 3 in [2] can be easily deduced from our Theorem 4. We note that the zero-coordinate sequence $Y(0,n)$, $n \in \mathbb{Z}$, of **Y** has been singled out by Miamee [6] in connection with a stationary dilation of a PC sequence. However this single coordinate does not have the full force of the process **Y** itself to recover X.

In the next theorem we observe that in the SαS case the regularity properties of a PD sequence X are also determined by the regularity properties of the associated T-variate strictly stationary sequence **Y**. The formulation of the concept of purely nondeterminizm for SαS sequences remains the same as for the second order case; one needs only to replace the L^2-norm by the $\| \ \|_\alpha$-norm.

THEOREM 6. Let $X=\{X(n): n \in \mathbb{Z}\}$ be a PD SαS stochastic sequence with $1<\alpha<2$ and a period T and let $\mathbf{Y}=\{Y(j,n): j \in \{0,...,T-1\}, n \in \mathbb{Z}\}$ be the T-variate strictly stationary sequence associated with X by the formula (3). Then the sequence X is purely nondeterministic if and only if the sequence **Y** is.

PROOF. Let $\mathbf{X}=\{X_k(n): k=0,1,...,T-1, n \in \mathbb{Z}\}$ where $X_k=\{X_k(n): n \in \mathbb{Z}\}$, $k=0,1,...,T-1$, are independent copies of the sequence X. Then **X** is a T-variate stochastic sequence. Let us show that **X** is purely nondeterministic if and only if the sequence X is purely nondeterministic. Suppose first that there is a nonzero $V \in M_{\mathbf{X}}(-\infty)$. Then

$V=\lim_{n \to \infty} \sum_{i=0}^{T-1} V_i^n$, where $V_i^n = \sum_{p \leq n} A_{p,i}^n X_i(p)$, and from (1) it follows that for every $i=0,...,T-1$ the sequence V_i^n converges to some element $V_i \in M_{X_i}(-\infty)$. Therefore $V=V_0+...+V_{T-1}$ with at least one V_i being nonzero.[i] The converse implication is trivial. Since the sequence $W(n)=\sum_{j=0}^{T-1} Y(j,n)\exp(-2\pi ijn/T)$, $n \in \mathbb{Z}$, defined by the formula (4), and the sequence X have the same finite dimensional distributions, W is purely nondeterministic if and only if X is purely nondeterministic. As shown above X is purely nondeterministic if and only if the T-variate sequence **X** is so. Therefore, in view of the obvious inclusions $M_W(k) \subseteq M_Y(k) \subseteq M_X(k+T-1)$, $k \in \mathbb{Z}$, we conclude that X is purely nondeterministic if and only if so is **Y**. $\square$

REFERENCES

[1] Cambanis, S. (1983). *Stable variables and processes.* Contributions to Statistics: Essays in Honor of Norman L. Johnson, ed. P.K. Sen, North Holland, New York, 63-79.

[2] Gladyshev, E.G. (1961). *Periodically correlated random sequences.* Soviet Math. Dokl. 2, 385-388.

[3] Hurd, H.L. (1974). *Stationarizing properties of random shift.* SIAM J. Appl. Math. 26, 203-211.

[4] Hurd, H.L., Mandrekar, V. (1991). *Spectral theory of periodically and quasi-periodically stationary SαS sequences.* Center for Stochastic Processes, Tech. Rep. No. 349, University of North Carolina, Chapel Hill.

[5] Masani, P. (1966). *Recent trends in multivariate prediction theory.* Multivariate Analysis, ed. P. R. Krishnaiah, Academic Press, New York and London, 351-382.

[6] Miamee, A.G. (1990). *Periodically correlated processes and their stationary dilations.* SIAM J. Appl. Math. 50, 1194-1199.

[7] Weron, A. (1984). *Stable processes and measures; a survey.* Probability Theory on Vector Spaces III, eds. D.Szynal and A.Weron, Lecture Notes in Mathematics 1080, Springer, 306-364.

A.Makagon
Dept. of Statistics and Probability,
Michigan State University,
East Lansing, MI 48824
and
Institute of Mathematics
Technical University of Wroclaw,
Wroclaw, 50-270 Poland

H.Salehi
Dept. of Statistics and Probability
Michigan State University,
East Lansing, MI 48824

Markov Property of
Measure-indexed Gaussian Random Fields[1]

V. MANDREKAR   and   SIXIANG ZHANG

Abstract. We give analytic characterizations of the RKHS of measure-indexed Gaussian random fields having Markov properties.

1. Introduction. The study of Markov property for the so-called multiparameter Lévy process was initiated by Lévy[7] and proved by McKean[11] by analytic techniques and by Molchan[12] by geometric techniques. In [15], a characterization of McKean Markov property for a certain class of multiparameter Gaussian processes was given and Kallianpur and Mandrekar[5] proved a characterization of germ field Markov property(GFMP) on open sets for generalized Gaussian random fields using the concept of dual field introduced by Molchan[12]. Künsch[6] strengthened the work of Pitt[15] and Mandrekar and Soltani[10] gave a generalization of [6] for generalized random fields. All the characterizations are in terms of the structure of the reproducing kernel Hilbert space(RKHS, [1]) of the covariance of the Gaussian processes. The main techniques used are elementary and use the geometry of RKHS.

In [2], Dynkin studied the Markov property for measure-indexed Gaussian random fields with the covariance given by the Green function of a symmetric R^d-valued Markov process. He studied their Markov property on all sets by using the additive functionals of symmetric R^d-valued Markov processes. Röckner[16] generalized Dynkin's work for the case of symmetric R^d-valued Markov processes associated with Dirichlet forms[3]. Our purpose here is to extend the basic characterizations([5], [12], [15], [6], [10]) of Gaussian random fields with Markov property on open sets to measure-indexed fields using only the geometric techniques. One can derive Markov properties on all sets for the case of Dynkin and Röckner by elementary results on the conditional independence as in [8] and [9]. As a consequence of our main results, Theorems 3.1 and 3.2 give a complete derivation of the main results of [6],[5]. The derivation of the main results of [16] from these is done in [18]. The details are in [19].

---

[1]Supported in part by ONR Grants: N000-14-91-J-1084, N000-14-85-K-0150

2. Preliminaries and Notations. Let $(\Omega, \mathcal{F}, P)$ be a complete probability space and $\mathcal{A}$, $\mathcal{B}$, $\mathcal{G}$ be sub-σ-fields of $\mathcal{F}$. We say that $\mathcal{A}$ is conditionally independent of $\mathcal{B}$ given $\mathcal{G}$ if $P(A \cap B \mid \mathcal{G}) = P(A \mid \mathcal{G})P(B \mid \mathcal{G})$ for all $A \in \mathcal{A}$ and $B \in \mathcal{B}$, and we write $\mathcal{A} \perp\!\!\!\perp \mathcal{B} \mid \mathcal{G}$. Let E be a locally compact Hausdorff space which is second countable (and hence metrizable); the metric on E is chosen so that all closed bounded sets are compact([21], p.8). Let $M(E)$ be a vector space of Radon measures on E with compact support. Here the support(supp(μ)) of a signed measure μ on E is defined to be the complement of the largest open set $O \subseteq E$ such that $|\mu|(O) = 0$ ($|\mu| = $ total variation measure of μ). Let $\underline{X} = \{X_\mu , \mu \in M(E)\}$ be a centered Gaussian random field defined on $(\Omega, \mathcal{F}, P)$ with covariance C. We assume C is bilinear on $M(E) \times M(E)$. Associated with $\underline{X}$ are the following σ-fields: given a set $S \subseteq E$, $F(S)$ is the σ-field generated by $\{X_\mu, \text{supp}(\mu) \subseteq S\}$ and all P-negligible sets of $\mathcal{F}$, denoted by $F(S) = \sigma\{X_\mu, \text{supp}(\mu) \subseteq S\}$, and $\Sigma(S) = \cap\{F(O): O \text{ open}, S \subseteq O\}$. We make the following assumptions on $M(E)$:

(A.1) $M(E)$ has the partition of unity property, namely for any $\mu \in M(E)$, if $\{O_1, O_2, ..,O_n\}$ is an open covering of supp(μ), then there are $\mu_1, ..., \mu_n$ in $M(E)$ with supp(μ_i) $\subseteq O_i$ and $\mu = \mu_1 + \cdots + \mu_n$.
(A.2) If f is a linear functional on $M(E)$ and the support of f (for short supp(f)) is contained in $A_1 \cup A_2$, where A_1 and A_2 are disjoint closed subsets of E, then $f = f_1 + f_2$ with f_1 and f_2 being linear functionals on $M(E)$ and supp(f_i) $\subseteq A_i$, $i = 1, 2$. The support of a linear functional f on $M(E)$ is defined as the complement of the largest open subset O of E such that $f(\mu) = 0$ for all $\mu \in M(E)$ with supp(μ) $\subseteq O$.

Under $(A.1)$ the support of a linear functional on $M(E)$ is well defined. The proof of the following lemma is not difficult and hence omitted.

Lemma 2.1: Under the Assumption $(A.1)$, we have the following.
(a) For every linear functional f on $M(E)$, supp(f)= complement of $\bigcup_i O_i$ where the union is over all open sets O_i such that $f(\mu) = 0$ for all $\mu \in M(E)$ with supp(μ) $\subseteq O_i$.
(b) If f is a linear functional on $M(E)$, then supp(f) is an empty set iff $f = 0$, i. e. $f(\mu) = 0$ for all $\mu \in M(E)$.
(c) If f_1 and f_2 are two linear functionals on $M(E)$, then
$$\text{supp}(f_1 + f_2) \subseteq \text{supp}(f_1) \bigcup \text{supp}(f_2).$$

The following lemma gives an assumption equivalent to $(A.2)$.

Lemma 2.2: Under $(A.1)$, $(A.2)$ is equivalent to the following $(A.2)'$:
(A.2)': If f is a linear functional on $M(E)$ and supp(f) $\subseteq A_1 \bigcup A_2$, where A_1 and A_2 are disjoint closed sets, then for any disjoint open sets

O_1 and O_2 with $A_i \subseteq O_i$, $i=1,2$, f can be decomposed into the sum of two linear functionals f_1 and f_2 with $supp(f_i) \subseteq O_i$, $i = 1,2$.

Proof. That $(A.2)$ implies $(A.2)'$ is obvious. To prove the converse, let O_1 and O_2 be disjoint open sets of E such that $O_i \supseteq A_i$, $i=1,2$. Then $f=f_1+f_2$ with f_i being linear functionals on $M(E)$ and $supp(f_i) \subseteq O_i$, $i = 1,2$. Choose another open set O_1' such that with $A_1 \subseteq O_1' \subseteq O_1$. Then $f=f_1'+f_2'$ with $supp(f_1') \subseteq O_1'$ and $supp(f_2') \subseteq O_2$, so $f_1 - f_1'=f_2'-f_2$. By Lemma 2.1(c) $supp(f_1 - f_1') \subseteq (supp(f_1)) \cup (supp(f_1')) \subseteq O_1$ and $supp(f_2' - f_2) \subseteq (supp(f_2')) \cup (supp(f_2)) \subseteq O_2$. Since $O_1 \cap O_2 = \emptyset$, $supp(f_1 - f_1')=\emptyset$. Then by Lemma 2.1(b), $f_1 - f_1'=0$. So $supp(f_1) \subseteq O_1'$. Hence $supp(f_1) \subseteq \cap \{O: \text{open}, A_1 \subseteq O \subseteq O_1\}=A_1$. Similarly we can show that $supp(f_2) \subseteq A_2$. $\square$

Using $(A.1)$ one can show the following lemma:

Lemma 2.3: $F(O \cup O') = F(O) \vee F(O')$, where O and O' are two open subsets of E.

Three different Markov Properties of $\underline{X}$ are defined as follows:

Definition 2.1: $\underline{X}$ has Markov Property I on an open subset S of E if for every open subset O of E with $\partial S \subseteq O$, $F(S) \perp\!\!\!\perp F(\overline{S}^c) \,|F(O)$.

Definition 2.2: (Germ Field Markov Property) $\underline{X}$ has Markov Property II on a subset S (not necessarily open) of E if $\Sigma(\overline{S}) \perp\!\!\!\perp \Sigma(\overline{S^c}) \,|\, \Sigma(\partial S)$.

Definition 2.3: $\underline{X}$ has Markov Property III on a subset S of E if $F(\overline{S}) \perp\!\!\!\perp F(\overline{S^c}) \,|\, F(\partial S)$.

We will explore some relationships between these Markov Properties.

Lemma 2.4: Let S be an open subset of E and O an open set containing ∂S. Then $F(S) \perp\!\!\!\perp F(\overline{S}^c) \,|\, F(O)$ iff $\Sigma(\overline{S}) \perp\!\!\!\perp \Sigma(S^c) \,|\, F(O)$.

Proof. The "if" part is easy because $F(S) \subseteq \Sigma(\overline{S})$ and $F(\overline{S}^c) \subseteq \Sigma(S^c)$. To prove the converse, using Lemma 2.3, $F(S) \vee F(O) = F(S \cup O) \supseteq \Sigma(\overline{S})$ and $F(\overline{S}^c) \vee F(O) = F(\overline{S}^c \cup O) \supseteq \Sigma(S^c)$. Then $\Sigma(\overline{S}) \perp\!\!\!\perp \Sigma(S^c) \,|\, F(O)$ for open set $O \supseteq \partial S$ by [9]. $\square$

We complete this section by giving three vector spaces $M(E)$ which satisfy assumptions $(A.1)$ and $(A.2)'$. These will be the indexing sets used in applications. For these we need the following lemma([20]).

Lemma 2.5: If E is a second countable locally compact Hausdorff space and $\{O_1, \ldots, O_n\}$ is an open covering of a closed set $A \subseteq E$, then there exist open sets $U_1, \ldots, U_n$ such that $\overline{U}_i \subseteq O_i$ and $\cup_{i=1}^{n} U_i \supseteq A$.

Examples: a) Let E be an open set in R^n and $C_0^\infty(E)$ be all infinitely differentiable functions on R^n with compact support in E. $M(E)=\{\mu,$ $d\mu/d\lambda = \varphi,\ \varphi \in C_0^\infty(E)\}$, where λ is Lebesgue measure on R^n. Clearly $M(E)$ is a vector space, $\text{supp}(\mu)=\text{supp}(\varphi)$, and it has the partition of unity as in [5]. To verify $(A.2)'$, let f be a linear functional on $M(E)$ and $\text{supp}(f) \subseteq A_1 \cup A_2$, where A_1 and A_2 are disjoint closed subsets of E. Let O_1 and O_2 be two disjoint open subsets such that $A_i \subseteq O_i$, $i=1,2$. Take O_i' disjoint such that $A_i \subseteq O_i' \subseteq O_i$ $(i=1,2)$ and $\bar{O}_1' \cap \bar{O}_2'=\emptyset$. Now with $O_3'=(A_1 \cup A_2)^c$, $\{O_1',O_2',O_3'\}$ is an open covering of E. Let $\varphi_i \in C^\infty(E)$, $\varphi_i \geq 0$, $\text{supp}(\varphi_i) \subseteq O_i'$ and $\Sigma_{i=1}^3 \varphi_i=1$. As $\varphi=\Sigma_{i=1}^3 \varphi\varphi_i$ and f is linear, $f(\varphi) = f(\varphi\varphi_1) + f(\varphi\varphi_2) + f(\varphi\varphi_3)$. Denote $f_i(\varphi) = f(\varphi\varphi_i)$. Since $f(\varphi\varphi_3) = 0$, we get for $\varphi \in C_0^\infty(E)$ with $\text{supp}(\varphi) \subseteq (\text{supp}(\varphi_i))^c$, $\varphi\varphi_i=0$ $(i=1,2)$. Hence $\text{supp}(f_i) \subseteq ((\text{supp}(\varphi_i)^c)^c= \text{supp}(\varphi_i) \subseteq O_i$ $(i=1,2)$.

b) Let E be a second countable locally compact Hausdorff space and $M(E)$ be a vector space of signed Radon measures with compact support on E satisfying the following: $\mu \in M(E) \Rightarrow \mu(A \cap \bullet) \in M(E)$ for any $A \in \mathfrak{B}(E)$, where $\mathfrak{B}(E)$ is the Borel σ-field of E. Suppose $\mu \in M(E)$, $\text{supp}(\mu) \subseteq \cup_{i=1}^n O_i$, where O_i's are open. Choose U_i as in Lemma 2.5. Then $\mu(\bullet) = \mu((\cup_{i=1}^n U_i) \cap \bullet) = \Sigma_{i=1}^n \mu(V_i \cap \bullet)$, where $V_i=U_i \cap U_1^c \cap \cdots \cap U_{i-1}^c$. Let $\mu_i(\bullet)=\mu(V_i \cap \bullet)$ $(i=1,\ldots,n)$. As $\text{supp}(\mu_i) \subseteq \bar{U}_i \subseteq O_i$, thus $(A.1)$ holds. Let $\text{supp}(f) \subseteq A_1 \cup A_2$, where A_1 and A_2 are disjoint closed sets. Choose open sets O_1 and O_2 with $A_i \subseteq O_i$ $(i=1,2)$ and $\bar{O}_1 \cap \bar{O}_2 = \emptyset$ with $O_3=(O_1 \cup O_2)^c$, we have $O_1 \cup O_2 \cup O_3 = E$ and the O_i's are disjoint. Then $\mu = \mu_1 + \mu_2 + \mu_3$ with $\mu_i =\mu(O_i \cap \bullet)$ $(i=1,2,3)$. Also $f(\mu) = f(\mu_1)+f(\mu_2)+f(\mu_3) = f_1(\mu)+f_2(\mu)+f_3(\mu)$ (say), with $f_3(\mu)=0$ for all $\mu \in M(E)$. If $\text{supp}(\mu) \subseteq \bar{O}_i$ $(i=1,2)$, $\mu_i=0$. Hence $f_i(\mu)= 0$ for $\mu \in M(E)$ with $\text{supp}(\mu) \subseteq \bar{O}_i^c$, i.e. $\text{supp}(f_i) \subseteq \bar{O}_i \subseteq O_i$ $(i=1,2)$ giving $(A.2)'$.

c) Let E be a second countable locally compact Hausdorff space and m be a fixed positive Radon measure on E with $\text{supp}(m) =E$. If we take $M(E)=\{\mu: d\mu/dm=\varphi$ and $\varphi \in C_0(E)\}$, where $C_0(E)$ is the space of all continuous functions on E with compact support, we get as in a) that $M(E)$ satisfies $(A.1)$ and $(A.2)$.

3. Main Theorems. Let $\underline{X}=\{X_\mu,\ \mu \in M(E)\}$ be a Gaussian random field and $H(S) = \overline{\text{sp}}\{X_\mu,\ \text{supp}(\mu) \subseteq S\}$, where the closure is in $L^2(\Omega,\mathfrak{F},P)$. Since $F(S) = \sigma\{H(S)\}$, we get by [9] that $\underline{X}$ has Markov property I on all open sets S iff

$$H(S \cup O) \cap H(\bar{S}^c \cup O) = H(O) \tag{3.1}$$

and

$$H(S \cup O)^\perp \perp H(\bar{S}^c \cup O)^\perp \tag{3.2}$$

for all open sets S and open sets $O \supseteq \partial S$.

Let C be the covariance of $\underline{X}$ and $K(C)$ the RKHS of C. Let Π be the linear isometry of $K(C)$ onto $H(X) = H(E)$ such that $\Pi(C(\,\cdot\,,\mu)) = X_\mu$. We now state our main theorem.

Theorem 3.1: Let E be a second countable locally compact Hausdorff space and $M(E)$ be a space of Radon measures on E with compact support satisfying $(A.1)$ and $(A.2)$. Then the centered Gaussian process $\{X_\mu \,,\ \mu \in M(E)\}$ has Markov Property I on all open sets iff the following (a) and (b) are satisfied.

(a) If f_1, $f_2 \in K(C)$ with $\operatorname{supp}(f_1) \cap \operatorname{supp}(f_2) = \emptyset$, then $(f_1, f_2)_{K(C)} = 0$.
(b) If $f \in K(C)$ and $f = f_1 + f_2$ where f_1 and f_2 are linear functionals on $M(E)$ with disjoint supports, then $f_1, f_2 \in K(C)$.

Proof. *Sufficiency:* Suppose that (a) and (b) hold. To verify (3.1) it is enough to prove that $H(S \cup O) \cap H(\bar{S}^c \cup O) \subseteq H(O)$ which is equivalent to

$$K(O)^\perp \subseteq K(S \cup O)^\perp \vee K(\bar{S}^c \cup O)^\perp. \tag{3.3}$$

Let $f \in K(O)^\perp$. Then $f(\mu) = (f, C(\,\cdot\,,\mu))_{K(C)} = 0$ if $\operatorname{supp}(\mu) \subseteq O$, hence $\operatorname{supp}(f) \subseteq O^c$. $S \cap O^c$ and $\bar{S}^c \cap O^c$ are disjoint closed sets whose union is O^c. By $(A.2)$, $f(\mu) = f_1(\mu) + f_2(\mu)$ with f_1 and f_2 being linear functionals of $M(E)$ and $\operatorname{supp}(f_1) \subseteq S \cap O^c$, $\operatorname{supp}(f_2) \subseteq \bar{S}^c \cap O^c$. By (b), $f_1, f_2 \in K(C)$. In addition $f_1 \in K((\bar{S} \cap O^c)^c)^\perp = K(\bar{S}^c \cup O^c)^\perp$, and $f_2 \in K((\bar{S}^c \cap O^c)^c)^\perp = K(\bar{S} \cup O^c)^\perp$. Note that $K(S \cup O)^\perp \subseteq \overline{\operatorname{sp}}\{f: f \in K(C)$ and $\operatorname{supp} f \subseteq \bar{S}^c\}$ and $K(\bar{S}^c \cup O)^\perp \subseteq \overline{\operatorname{sp}} \{f: f \in K(C)$ and $\operatorname{supp} f \subseteq S\}$ giving (3.2) by (a) and $\Pi(K(D)^\perp) = H(D)^\perp$ for open set D.

Necessity: Suppose that (3.1) and (3.2) hold. Let $f_1, f_2 \in K(C)$ have disjoint supports. Then there exists an open set S such that $\operatorname{supp}(f_1) \subseteq S$ and $\operatorname{supp}(f_2) \subseteq \bar{S}^c$. Then $O = [\operatorname{supp}(f_1) \cup \operatorname{supp}(f_2)]^c$ is an open set containing ∂S. Since $S \cup O \subseteq (\operatorname{supp}(f_2))^c$ and $\bar{S}^c \cup O \subseteq (\operatorname{supp}(f_1))^c$, $f_1 \in K(\bar{S}^c \cup O)^\perp$ and $f_2 \in K(S \cup O)^\perp$. Hence $(f_1, f_2)_{K(C)} = 0$ by (3.2) which proves (a). To prove (b), assume $f \in K(C)$ and $f = f_1 + f_2$, where f_1 and f_2 are linear functionals on $M(E)$ and have disjoint supports. Choose an open set S such that $\operatorname{supp}(f_1) \subseteq S$ and $\operatorname{supp}(f_2) \subseteq \bar{S}^c$. Letting $O = [\operatorname{supp}(f_1) \cup \operatorname{supp}(f_2)]^c$, by (3.1) and (3.2) we have

$$K(X) = K(S \cup O)^\perp \oplus K(\bar{S}^c \cup O)^\perp \oplus K(O).$$

Since $\operatorname{supp}(f) \subseteq \operatorname{supp}(f_1) \cup \operatorname{supp}(f_2) = O^c$, $f \in K(O)^\perp$. Hence $f = f_1'$ $+ f_2'$ where $f_1' \in K(\bar{S}^c \cup O)^\perp$ and $f_2' \in K(S \cup O)^\perp$. Moreover $\operatorname{supp}(f_1') \subseteq (\bar{S}^c \cup O)^c = \bar{S} \cap O^c = \bar{S} \cap [\operatorname{supp}(f_1) \cup \operatorname{supp}(f_2)] = [\bar{S} \cap \operatorname{supp}(f_1)] \cup [\bar{S} \cap \operatorname{supp}(f_2)] = \bar{S} \cap \operatorname{supp}(f_1) \subseteq \operatorname{supp}(f_1)$. Similarly $\operatorname{supp}(f_2') \subseteq \operatorname{supp}(f_2)$. Now $f = f_1 + f_2 = f_1' + f_2'$, $f_1 - f_1' = f_2' - f_2$ with $\operatorname{supp}(f_1 - f_1') \cap \operatorname{supp}(f_2' - f_2) = \emptyset$ which implies $f_1 = f_1'$ and $f_2 = f_2'$ by virtue of Lemma 2.1(b). Hence $f_1, f_2 \in K(C)$. $\square$

If we only consider the Markov Property I on all open subsets which are bounded or have bounded complements, then we have the following theorem similar to Theorem 3.1.

Theorem 3.2: Let E and $M(E)$ be as in Theorem 3.1. Then $\{X_\mu, \mu \in M(E)\}$ has Markov Property I on all open subsets of E which are bounded or have bounded complements iff the following (a) and (b) hold.
(a) If $f_1, f_2 \in K(C)$, $\text{supp}(f_1) \cap \text{supp}(f_2) = \emptyset$ and at least one of $\text{supp}(f_i)$ is compact, then $(f_1, f_2)_{K(C)} = 0$.
(b) If $f \in K(C)$, $f = f_1 + f_2$, where both f_1 and f_2 are linear functionals of $M(E)$ with $\text{supp} f_1 \cap \text{supp} f_2 = \emptyset$, and at least one of the $\text{supp} f_i$ $(i=1,2)$ is compact, then $f_1, f_2 \in K(C)$.

Proof. All the arguments in the the proof of Theorem 3.1 go through, except when one of $\text{supp} f_i$ $(i=1,2)$ is compact, one can choose a bounded open set S to cover the compact one and $\bar{S}^c$ to cover the other. $\square$

The property in Theorem 3.1(a) is called local property of $K(C)$ and in Theorem 3.2(a) is called compact local property of $K(C)$.

We extend the concept of the dual (or biorthogonal in the terminology of [17]) process introduced in [12] to the measure-indexed field $\{X_\mu, \mu \in M(E)\}$. Let $G(E)$ be a subset of $C_0(E)$ ($G(E)$ need not be a linear subspace of $C_0(E)$) and $\hat{X} = \{\hat{X}_g, g \in G(E)\}$ be a Gaussian field defined on the same probability space as $\underline{X} = \{X_\mu, \mu \in M(E)\}$. Then we have the following:

Definition 3.1: The Gaussian field $\hat{X} = \{\hat{X}_g, g \in G(E)\}$ is called a biorthogonal field of $\underline{X} = \{X_\mu, \mu \in M(E)\}$ if $H(X) = H(\hat{X})$, and $E(\hat{X}_g X_\mu) = \int_E g d\mu$ for every $g \in G(E)$ and $\mu \in M(E)$, where $H(X)$ and $H(\hat{X})$ are the linear subspaces of $L_2(\Omega. \mathfrak{F}, P)$ generated by $\{X_\mu, \mu \in M(E)\}$ and $\{\hat{X}_g, g \in G(E)\}$ respectively.

For any $g \in G(E)$ let $f_g(\mu) = \int_E g d\mu$, $\mu \in M(E)$. Assuming that the biorthogonal field $\hat{X}$ exists, we clearly have $f_g \in K(C)$. For any open subset D of E, we define subspaces $M(D)$ and $\hat{M}(D)$ of $K(C)$ as follows:
$$M(D) = \overline{\text{sp}}\{f, f \in K(C) \text{ and } \text{supp}(f) \subseteq D\}, \quad (3.4)$$
$$\hat{M}(D) = \overline{\text{sp}}\{f_g, g \in G(E) \text{ and } \text{supp}(g) \subseteq D\}, \quad (3.5)$$
where $\text{supp}(g)$ is the closure of $\{e, g(e) \neq 0\}$ in E. We observe that $\hat{M}(D) \subseteq M(D) \subseteq K(D^c)^\perp$ for every open set D.

Definition 3.2: $G(E)$ has the partition of the unity property if for every $g \in G(E)$, and $O_1, \ldots, O_n$ open sets covering $\text{supp}(g)$, we have $g = \sum_{i=1}^n g_i$ with $g_i \in G(E)$ and $\text{supp}(g_i) \subseteq O_i$ $(i=1, \ldots, n)$.

Lemma 3.1: If $G(E)$ has the partition of unity property, then $\hat{M}(D_1 \cup D_2) = \hat{M}(D_1) \vee \hat{M}(D_2)$ for any open sets D_1 and D_2.

Proof. Let $g \in G(E)$ with $\text{supp}(g) \subseteq D_1 \cup D_2$. Then $g = g_1 + g_2$ with $g_i \in G(E)$ and $\text{supp}(g_i) \subseteq D_i$ $(i=1,2)$. Hence $f_g(\mu) = f_{g_1}(\mu) + f_{g_2}(\mu)$ for $\mu \in M(E)$. But $f_{g_i} \in \hat{M}(D_i)$ $(i=1,2)$, hence $f_g \in \hat{M}(D_1) \vee \hat{M}(D_2)$, which implies $\hat{M}(D_1 \cup D_2) = \hat{M}(D_1) \vee \hat{M}(D_2)$. $\square$

Theorem 3.3: Let $\hat{\underline{X}}$ be a biorthogonal field of $\underline{X}$ such that $G(E)$ has the partition of unity property and $\hat{M}(D) = M(D)$ for all open sets D. Then condition (a) of Theorem 3.1 implies condition (b) of Theorem 3.1.

Proof. If condition (a) of Theorem 3.1 holds, then $M(D_1) \perp M(D_2)$ for every two disjoint open sets D_1 and D_2. By Lemma 3.1 and the assumption $M(D_1 \cup D_2) = \hat{M}(D_1 \cup D_2) = \hat{M}(D_1) \vee \hat{M}(D_2) = M(D_1) \oplus M(D_2)$. Let $f \in K(C)$ with $f = f_1 + f_2$, where f_1, f_2 are linear functionals on $M(E)$ with $\text{supp}(f_1) \cap \text{supp}(f_2) = \emptyset$. Then we can choose open sets D_1, D_2 such that $\text{supp}(f_i) \subseteq D_i$ $(i=1,2)$, and $\bar{D}_1$, $\bar{D}_2$ are disjoint. Let $D = D_1 \cup D_2$. Then $\text{supp}(f) \subseteq D$, hence $f \in M(D)$. We can write

$$f = Proj_{M(D)} f = Proj_{M(D_1)} f + Proj_{M(D_2)} f = f_1' + f_2' \qquad \text{(say)},$$

where f_1' as an element of $M(D_1)$ is the limit of a sequence $\{\tilde{f}_n\}$ in $K(C)$ with $\text{supp}(\tilde{f}_n) \subseteq D_1$. Hence $f_1' = \lim_n \tilde{f}_n(\mu) = 0$ for any $\mu \in M(E)$ with $\text{supp}(\mu) \subseteq \bar{D}_1^c$, and so $\text{supp}(f_1') \subseteq \bar{D}_1$. Similarly $\text{supp}(f_2') \subseteq \bar{D}_2^c$. Now $f = f_1 + f_2 = f_1' + f_2'$ gives $f_1 - f_1' = f_2' - f_2$. But $\text{supp}(f_1 - f_1') \subseteq \bar{D}_1$, and $\text{supp}(f_2 - f_2') \subseteq \bar{D}_2$, and $\bar{D}_1 \cap \bar{D}_2 = \emptyset$ imply $\text{supp}(f_1 - f_1') = \text{supp}(f_2' - f_2) = \emptyset$. Hence $f_i = f_i' \in K(C)$, $i = 1,2$. $\square$

The following can be proved as above.

Theorem 3.4: Let $\hat{\underline{X}}$ be a biorthogonal field of $\underline{X}$ such that $G(E)$ has the partition of unity property and $\hat{M}(D) = M(D)$ for all open sets which are bounded or have bounded complements. Then condition (a) of Theorem 3.2 implies condition (b) of Theorem 3.2.

The following corollary is an immediate consequence of Theorems 3.1 and 3.3.

Corollary 3.1: Under the assumptions of Theorem 3.3, $\underline{X}$ has Markov Property I on all open sets iff $K(C)$ has the local property.

A similar corollary can be stated with the compact local property for $K(C)$ using Theorems 3.2 and 3.4.

By Corollary 3.1, condition $M(D) = \hat{M}(D)$ allows us to check the Markov property using the local property of $K(C)$. We now give a condition on $M(E)$, $G(E)$ and the norm in $K(C)$ in order that $M(D) = \hat{M}(D)$ for the class of all open sets or the class of all open sets which are bounded or have bounded complements. The condition is motivated by([17], p.108), where separating function is defined.

Assumption 3.1: (a) For every closed set A, there is a separating function $w \in C(E)$ such that $wg \in G(E)$, $\int . wd\mu \in M(E)$ for all $g \in G(E)$ and all $\mu \in M(E)$. (b) is exactly as above except that closed is replaced by compact.

Remark 3.1: In the case $E \subseteq R^n$ open, $G(E)=C_0^\infty(E)$ and $M(E)=\{\varphi d\lambda$, $\varphi \in C_0^\infty(E)\}$, Assumption 3.1(a) is satisfied with $w \in C^\infty(E)$ and (b) is satisfied with $w \in C_0^\infty(E)$. In the case $G(E)=C_0(E)$ and $M(E)=\{\varphi dm$, $\varphi \in C_0(E)\}$, where m is a fixed positive Radon measure on a second countable locally compact Hausdorff space E with supp$(m)=E$, both assumptions are satisfied with some $w \in C_0(E)$.

We now state a lemma whose proof is exactly as in Rozanov ([17], p.108) and so is omitted.

Lemma 3.3: Let $\underline{X}$ be a centered Gaussian random field with a biorthogonal field $\hat{X}$ such that $G(E)$ has the partition of unity property. If assumption 3.1(a) (resp.(b)) is satisfied and furthermore, w can be chosen to satisfy the following inequality:

$$E \mid \hat{X}_{wg} \mid^2 \le L_w E \mid \hat{X}_g \mid^2 \quad \text{for all } g \in G(E),$$

where L_w is a constant depending only on w and the corresponding closed set A, then $M(D) = \hat{M}(D)$ for all open sets D (resp. all open sets D which are bounded or have bounded complements).

Using Lemma 3.3 and Corollary 3.1 we get

Theorem 3.5: Let $\underline{X}$ be a centered Gaussian random field with a biorthogonal field $\hat{X}$ such that $G(E)$ has the partition of unity property. If Assumption 3.1(a) (resp. (b)) is satisfied, and furthermore, w satisfies the condition given in Lemma 3.3 , then the following are equivalent:
(a) $K(C)$ has local (resp. compact local) property,
(b) $\hat{M}(D_1) \perp \hat{M}(D_2)$ for all disjoint open (resp. bounded open) sets D_1 and D_2.

We note that in the case $M(E)$ is as in Example(a) and $G(E)$ $=C_0^\infty(E)$, $E \subseteq R^n$ open, then $\{\hat{X}_\varphi, \varphi \in C_0^\infty(E)\}$ is a generalized Gaussian random field in the sense of [4] and condition $\hat{M}(D_1) \perp \hat{M}(D_2)$ gives us that it has independent values at every point. In view of Remark 3.1, this result includes the main results of [5] and [12].

Let E be a second countable locally compact Hausdorff space and $\{\xi_t , t \in E\}$ be a mean zero Gaussian process. We say that $\{\xi_t, t \in E\}$ has the Markov Property on $A \subseteq E$ if $F(\bar{A}) \perp\!\!\!\perp F(A^c) \mid \Sigma(\partial A)$. Assume that $R(s,t)=E(\xi_t \xi_s)$ is continuous. Choose a positive

Radon measure m on E with supp(m) $=E$. Take $M(E)=\{\varphi dm,$ $\varphi \in C_0(E)\}$. From Example(c), $M(E)$ satisfies Assumptions $(A.1)$ and $(A.2)$. Define $X_\mu= \int_E \xi_t d\mu$, $\mu \in M(E)$. Then the Markov property of $\{\xi_t,\ t \in E\}$ for all open sets is equivalent to the Markov Property I of $\{X_\mu,\ \mu \in M(E)\}$ for all open sets. Let $K(C_X)$ and $K(C_\xi)$ be the RKHS's of $\{X_\mu,\ \mu \in M(E)\}$ and $\{\xi_t,\ t \in E\}$ respectively. Define J: $K(C_\xi) \to K(C_X)$ as $(Jf)(\mu)= \int_E f(t)d\mu$, $f \in K(C_\xi)$ and $\mu \in M(E)$. Then J is an isometric map between $K(C_\xi)$ and $K(C_X)$ and furthermore supp(Jf)=supp(f). We can obtain the following extension of Künsch([6], Theorem 5.1) and Pitt([15], Theorem 3.3):

$\{\xi_t,\ t \in E\}$ has Markov Property on all open sets iff

(a) $(f_1,f_2)_{K(C_\xi)}=0$ if f_1 and f_2 are in $K(C_\xi)$ with disjoint support, and

(b) $f \in K(C_\xi)$ such that $f=f_1+f_2$, where f_1 and f_2 are continuous and have disjoint supports implies $f_1,f_2 \in K(C_\xi)$.

Similarly we can show that $\{\xi_t,\ t \in E\}$ has the Markov Property on all open sets which are bounded or have bounded complements iff the above (a) and (b) hold when at least one of suppf_1 or suppf_2 is compact.

In view of Remark 3.1, we get under Assumption 3.1 that the analog of Theorem 3.5 holds. In case $E=R^n$ this theorem is used in the study of stationary random fields([17]). In many applications one encounters E more general than R^n. The study of such fields is in progress and will appear in future work of the authors.

References:

[1]   Aronszajn, N.(1950). Theory of Reproducing Kernels. *Trans. Amer. Math. Soc.* **68** 337-404.

[2]   Dynkin, E., B.(1980). Markov Processes and Random Fields. *Bull. Amer. Math. Soc.* **3** 975-999.

[3]   Fukushima, M. *Dirichlet Forms and Markov Processes.* North Holland, Amsterdam, 1980.

[4]   Gelfand, I. M. and Vilenkin, V. *Generalized Functions.* Academic Press, New York, 1964.

[5]   Kallianpur, G. and Mandrekar, V. (1974). The Markov Property of Generalized Gaussian Random Fields. *Ann. Inst.Fourier (Grenoble)* **24** 143-167.

[6]   Künsch, H.(1979). Gaussian Markov Random Fields. *J. Fac. Sci. Univ. Tokyo Sect. IA Math* **26** 53-73.

[7]   Lévy, P.(1956). A Special Problem of Brownian Motion and General
      Theory of Gaussian Random Functions. *Proc. Third Berkeley
      Symp. on Math. Stat. Prob.* Vol.2 133-175.
[8]   Mandrekar, V., Germ-field Markov Property for Multiparameter
      Processes. *In " Seminaire de Probabilités X"*, Lecture Notes in
      Math., no. 511, Springer, Berlin, 1976. 78-85.
[9]   Mandrekar, V., Markov Properties of Random Fields. *in
      "Probabilistic Analysis and Related Topics"* (A.T. Bharucha-Reid,
      ed.) Vol.3, Academic Press, New York, 1983. 161-193.
[10] Mandrekar, V. and Soltani, A. R.(1982). Markov Property for
      Ultraprocess. Technical Report No.5 , Center for Stochastic
      Processes, Univ. of North Carolina (Chapel Hill).
[11] McKean, H. P. Jr. (1963). Brownian Motion with Several
      Dimensional Time. *Theor. Prob. Appl.* **8** 335-354.
[12] Molchan, G., M. (1971). Characterization of Gaussian Fields with
      Markov Property. *Soviet Math Dokl.* **12** 563-567.
[13] Nelson, E. (1979). Construction of Quantum Fields From Markov
      Fields. *J. Funct. Anal.* **12** 97-112.
[14] Neveu, J. *Discrete Parameter Martingales.* North-Holland,
      Oxford, 1975.
[15] Pitt, L. D. (1971). A Markov Property for Gaussian Processes
      with a Multidimensional Parameter. *Arch. Rat. Mech. Anal.* **43**
      367-391.
[16] Röckner, M. (1985). Generalized Markov Fields and Dirichlet
      Forms. *Acta Appl. Math.* **3** 285-311.
[17] Rozanov, Yu. A. *Markov Random Fields.* Springer, Berlin, 1982.
[18] Zhang, S. *Markov Properties of Measure-indexed Gaussian
      Random Fields.* Ph.D. Thesis, Dept. of Stat. and Prob., Michigan
      State Univ., 1990.
[19] Zhang, S.(1991). Measure-indexed Gaussian Random Fields
      Related to Potential Theory (preprint).
[20] Schwartz, L. *Analyse Mathématique,* vol. 1, Hermann, Paris, 1967.
[21] Blumenthal, R. M. and Getoor, R. K., *Markov Processes and
      Potential Theory,* Academic Press, New York, 1968.

               V. Mandrekar   Sixiang Zhang
Department of Statistics and Probability   Department of Mathematics
        Michigan State University   Gannon University
        East Lansing, MI 48824   Erie, PA 16541

Relative Entropy as a Countably-Additive Measure

P. R. MASANI

Abstract. Let P be a countably additive probability measure, and μ be any countably additive $[0,\infty]$-valued measure, both on a σ-algebra $\mathcal{A}$ over a space Ω. We define a $(P,\mu$ dependent) countable additive measure H on the δ-ring $\mathcal{A}_\mu$ of sets $A \in \mathcal{A}$ such that $\mu(A) < \infty$, with values in $(-\infty,\infty]$, with the property that in case $\Omega \in \mathcal{A}_\mu$, $H(\Omega)$ is the total entropy of P relative to μ. We study the extension of H beyond $\mathcal{A}_\mu$, and its Lebesgue decomposition with respect to μ.

1. Introduction

The concept of "the entropy of a probability measure P" is a misnomer: there is always another measure μ in the background. In the discrete case considered by Shannon, μ is the cardinality measure [9, p. 19]; in the continuous case considered by both Shannon and Wiener, μ is the Lebesgue measure [9, p. 54] and [12, pp. 61, 62]. *All entropies are relative entropies of P with respect to some measure* μ, as Shannon and Wiener both emphasized, cf. [9, pp. 57, 58] and [12, pp. 61, 62].

The important paper by Gelfand, Komogorov and Yaglom [3] called attention to the case in which the underlying μ is an arbitrary probability measure. This case was studied independently (c. 1960) by Kallianpur [5] in the United States and Pinsker [7] in the Soviet Union. They took countably additive (CA) probability measures P, μ on a σ-algebra $\mathcal{A}$ over an arbitrary set Ω, and taking Π_Ω to be the class of all partitions π of Ω into a finite number of disjoint ($\|$) cells Δ in $\mathcal{A}$, they defined $H_\pi(P|\mu)$ and $H(P|\mu)$, by

$$(1.1) \quad \begin{cases} \forall \pi \in \Pi_\Omega, \ H_\pi(P|\mu) := \sum_{\Delta \in \pi+} \left\{ \log \frac{P(\Delta)}{\mu(\Delta)} \right\} \cdot P(\Delta) \\ \& \qquad\qquad H(P|\mu) := \sup_{\pi \in \Pi_\Omega} H_\pi(P|\mu). \end{cases}$$

Here, $\pi+ = \{\Delta : \Delta \in \pi \ \& \ \mu(\Delta) > 0$ or $P(\Delta) > 0\}$. The number $H(P|\mu)$ is called *the total entropy of P relative to* μ. A basic result is that

(1.2) $H(P|\mu) \in [0,\infty].$

However, *the underlying measure* μ *cannot always be taken to be a probability measure*. Both Shannon and Wiener took the entropy of a probability P over $\mathbf{R}$ relative to Lebesgue measure μ over $\mathbf{R}$. Indeed, the probability imposition on μ excludes even the elementary case considered by Shannon, as the following easily proved proposition shows:

1.3 Prop. *Let* Ω *be finite with 2 or more members,* $\mathcal{A} = 2^\Omega$ *and* $\forall \omega \in \Omega$, $P\{\omega\} > 0$, *and let S be the Shannon entropy defined by:*

$$S = \sum_{\omega \in \Omega} (\log P\{\omega\}) \cdot P\{\omega\}.$$

Then for all probability measures μ *on* $\mathcal{A}$, $S < H(P|\mu)$.

The inequality in 1.3 can be turned into an equality by removing the probability restraint on μ, i.e. replacing it by the cardinality measure μ_0 on $\Omega = \{1,...,n\}$, so that each $\mu_0\{k\} = 1$, for $k \in \Omega$.[1] If $\Omega = \mathbf{N}_+$, the sets of all positive integers, the required measure μ_0 will, like Lebesgue measure, be $[0,\infty]$-valued. In the general case in which μ is any $[0,\infty]$ CA measure on $\mathcal{A}$, $H(P|\mu)$ can assume negative values in contrast to (1.2).

The definitions of $H_\pi(P|\mu)$, $H(P|\mu)$ in (1.1) break down when μ is not a bounded measure on $\mathcal{A}$. This paper will provide a universal definition of $H(P|\mu)$ valid for all $[0,\infty]$-valued measures μ. The few who have dealt with $[0,\infty]$-valued μ, for instance Rosenblatt-Roth [8] and Cziszar [1],[2] have assumed that μ is σ-finite, and have tailored their definitions accordingly. This assumption is satisfied in most applications, but is theoretically unnecessary.

The main purpose of this paper is to show that the subject admits considerable widening when we interpret $H(P|\mu)$, not as a *number* depending on P and μ, but rather as a (P,μ) dependent *set-function* on the δ-ring $\mathcal{A}_\mu$ of subsets A of $\mathcal{A}$ for which $\mu(A) < \infty$, cf. Def. 2.3 below. The entropy measure $H := H(P|\mu)$ so obtained turns out to be countably additive on the δ-ring $\mathcal{A}_\mu$ to $(-\infty,\infty]$, (Thm. 2.9). By taking its total variation $|H|$ and its Hahn-Jordan decomposition, one can show that H extends to a CA measure $\bar{H}$ on a larger family $\mathcal{P}$, (Thm. 3.9). This family is a pre-ring, which is not a ring in general, (Lma. 3.8). $\bar{H}$ does not therefore admit an extension beyond $\mathcal{P}$.

It turns out that the Lebesgue decomposition of P into parts P_a and P_b that are absolutely continuous and singular with respect to μ, yields a corresponding decomposition of the entropy measure $\bar{H}$ on $\mathcal{P}$ into parts H_a, H_b that are absolutely continuous and singular with respect to μ, (Thms. 2.13 and 3.12).

---

[1] We obviously have $S = H(P|\mu_0)$, on noting that the partition $\pi_0 = \{\{1\},...,\{n\}\}$ has no proper refinements, that $H_\pi(P|\pi)$ increases monotonically with the refinement of π, and that $S = H_{\pi_0}(P|\mu)$.

[2] The writer is grateful to Professor A. Barron for this reference.

Moreover, the classical theorems, due to Kallianpur and others, on the total entropy of P relative to a probability measure μ (which in our notation would read $H(P|\mu)(\Omega)$), admit for arbitrary μ full-fledged extensions to $\bar{H}(P|\mu)(E)$ when $E \in \mathcal{P}$, and extensions to $\bar{H}(P|\mu)(\Omega)$ when $\Omega \in \mathcal{P}$, cf. 2.4(c), 3.11, 3.12. (We can give examples of P and μ, for which $\Omega \in \mathcal{P}$ and $\Omega \notin \mathcal{P}$.) When P has a density f with respect to μ, we can show that the entropy measure $\bar{H}$ itself has the density $\varphi{\circ}f$ with respect to μ, where $\varphi(x) := x \log x, x > 0$, (Thm. 4.16).

The measure $H(P|\mu)$ turns out to be invariant under all μ-measure preserving transformations of the space Ω, (§5). This result shows that the entropy in non-discrete situations is no less intrinsic than in the discrete—an issue raised by Vallée [10].

This work, stimulated by Kallianpur's paper [5], also bears the impress of useful conversations with him. It is therefore a pleasure to submit this paper. Unfortunately, space limitations will not allow us to give proofs. These should appear in 1992 in papers in *J. Comp. & App. Math.*

2. The relative entropy measure on the fundamental δ-ring

In the rest of this paper we shall adhere to the following notation:

2.1 Notation.
(a) $\mathcal{A}$ is a σ-algebra over Ω,

P is a CA probability or (occasionally) subprobability measure on $\mathcal{A}$,

μ is a CA measure on $\mathcal{A}$ to $[0,\infty]$, in symbols $\mu \in CA(\mathcal{A}, [0,\infty])$,

$\mathcal{A}_\mu := \{A : A \in \mathcal{A} \ \& \ \mu(A) < \infty\}$,

$\mathcal{A}_\mu{+} := \{A : A \in A_\mu \ \& \ \mu(A) > 0 \text{ or } P(A) > 0\}$.

(b) $\forall A \in \mathcal{A}_\mu, \ \Pi_A := \{\pi : \pi \text{ is a partition of } A \text{ into a finite number of } \| \text{ sets}$
$$\Delta \in \mathcal{A}_\mu, \text{ i.e. } \bigcup_{\Delta \in \pi} \Delta = A\}.$$

For $\pi', \pi \in \Pi_A$, we say that π' is a *refinement* of π, and write $\pi' < \pi$,
$$\text{iff } \forall \Delta' \in \pi', \ \exists \Delta' \in \pi \ni \Delta' \subseteq \Delta.$$

(c) $\forall A \in \mathcal{A}_\mu \ \& \ \forall \pi \in \Pi_A, \ \pi{+} := \pi^+ \cup \pi_+$, where
$$\pi^+ := \{\Delta : \Delta \in \pi \ \& \ P(\Delta) > 0\}, \quad \pi_+ := \{\Delta : \Delta \in \pi \ \& \ \mu(\Delta) > 0\}.$$

(d) $\forall \Delta \in \pi{+}, \ \rho(\Delta) := \dfrac{P(\Delta)}{\mu(\Delta)}, \quad Q(\Delta) := \{\log \rho(\Delta)\}{\cdot}P(\Delta)$.

We note that for $\Delta \in \pi \backslash \pi{+}, P(\Delta) = 0 = \mu(\Delta)$, and $\rho(\Delta) = 0/0$ is meaningless. However, since P and μ are FA, and the function $\varphi : \varphi(x) := x \log x, x \in \mathbf{R}_+$ has $-1/e$ as its absolute minimum, it follows that:

$$(2.2) \begin{cases} \forall A \in \mathcal{A}_\mu{+} \ \& \ \forall \pi \in \Pi_A, \quad \pi{+} \text{ is non-void,} \\[2pt] \qquad P(A) = \sum_{\Delta \in \pi{+}} P(\Delta) \ \& \ \mu(A) = \sum_{\Delta \in \pi{+}} \mu(\Delta), \\[2pt] \forall \Delta \in \pi{+}, \ \rho(\Delta) \in [0,\infty] \ \& \ -(1/e)\mu(A) \le -(1/e)\,\mu(\Delta) \le Q(\Delta) \le \infty. \end{cases}$$

Likewise, for $A \in \mathcal{A} \backslash \mathcal{A}_\mu +$, both π^+ and π_+ are void and so therefore is $\pi+$. These facts suggest the adoption of the following generalization of (1.1):

2.3 Def. Let $A \in \mathcal{A}_\mu$. Then[3] $H(A) = \sup\limits_{\pi \in \Pi_A} H_\pi(A)$, where

$$\forall \pi \in \Pi_A, \quad H_\pi(A) = \begin{cases} \sum\limits_{\Delta \in \pi+} Q(\Delta), & A \in \mathcal{A}_\mu +, \\ 0, & A \in \mathcal{A}_\mu \backslash \mathcal{A}_\mu +. \end{cases}$$

An easy consequence of (2.2) and this definition is the following useful triviality:

2.4 Triv. (a) $\forall A \in \mathcal{A}_\mu +, \quad -(1/e)\mu(A) \le Q(A) \le H(A) \le \infty.$

(b) $\forall A, B \in \mathcal{A}_\mu, \quad H(A) = \infty \ \& \ A \subseteq B \ \Rightarrow \ H(B) = \infty.$

(c) $\forall A \in \mathcal{A}_\mu, \quad H(A) < \infty \ \Rightarrow \ P \underset{A}{\overset{\prec\prec}{}} \mu,$
where $P \underset{A}{\overset{\prec\prec}{}} \mu$ means: $\Delta \in \mathcal{A}_\mu \cap 2^A \ \& \ \mu(\Delta) = 0 \ \Rightarrow \ P(\Delta) = 0.$

To show the countable additivity of the measure $H(\cdot)$, we must consider countable partitions of A in $\mathcal{A}_\mu$. Let

$$(2.5) \begin{cases} \forall A \in \mathcal{A}_\mu, \ \overline{\Pi}_A := \{\pi: \pi \text{ is a partition of } A \text{ into a finite or} \\ \qquad \qquad \qquad \text{countable number of } \| \text{ cells in } \mathcal{A}_\mu \}. \end{cases}$$

For $A \in \mathcal{A}_\mu +$, and $\pi \in \overline{\Pi}_A$, the series $\sum_{\Delta \in \pi+} Q(\Delta)$ is infinite, and the question of its unconditional convergence or unconditional divergence arises. But this matter is easily settled: by 2.4(a), each $Q(\Delta) \ge -(1/e)\mu(\Delta)$; hence the sum of the negative terms of the series $\sum_{\Delta \in \pi+} Q(\Delta)$ is bounded below by $-(1/e)\mu(A)$ and cannot be $-\infty$. Thus, $\forall A \in \mathcal{A}_\mu$ and $\forall \pi \in \overline{\Pi}_A$, the series

$$H_\pi^*(A) := \sum_{\Delta \in \pi+} Q(\Delta)$$

converges unconditionally to some number in $[-(1/e)\mu(A), \infty)$, or it diverges unconditionally to $+\infty$. This allows us to show that the definition of entropy measure given in 2.3 is unaffected by the replacement of finite partitions by countable partitions:

2.6 Lma. Let $A \in \mathcal{A}_\mu +$. Then $H(A) = \sup\limits_{\pi \in \overline{\Pi}_A} H_\pi^*(A).$

So far no use has been made of the concavity of the log function on $(0, \infty)$. Its use yields a sharpening of Triv. 2.4(a), and thence the useful result that $H_\pi(A)$ and $H_\pi^*(A)$ do not diminish under the refinement of π in Π_A:

---

[3] The full symbol for $H(A)$ should be $\{H(P|\mu)\}(A)$ or $H_{P|\mu}(A)$. But since the measures P and μ are fixed, the abbreviation $H(A)$ suffices.

(2.7) $\forall A \in \mathcal{A}_\pi^+, \quad \pi, \pi' \in \Pi_A \ \& \ \pi' \prec \pi \ \Rightarrow \ H_\pi(A) \le H_{\pi'}(A).$

(2.8) $\forall A \in \mathcal{A}_\pi^+, \quad \overline{\pi}, \overline{\pi}' \in \overline{\Pi}_A \ \& \ \overline{\pi}' \prec \overline{\pi} \ \Rightarrow \ H_{\overline{\pi}}(A) \le H_\pi^*(A).$

These results culminate in the following theorem:

2.9 Thm. $H(\cdot)$ *is a* CA *measure on the* δ-*ring* $\mathcal{A}_\mu$ *to* $(-\infty, \infty]$.

For the measures at hand, defined on δ–rings, the relevant notions of global and local absolute continuity and singularity (symbolized by $\ll$, $\overset{loc}{\ll}$ and $\perp\!\!\!\perp$) due to von Neumann [11, p. 197], and other cognate ideas are defined as follows:

2.10 Def. Let $\varphi \ne \mathcal{F} \subseteq 2^\Omega$, $\mathcal{F}^{loc} := \{A : A \subseteq \Omega \ \& \ \forall F \in \mathcal{F}, A \cap F \in \mathcal{F}\}$, X be a normed vector space or $(-\infty, \infty]$, and ξ be a function on $\mathcal{F}$ to X. We say that

(a) C is a *carrier* of ξ iff $C \in \mathcal{F}^{loc} \ \& \ \forall F \in \mathcal{F}, \xi(C \cap F) = \xi(F)$.

(b) N is a ξ-*negligible* iff $N \in \mathcal{F}^{loc} \ \& \ \forall F \in \mathcal{F}, \xi(N \cap F) = 0$.

$\mathcal{N}_\xi$ denotes the family of ξ-negligible subsets of Ω.

2.11 Def. Let (i) $\mathcal{R}$ be a ring over Ω, and X be as in Def. 2.10, (ii) μ and ξ be CA measures on $\mathcal{R}$ to $[0, \infty]$ and X, respectively. We say that

(a) ξ is *absolutely continuous with respect to* μ,, and write $\xi \ll \mu$, iff
$$\forall A \in \mathcal{R} \ \& \ \forall \varepsilon > 0, \quad \exists \delta_{A,\varepsilon} > 0 \ \ni$$
$$R \in \mathcal{R} \cap 2^A \ \& \ \mu(R) < \delta_{A,\varepsilon} \ \Rightarrow \ |\xi(A)| < \varepsilon.$$

(b) ξ is *locally absolutely continuous with respect to* μ and write $\xi \overset{loc}{\ll} \mu$, iff $\forall A \in \mathcal{R} \ \ni |\xi|(A) < \infty \ \& \ \forall \varepsilon > 0, \ \exists \delta_{A,\varepsilon} > 0 \ \ni$
$$R \in \mathcal{R} \cap 2^A \ \& \ \mu(R) < \delta_{A,\varepsilon} \ \Rightarrow \ |\xi(A)| < \varepsilon.\ ^{4}$$

(c) ξ and μ are *mutually singular*, and write $\xi \perp\!\!\!\perp \mu$, iff there exist disjoint sets $C_1, C_2 \in \mathcal{R}^{loc}$ such that C_1 is a carrier of ξ and C_2 is a carrier of μ.

It is easily seen that for a δ-ring $\mathcal{R}$, $\mathcal{R}^{loc}$ is a σ-algebra, and $\mathcal{R} \subseteq \sigma\text{-alg}(\mathcal{R}) \subseteq \mathcal{R}^{loc}$. For the δ-ring $\mathcal{A}_\mu$ of (2.1)(a) in particular, $\mathcal{A}_\mu \subseteq \sigma\text{-alg}(\mathcal{A}_\mu) \subseteq \mathcal{A} \subseteq \mathcal{A}_\mu^{loc}$. Each of these inclusions can be proper, as can be seen by taking $\Omega = \mathbf{R}$, and μ to be the cardinality measure on the σ-algebra $\mathcal{A}$ of Borel subsets of $\mathbf{R}$. With regard to absolute continuity, we have:

2.12 Prop. (a) $H \overset{loc}{\ll} \mu \ \Rightarrow \ \mathcal{N}_\mu \subseteq \mathcal{N}_H \ \Leftrightarrow \ P \ll \mu.$

(b)   Range $H \subseteq \mathbf{R} \ \Rightarrow \ P \ll \mu \ \& \ H \ll \mu.$

(c) $P \perp\!\!\!\perp \mu \ \Rightarrow \ H \perp\!\!\!\perp \mu \ \Rightarrow \ $ Range $H = $ *the binary set* $\{0, \infty\}$.

The probability P has a Lebesgue decomposition with sub-probability

---

[4] We write $\overset{loc}{\ll}$ rather than $\underset{loc}{\ll}$ a to stress that "locally" here refers to sets that are "finite" with respect to ξ, not μ.

components P_a, P_b that are absolutely continuous and singular with respect to μ, so that $P = P_a + P_b$, $P_a \ll \mu$ and $\mu \perp\!\!\!\perp P_b \perp\!\!\!\perp P_a$. The next theorem asserts that the entropy measures of these components yield the corresponding components in the Lebesgue decomposition of H itself with respect to μ. Briefly, there is a concordance between the two Lebesgue decompositions:

2.13 Concordance Thm. *Let* H, H_a, H_b *be the entropy measures of* $\mathcal{A}_\mu$ *of* P, P_a, P_b *relative to* μ, *respectively. Then*

$$H = H_a + H_b, \quad H_a \overset{loc}{\ll} \mu \quad \& \quad \mu \perp\!\!\!\perp H_b \perp\!\!\!\perp H_a.$$

3. Extension of the relative entropy measure beyond the fundamental δ-ring

To investigate how far beyond the δ-ring $\mathcal{A}_\mu$ the entropy measure $H(\cdot)$ can be extended, define the families of "positive" and "negative" subsets of the δ-ring $\mathcal{A}_\mu$ by:

$$(\mathbf{3.1}) \quad \begin{cases} \mathcal{A}_\mu^+ := \{ E: E \in \mathcal{A}_\mu \ \& \ \forall A \in \mathcal{A}_\mu, \ H(E \cap A) \geq 0 \} \\ \mathcal{A}_\mu^- := \{ E: E \in \mathcal{A}_\mu \ \& \ \forall A \in \mathcal{A}_\mu, \ H(E \cap A) \leq 0 \}. \end{cases}$$

These are ideals of the σ-algebra $\mathcal{A}_\mu^{loc}$. From the classical Hahn-Jordan decomposition theory, cf. Halmos [4], we conclude that

$$(\mathbf{3.2}) \quad \begin{cases} \forall A \in \mathcal{A}_\mu, \ \exists H\text{-essentially unique sets } A^+ \ \& \ A^- \ni \\ A^+ \in \mathcal{A}_\mu^+, \ A^- \in \mathcal{A}_\mu^-, \ A^+ \parallel A^- \ \& \ A^+ \cup A^- = A. \end{cases}$$

This allows us to define H^+, H^- unequivocally on $\mathcal{A}_\mu$ by:

$$(\mathbf{3.3}) \qquad \forall A \in \mathcal{A}_\mu, \ H^+(A) := H(A^+), \quad H^-(A) := -H(A^-).$$

The properties of the set-functions H^+, H^- are given in the next theorem.

3.4 Thm. (a) $H^+ \in CA(\mathcal{A}, [0,\infty])$, $H^- \in CA(\mathcal{A}, [0,\infty))$.

(b) $H = H^+ - H^-$ & $|H| = H^+ + H^-$ on $\mathcal{A}_\mu$.

(c) $\forall A \in \mathcal{A}_\mu$, $-(1/e)\mu(A) \leq -H^-(A) \leq H(A) \leq H^+(A) \leq \infty$.

Recalling that for $\xi \in CA(\mathcal{R}, X)$, $\mathcal{R}$ a ring, $|\xi| \in CA(\mathcal{R}^{loc}, [0,\infty])$, cf. Dinculeanu [2, p. 35, No. 9], we see from 3.4(a) and Masani-Niemi [6, 2.4(b)] that

$$(\mathbf{3.5}) \qquad H^+, H^- \in CA(\mathcal{A}_\mu^{loc}, [0,\infty]) \ \& \ |H| = |H^+| + |H^-| \text{ on } \mathcal{A}_\mu^{loc}.$$

The last equality is an extension of the second equality in 3.4(b). Obviously, we could imitate the first equality in 3.4(b) to extend H beyond $\mathcal{A}_\mu$ by defining

$$\forall A \in \mathcal{A}_\mu^{loc}, \ \bar{H}(E) := |H^+|(E) - |H^-|(E),$$

except in the case where both $|H^+|(E)$, $|H^-|(E)$ are ∞. This suggests introducing in $\mathcal{A}_\mu^{loc}$ a subfamily $\mathcal{P}$, and a measure $\bar{H}$ on $\mathcal{P}$ defined by:

$$(3.6) \quad \begin{cases} \mathcal{P} := \{E: E \in \mathcal{A}_\mu^{loc} \ \& \ |H^+|(E) < \infty \text{ or } |H^-|(E) < \infty\} \\ \forall A \in \mathcal{P}, \quad \bar{H}(E) := |H^+|(E) - |H^-|(E). \end{cases}$$

The structure of the family $\mathcal{P}$ is given in the following lemma:

3.7 Lma. (a) $\mathcal{P}$ is a pre-ring in $\mathcal{A}_\mu^{loc}$ with the ideal property
$$E \in \mathcal{P} \ \& \ F \in \mathcal{A}_\mu^{loc} \ \Rightarrow \ E \cap F \in \mathcal{P}.$$

(b) $\mathcal{A}_\mu \subseteq \mathcal{P} \ \& \ \mathcal{P}^{loc} = \mathcal{A}_\mu^{loc}$.

It is clear, however, that if E, F in $\mathcal{P}$ fall in the different components of $\mathcal{P}$ given in (3.6), i.e. if

$$|H^+|(E) < \infty = |H^-|(E) \ \& \ |H^-|(F) < \infty = |H^+|(F),$$

then $E \cup F \notin \mathcal{P}$. Thus, almost never is $\mathcal{P}$ a ring or a lattice. However, the ideal property given in 3.7(a) endows $\mathcal{P}$ with the following property, absent in the familiar pre-rings of intervals encountered in elementary measure theory:

3.8 Cor. (a) $\forall \Omega_0 \in \mathcal{P}$, $\mathcal{P} \cap 2^{\Omega_0}$ is a σ-algebra over Ω_0.

(b) If $\Omega \in \mathcal{P}$, then $\mathcal{P} = \mathcal{A}_\mu^{loc} = a$ σ-algebra over Ω.

To turn to the properties of the extension $\bar{H}(\cdot)$ introduced in (3.6), it follows at once from the countable additivity of $|H^+|$, $|H^-|$ on $\mathcal{A}_\mu^{loc}$, cf. (3.5) and from the facts that on $\mathcal{A}_\mu$, $|H^+| - |H^-| = H^+ - H^- = H$, and $\mathcal{A}_\mu \subseteq \mathcal{P}$, cf. 3.4(b), 3.7(b), that

3.9 Thm. $\bar{H}$ is a CA measure on the pre-ring $\mathcal{P}$ to $[-\infty, \infty]$,[5] and $H \subseteq \bar{H}$.

It is obvious that since the ring $\mathcal{R}$ generated by $\mathcal{P}$ will have sets R for which $|H^+|(R) = |H^-|(R) = \infty$, that $\bar{H}$ has no extension to $\mathcal{R}$. Thus, the pre-ring $\mathcal{P}$ is the maximal domain for the existence of the relative entropy measure $\bar{H}(\cdot)$. Since, in general, $\Omega \notin \mathcal{P}$, we cannot speak of "the total entropy of P relative to μ". When, however, $\Omega \in \mathcal{P}$, we can define *the total entropy* to be $\bar{H}(\Omega)$.

The results 2.4(b),(c) on $\mathcal{A}_\mu$ extend to $\mathcal{P}$.

$$(3.10) \qquad \forall E, F \in \mathcal{P}, \ \bar{H}(E) = \left\{ {}^{+\infty}_{-\infty} \ \& \ E \subseteq F \ \Rightarrow \ \bar{H}(F) = \left\{ {}^{+\infty}_{-\infty}. \right.\right.$$

---

[5] The reader should note the definition of countable additivity of ξ on a pre-ring $\mathcal{P}$: only when $A_k \in \mathcal{P}$ are $\|$ and $A = \bigcup_{k=1}^\infty A_k \in \mathcal{P}$, is it required that the series $\sum_{k=1}^\infty \xi(A_k)$ converge unconditionally to $\xi(A)$.

$$(3.11) \begin{cases} \text{(a)} \ \Omega_0 \in \mathcal{P} \ \& \ \bar{H}(\Omega_0) < \infty \ \Rightarrow \ P \underset{\Omega_0}{\ll\!\!<} \mu; \\[2ex] \text{(b)} \ \Omega \in \mathcal{P} \ \& \ \bar{H}(\Omega) < \infty \ \Rightarrow \ P \ll\!\!< \mu \quad \text{(cf. 2.11).} \end{cases}$$

The result 2.4(c) subsumes Lemma 1 in Kallianpur [5, §2]. The result (3.11)(b) is a full-fledged extension of this to the pre-ring $\mathcal{P}$, i.e. to the maximal domain on which the relative entropy measure $\bar{H}(\cdot)$ is defined.

Finally, the concordance of the Lebesgue decompositions of P and H with respect to μ, established in Thm. 2.13, extends to $\mathcal{P}$:

3.12 Concordance Thm. *With the notation* H, H_a, H_b *introduced in Thm. 2.13, let* (i) $\mathcal{P}$ *and* $\bar{H}$ *be the pre-ring and extension of* $\bar{H}$ *defined in (3.6), and* $\mathcal{P}_a$ *and* $\overline{(H_a)}$*, and* $\mathcal{P}_b$ *and* $\overline{(H_b)}$ *be defined analogously for* H_a *and* H_b*,*
(ii) $\bar{H}_a := \overline{(H_a)}$ *on* $\mathcal{P}_a$*, and* $\bar{H}_b := \overline{(H_b)}$ *on* $\mathcal{P}_b$*, for brevity.*
Then (a) $\quad \mathcal{P} \subseteq \mathcal{P}_a \cap \mathcal{P}_b;$

(b)[6] $\quad \bar{H} = \bar{H}_a + \bar{H}_b, \quad \mathcal{N}_\mu \subseteq \mathcal{N}_{\bar{H}_a} \ \& \ \mu \perp\!\!\!\perp \bar{H}_b \perp\!\!\!\perp \bar{H}_b.$

4. Probability measures with densities

In this section we will make the following assumptions:

4.1 Assumptions. (a) P has a density f with respect to μ, i.e.

$$\exists f \in L_1(\Omega, \mathcal{A}, \mu; \mathbf{R}_{0+}) \ \ni \ \forall E \in \mathcal{A}, \ P(E) = \int_E f(\omega)\mu(d\omega).$$

(b) $\quad \varphi(x) := x \log x, \quad x \in [0, \infty).$

Under these assumptions, the entropy measure $\bar{H}(\cdot)$ itself can be shown to have a density, viz. $\varphi \circ f$, which is integrable on sets in the δ-ring:

$$\mathcal{D}_H := \{E : E \in \mathcal{A}_\mu^{\text{loc}} \ \& \ |H|(E) < \infty\}.$$

We require nets of averaging operators for arbitrary μ to do what martingales accomplish when μ is a probability. Write:

$$L_{1,\mu} = L_1(\Omega, \mathcal{A}, \mu; \mathbf{R}); \quad \forall f \in L_{1,\mu}, \ |f|_{1,\mu} := \int_\Omega |f(\omega)|\mu(d\omega);$$

$$S(\mathcal{A}_\mu, \mathbf{R}) := \{s : s \text{ is a } \mathcal{A}_\mu \text{ simple function on } \Omega \text{ to } \mathbf{R}\}.$$

Then $S(A_\mu, \mathbf{R})$ is an everywhere dense linear manifold in the Banach space $L_{1,\mu}$.

4.2 Def. (Averaging operator over A). $\forall A \in \mathcal{A}_\mu^+ \ \& \ \forall \pi \in \Pi_A, \ S_\pi^A$ is the operator on $L_{1,\mu}$ to $S(\mathcal{A}_\mu, \mathbf{R})$ defined by

---

[6] Since $\mathcal{P}$ is not a ring, and $\bar{H}$ on $\mathcal{P}$ does not extend to ring($\mathcal{P}$), the symbols $\ll\!\!<$, $\ll\!\!<^{\text{loc}}$ defined in 2.11 are not usable.

$$S_\pi^A(f) := \sum_{\Delta \in \pi_+} \left\{ \frac{1}{\mu(\Delta)} \int_\Delta |f(\omega)| \mu(d\omega) \right\} \chi_\Delta, \quad \forall f \in \mathcal{L}_{1,\mu}.$$

The basic properties of the operators S_π^A are given by the following proposition:

4.3 Prop. $\forall A \in \mathcal{A}_\mu + \ \& \ \forall \pi \in \Pi_A,\ S_\pi^A$ *is a linear contraction of norm 1 on* $\mathcal{L}_{1,\mu}$ *into itself, with* range $\subseteq \mathcal{S}(\mathcal{A}_\mu, \mathbf{R}) \subseteq \mathcal{L}_{1,\mu}$.

4.4 Thm. $\forall A \in \mathcal{A}_\mu +,\ \underset{\pi \downarrow}{\text{slim}}\ S_\pi^A = \chi_A$ *in the* $\mathcal{L}_{1,\mu}$ *topology.* (*Here* $\mu \in \Pi_A$.)

Since $\mathcal{L}_{1,\mu}$ is a metric space and therefore first countable, this result has the following sequential version:

4.5 Cor. *Let* $f \in \mathcal{L}_{1,\mu}$. *Then* $\forall A \in \mathcal{A}_\mu +,$ *there exists a sequence* $(\pi_k)_{k=1}^\infty$ *in* Π_A *such that* $\pi_{k+1} \prec \pi_k \ \& \ \underset{k \to \infty}{\lim} S_{\pi_k}^A(f) = \chi_A f$ *in the* $\mathcal{L}_{1,\mu}$ *topology.*

Appealing to the principle that a mean-convergent sequence contains a subsequence converging almost everywhere, we now extract from $(\pi_k)_{k=1}^\infty$ a sequence $(\pi_{k_i})_{i=1}^\infty$ that satisfies the condition of a.e. convergence for the mean convergent sequence $(S_{\pi_k}^A(f))_{k=1}^\infty$. For notational brevity we shall use the same symbol for this subsequence. We thus conclude:

4.6 Cor. *Let* $f \in \mathcal{L}_{1,\mu}$. *Then* $\forall A \in \mathcal{A}_\mu +,$ *there exists a sequence* $(\pi_k)_{k=1}^\infty$ *in* $\Pi_A,$ *and a set* $N \in \mathcal{A}$ *such that* $\mu(N) = 0$ (*the* π_k *and* N *dependent on* f) *such that* $\pi_{k+1} \prec \pi_k$ *and*

(a) $\forall \omega \in \Omega \backslash N,\ \underset{k \to \infty}{\lim}\ [S_{\pi_k}^A(f)](\omega) = \chi_A(\omega) f(\omega);$

(b) $\forall$ *continuous functions* ψ *on a closed interval* I *of* $\mathbf{R}$ *which includes* (Range f) $\cup \{0\}$ *and such that* $\psi(0) = 0,$

$$\forall \omega \in \Omega \backslash N,\ \underset{k \to \infty}{\lim}\ \psi[\{S_{\pi_k}^A(f)\}(\omega)] = \chi_A(\omega) \psi\{f(\omega)\}.$$

The importance of Cor. 4.6(b) stems from the circumstance that when f satisfies condition 4.1(a), then $\forall A \in \mathcal{A}_\mu + \ \& \ \forall \pi \in \Pi_A,$

(4.7) $$S_\pi^A(f) = \sum_{\Delta \in \pi_+} \frac{P(\Delta)}{\mu(\Delta)} \chi_\Delta = \sum_{\Delta \in \pi+} \rho(\Delta) \chi_\Delta, \quad \text{cf. 2.1(d),}$$

for now $\pi_+ = \pi+$, since $P \ll \mu$ and $\pi^+ \subseteq \pi_+$. It obviously follows that for the function φ on $[0,\infty)$ given by 4.1(b),

(4.8) $$\varphi \circ S_\pi^A(f) = \sum_{\Delta \in \pi+} \varphi\{\rho(\Delta)\} \chi_\Delta.$$

An integration on A with respect to μ shows at once that

(4.9) $\forall A \in \mathcal{A}_\mu+ \ \& \ \forall \pi \in \Pi_A, \quad \int_A \varphi[\{S^A_\pi(f)\}(\omega)]\mu(d\omega) = H_\pi(A).$

With the aid of these results, we can show that

(4.10) $\forall A \in \mathcal{A}_\mu+, \quad -(1/e)\mu(A) \leq \int_A \varphi\{f(\omega)\}\mu(d\omega) \leq H(A),$

(4.11) $\forall A \in \mathcal{A}_\mu+, \qquad H(A) \leq \int_A \varphi\{f(\omega)\}\mu(d\omega),$

where in the last, we appeal to the convexity of φ. Since for $A \in \mathcal{A}_\mu \backslash \mathcal{A}_\mu+$, both $H(A)$ and the integral are zero. We arrive at the following theorem:

4.12 Thm. $\forall A \in \mathcal{A}_\mu, \ H(A) = \int_A \varphi\{f(\omega)\}\mu(d\omega).$

To extend the last equality to sets $E \in \mathcal{P}$, consider the measure M defined on $\mathcal{A}_\mu$ by

(4.13) $M(A) := \int_A \varphi\{f(\omega)\}\mu(d\omega), \quad A \in \mathcal{A}_\mu.$

Since each $M(A) \geq -(1/e)\mu(A)$, we see that, like H, M is a CA measure on $\mathcal{A}_\mu$ to $(-\infty, \infty]$. We can therefore do for M what we did for H in §3, viz. take its Hahn-Jordan components M^+, M^- and then their total variations $|M^+|, |M^-|$. Moreover, we can define the analogue $\mathcal{Q}$ of the family $\mathcal{P}$ of (3.6) by

$$Q := \{E : E \in \mathcal{A}_\mu^{loc} \ \& \ |M^+|(E) < \infty \ \text{or} \ |M^-|(E) < \infty\},$$

and then define the analogue $\bar{M}$ of $\bar{H}$ by

$$\forall E \in \mathcal{Q}, \quad \bar{M}(E) := |M^+|(E) - |M^-|(E).$$

Since from (4.13) it follows classically that $\forall E \in \mathcal{A}_\mu^{loc}$,

(4.14) $|M^+|(E) = \int_E \left[\varphi\{f(\omega)\}\right]^+ \mu(d\omega), \quad |M^-|(E) = \int_E \left[\varphi\{f(\omega)\}\right]^- \mu(d\omega),$

we see that

$$\forall E \in \mathcal{Q}, \quad \bar{M}(E) = \int_E \left(\left[\varphi\{f(\omega)\}\right]^+ - \left[\varphi\{f(\omega)\}\right]^-\right)\mu(d\omega)$$

(4.15) $= \int_E \varphi\{f(\omega)\}\mu(d\omega) \in [-\infty, \infty].$

However, Thm. 4.12 tells us that $M = H$ on $\mathcal{A}_\mu$. It follows at once that $M^+ = H^+$ and $M^- = H^-$ on $\mathcal{A}_\mu$, and thence (cf. Masani-Niemi [6, 2.4(b)], that $|M^+| = |H^+|$ and $|M^-| = |H^-|$ on $\mathcal{A}_\mu^{loc}$. This in turn entails that $Q = \mathcal{P}$ and $\bar{M} = \bar{H}$. By combining 4.12, (4.13) and (4.14), we thus get:

4.16 Thm. (a) $\forall E \in \mathcal{A}_\mu^{loc}, \ E \in \mathcal{P}$ iff

$$|H^+|(E) = \int_E \left[\varphi\{f(\omega)\}\right]^+ \mu(d\omega) < \infty \ \text{or} \ |H^-|(E) = \int_E \left[\varphi\{f(\omega)\}\right]^- \mu(d\omega) < \infty.$$

(b) $\forall E \in \mathcal{P}, \quad \bar{H}(E) = \int_E \varphi\{f(\omega)\}\mu(d\omega) \in [-\infty,\infty]$.

(c) $\forall E \in \mathcal{D}_H, \quad \bar{H}(E) = \int_E \varphi\{f(\omega)\}\mu(d\omega) \in \mathbf{R}$.

By virtue of (a), the results (b) and (c) are the best possible.

5. Other topics

To see how the measures H and $\bar{H}$ are affected by a transformation T of the space Ω, assume that T is a one-one function on Ω onto Ω and T is $\mathcal{A}, \mathcal{A}$ measurable, i.e. $\forall E \in \mathcal{A}, T^{-1}(E) \in \mathcal{A}$. It then follows quite easily that

5.1 Triv. If T is μ-measure preserving, then

$$H(P|\mu) = H(P|\mu \circ T^{-1}) \text{ on } \mathcal{A}_\mu \quad \& \quad \bar{H}(P|\mu) = \bar{H}(P|\mu \circ T^{-1}) \text{ on } \mathcal{P}.$$

R. Vallée [10, p.405] has remarked that the Shannon-Wiener entropy for probabilities over $\Omega = \mathbf{R}$ is not invariant under all measurable automorphisms of Ω, and consequently lacks the intrinsicality of the Shannon entropy in the discrete case, which is so invariant. Both cases, however, are subject to the universal result 5.1. Their apparent divergence stems from the special circumstance that *all* one-one transformations on Ω onto Ω preserve the cardinality measure, i.e. preserve the underlying measure μ in the discrete case.

The entropy measure has nice properties on the product space. For instance, if P_i, μ_i are on $\mathcal{A}_i$, $i = 1,2$ and $\mathcal{A} := \sigma\text{-alg}(\mathcal{A}_1 \times \mathcal{A}_2)$. Then on the δ-ring, $\mathcal{A}_{\mu_1 \times \mu_2}$, we have

$$H(P_1 \times P_2 | \mu_1 \times \mu_2) = H(P_1|\mu_1) \times P_2 + P_1 \times H(P_2|\mu_2).$$

In case $\mu_i(\Omega_i) < \infty$, we get the known result for the total entropy:

$$H(P_1 \times P_2 | \mu_1 \times \mu_2)(\Omega_1 \times \Omega_2) = H(P_1|\mu_1)(\Omega_1) + H(P_2|\mu_2)(\Omega_2).$$

References

[1]   Cziszar, I., On generalized entropy, *Studia Scientarum Math. Hungarica* **4**, 1969, 401–419.

[2]   Dinculeanu, N., *Vector Measures*, Pergamon, Oxford, 1953.

[3]   Gelfand, I. M., Kolmogorov, N. A. and Yaglom, A. M., On the general definition of the amount of information, *DAN USSR*, 111, 4, 1956, 745–748. (In Russian.)

[4]   Halmos, P. R., *Measure Theory*, Van Nostrand, New York, 1950.

[5]   Kallianpur, G., On the amount of information contained in a σ-field, *Contributions to Probability and Statistics: Essays in Honor of Harold Hotelling* (Eds. I. Olkin and S. G. Ghurye), Stanford Univ. Press, Stanford, 1960, 265–273.

[6]   Masani, P. and Niemi, H., The integration of Banach space valued
        measures and the Tonelli-Fubini theorems. Part I: Scalar-valued measures
        on δ-rings, *Advances in Mathematics* **73**, 1989, 204-241.
[7]   Pinsker, M. S., *Information and Information Stability of Random
        Variables and Processes*, (Russian Ed., 1960), Holden Day, San
        Francisco, 1964.
[8]   Rosenblatt-Roth, M., The concept of entropy in probability theory and its
        application in the theory of information transmission through
        communication channels, *Theory of Probability and its Applications*,
        Vol. IX, No. 2, 1964, 212–235.
[9]   Shannon, C. E. and Weaver, W., *The Mathematical Theory of
        Communication*, Univ. of Illinois Press, Urbana, 1949.
[10]   Vallée, R., Information entropy and state observation of a dynamical
        system, *Uncertainty in Knowledge-Based Systems* (Eds. B. Bouchon and
        R. R. Yager), Springer-Verlag, Berlin, 1987, 403–405
[11]   Von Neumann, J., *Functional Operators*, I, Princeton Univ. Press,
        Princeton, NJ, 1950.
[12]   Wiener, N., *Cybernetics*, 2nd edition, MIT Press, 1961.

Department of Mathematics and Statistics
University of Pittsburgh
Pittsburgh, PA 15260, USA

Probability Bounds, Multivariate Normal Distribution and An Integro-Differential Inequality for Random Vectors

B.L.S. PRAKASA RAO

Abstract

In the light of an inequality derived by Chernoff (1981), a characterization of the normal distribution was obtained by Borovkov and Utev (1983). Prakasa Rao and Sreehari (1986) derived a multivariate analogue characterizing the multivariate normal distribution. A bound is obtained for the variation between the probability distribution of a random vector with mean zero and a finite covariance matrix $\sum$ and the corresponding multivariate normal distribution with mean zero and the same covariance matrix $\sum$. As applications, characterization of a multivariate normal distribution due to Prakasa Rao and Sreehari (1986) is derived and a multivariate limit theorem is given. Results obtained extend the work of Utev (1989). Inter alia, an integro-differential inequality valid for random vectors with finite covariance matrix is obtained.

1 Introduction

Suppose ξ is a random variable with the standard normal distribution. It is easy to check that

$$E[\xi \; g(\xi)] = E[g'(\xi)] \tag{1.1}$$

for any differentiable function g with $E|g'(\xi)| < \infty$. Stein (1973) proved that, if (1.1) holds for all differentiable functions g for a random variable ξ, then ξ has to be standard normal. This characterization theorem has been generalized to one parameter exponential families of both discrete and continuous type in Prakasa Rao (1979). It is easy to check that ξ has $N(0, \sigma^2)$, that is, the normal distribution with mean 0 and variance σ^2 iff for all differentiable functions h with $E|h'(\xi)| < \infty$,

$$E[\xi \; h(\xi)] = \sigma^2 E[h'(\xi)] \, . \tag{1.2}$$

Suppose $\boldsymbol{\xi}$ is $N_k(\boldsymbol{O}, \Sigma)$ where Σ is positive definite. Here $N_k(\boldsymbol{O}, \Sigma)$ denotes the k-variate normal distribution with mean vector $\boldsymbol{O}$ and covariance matrix Σ. It is well known that $\boldsymbol{\xi}$ is $N_k(\boldsymbol{O}, \Sigma)$ iff $\boldsymbol{\lambda}^T \boldsymbol{\xi}$ is $N(0, \boldsymbol{\lambda}^T \Sigma \boldsymbol{\lambda})$ for all $\boldsymbol{\lambda} \in$

R^k. In view of this fact, it is easy to check that ξ is $N_k(O, \Sigma)$ iff for all differentiable functions h,

$$E[\lambda^T \xi \, h(\lambda^T \xi)] = \lambda^T \Sigma \lambda \, E[h'(\lambda^T \xi)], \lambda \in R^k \, . \tag{1.3}$$

Chernoff (1981) proved that if ξ is $N(0, \sigma^2)$, then for any differentiable function g,

$$\text{Var}[g(\xi)] \leq E[g'(\xi)]^2 \, \text{Var}(\xi) \, . \tag{1.4}$$

Let ξ be any random variable with mean zero and $0 < \text{Var} \, \xi < \infty$. Define

$$U_\xi = \sup_{g \in \zeta} \frac{\text{Var}[g(\xi)]}{\text{Var} \, \xi \, E[g'(\xi)]^2} \tag{1.5}$$

where ζ is the class of all absolutely continuous functions for which $E[g(\xi)]^2 < \infty$ and $0 < E[g'(\xi)]^2 < \infty$. It is clear that $U_\xi \geq 1$ from (1.5). Infact, for linear functions $g(\cdot)$, $\text{Var} \, [g(\xi)] = \text{Var} \, [\xi] \, E[g'(\xi)]^2$. Borovkov and Utev (1983) prove that $U_\xi = 1$ iff ξ is $N(0, \sigma^2)$ for some σ^2. This suggests using $(U_\xi - 1)$ as a measure of the deviation of the distribution of ξ from normality. For any two probability measures P_1 and P_2 on the real line, define

$$\rho(P_1, P_2) = \sup_{A \in \mathcal{B}} |P_1(A) - P_2(A)| \tag{1.6}$$

where $\mathcal{B}$ is the σ-algebra of Borel subsets of R. Let Φ be the probability measure on R corresponding to the standard normal distribution. Utev (1989) proved the following theorem.

Theorem 1.1 : (Utev (1989)) : Suppose $E\xi = 0$ and $\text{Var} \, \xi = 1$. Then

$$\rho(P_\xi, \Phi) \leq 3[U_\xi - 1]^{1/2} \tag{1.7}$$

where P_ξ is the probability measure corresponding to ξ.

Chen (1982) generalized Chernoff's inequality, given by (1.4), to the multivariate normal distribution. He proved that if $\boldsymbol{\xi}$ is a k-variate normal random vector with mean O and a positive definite covariance matrix Σ, then for any totally differentiable mapping $g(\cdot)$ from R^k into R,

$$\text{Var}[g(\boldsymbol{\xi})] \leq E[\nabla g(\boldsymbol{\xi})^T \Sigma \nabla g(\boldsymbol{\xi})] \tag{1.8}$$

where $\nabla g(\boldsymbol{\xi})$ denotes the gradient of $g(\cdot)$. Let $\boldsymbol{\xi}$ be a random vector with mean zero vector and a positive definite covariance matrix Σ. Define

$$U_{\boldsymbol{\xi}} = \sup_{g \in \zeta} \frac{\text{Var}[g(\boldsymbol{\xi})]}{E[\nabla g(\boldsymbol{\xi})^T \Sigma \nabla g(\boldsymbol{\xi})]} \tag{1.9}$$

where ζ is the class of all totally differentiable functions g on R^k such that $0 < E[\nabla g(\xi)^T \Sigma \nabla g(\xi)] < \infty$. It can be checked that $U_\xi \geq 1$ from (1.9). Prakasa Rao and Sreehari (1986) proved that $U_\xi = 1$ iff ξ is $N_k(O, \Sigma)$.

Our aim in this paper to obtain a bound analogous to (1.7) for k-dimensional random vectors ξ with mean O and a positive definite covariance matrix Σ. We obtain a bound on

$$\sup_{\lambda \in R^k, W \in \mathcal{B}} |Pr(\lambda^T \xi \in W) - Pr(\lambda^t Z \in W)| \qquad (1.10)$$

where Z is $N_k(O, \Sigma)$ and $\mathcal{B}$ is the σ-algebra of Borel subsets of R.

2 An Integro–Differential Inequality

Let ξ be a random vector with covariance matrix Σ_0. Define, for any positive definite matrix Σ and σ-finite measure ν on $(R^k, \mathcal{B}_k)$ satisfying $\nu(\{0\}) = 0$,

$$U_\xi(\Sigma, \nu) = \sup_{g \in \zeta} \frac{\text{Var}[g(\xi)]}{E[\nabla g(\xi)^T \Sigma \nabla g(\xi)] + \int_{R^k} E[\Delta_\eta g(\xi)]^2 \nu(d\eta)} \qquad (2.1)$$

where $\Delta_\eta g(\xi) = g(\xi + \eta) - g(\xi)$, ζ is the class of all totally differentiable functions g on R^k such that $0 \leq \text{Var}[g(\xi)] < \infty$ and

$$0 < E[\nabla g(\xi)^T \Sigma \nabla g(\xi)] + \int_{R^k} E[\Delta_\eta g(\xi)]^2 \nu(d\eta) < \infty . \qquad (2.2)$$

It is easy to see that $U_\xi(\Sigma, \nu) \geq 1$. Let

$$\tilde{g}(\xi) = \lambda^T \xi + \mu\, g(\xi) \qquad (2.3)$$

where $\mu \in R$. Then

$$\nabla \tilde{g} = \lambda + \mu \nabla g . \qquad (2.4)$$

Note that

$$\text{Var}[\tilde{g}(\xi)] \leq U_\xi(\Sigma, \nu)\{E[(\lambda^T + \mu \nabla g^T)\Sigma(\lambda + \mu \nabla g)]$$

$$+ \int_{R^k} E[\lambda^T \eta + \mu \Delta_\eta g(\xi)]^2 \nu(d\eta)\} \qquad (2.5)$$

from the definition of $U_\xi(\Sigma, \nu)$ and

$$\begin{aligned} \text{Var}[\tilde{g}(\xi)] &= \text{Var}[\lambda^T \xi + \mu\, g(\xi)] \\ &= \lambda^T \Sigma_0 \lambda + 2\mu\, \text{Cov}(\lambda^T \xi,\, g(\xi)) + \mu^2 \text{Var}[g(\xi)] . \end{aligned} \qquad (2.6)$$

Hence
$$\lambda^T \Sigma_0 \lambda + 2\mu \text{ Cov }(\lambda^T \xi, g(\xi)) + \mu^2 \text{ Var}[g(\xi)]$$

$$\begin{aligned}
\leq \quad & U\{\lambda^T \Sigma \lambda + 2\mu \, E[\lambda^T \Sigma \nabla g] + \mu^2 E[\nabla g^T \Sigma \nabla g] \\
& + \int_{R^k} (\lambda^T \eta)^2 \nu(d\eta) + 2E \int_{R^k} \mu(\lambda^T \eta)(\Delta_\eta g(\xi)) \nu(d\eta) \\
& + \mu^2 E(\int_{R^k} (\Delta_\eta g(\xi))^2 \nu(d\eta))\}
\end{aligned} \tag{2.7}$$

with the obvious notations for $U, \nabla g$ etc.

Let Σ be chosen so that

$$\lambda^T \Sigma_0 \lambda = \lambda^T \Sigma \lambda + \int_{R^k} (\lambda^T \eta)^2 \nu(d\eta). \tag{2.8}$$

Then

$$\lambda^T \Sigma_0 \lambda + 2\mu \text{ Cov}(\lambda^T \xi, g(\xi)) + \mu^2 \text{Var}[g(\xi)]$$

$$\begin{aligned}
\leq \quad & U\{\mu^2 E[\nabla g^T \Sigma \nabla g] + 2\mu E[\lambda^T \Sigma \nabla g] \\
& + \lambda^T \Sigma_0 \lambda + 2\mu \, E(\int_{R^k} (\lambda^T \eta)(\nabla_\eta g(\xi)) \nu(d\eta)) \\
& + \mu^2 E(\int_{R^k} (\nabla_\eta g(\xi))^2 \nu(d\eta))\}
\end{aligned} \tag{2.9}$$

and hence

$$\mu^2 \{\text{Var}(g(\xi)) - UE[\nabla g^T \Sigma \nabla g] - U \int_{R^k} E(\nabla_\eta g(\xi))^2 \nu(d\eta)\}$$

$$+2\mu\{E[(\lambda^T \xi - E(\lambda^T \xi))g(\xi)] - UE[\lambda^T \Sigma \nabla g] - UE(\int_{R^k} (\lambda^T \eta)(\nabla_\eta g(\xi)) \nu(d\eta))\}$$

$$\leq \lambda^T \, \Sigma_0 \lambda \, (U - 1) \, . \tag{2.10}$$

Let us rewrite the above inequality in the form

$$\mu^2 S + 2\mu W \leq \lambda^T \Sigma_0 \lambda \, (U - 1) \equiv C(say) \tag{2.11}$$

with the obvious notation. Note that $S \leq 0$ by the definition of U. Since the relation (2.11) holds for all $\mu \in R$, it follows that

$$W^2 + SC \leq 0 \tag{2.12}$$

or equivalently

$$
\begin{aligned}
W^2 &\leq (U-1)\,\boldsymbol{\lambda}^T\Sigma_0\boldsymbol{\lambda}\,|S| \\
&\leq (U-1)\,\boldsymbol{\lambda}^T\Sigma_0\boldsymbol{\lambda}\,\{UE[\nabla g^T\Sigma\nabla g] \\
&\qquad\qquad +U\int_{R^k}E(\nabla_{\boldsymbol{\eta}}g(\boldsymbol{\xi}))^2\nu(d\boldsymbol{\eta})-\mathrm{Var}(g(\boldsymbol{\xi}))\} \\
&\leq U(U-1)\,\boldsymbol{\lambda}^T\Sigma_0\,\boldsymbol{\lambda}\{UE[\nabla g^T\Sigma\nabla g] \\
&\qquad\qquad +\int_{R^k}E(\nabla_{\boldsymbol{\eta}}g(\boldsymbol{\xi}))^2\nu(d\boldsymbol{\eta})\}\,.
\end{aligned}
\tag{2.13}
$$

This result can now be stated as follows .

Theorem 2.1: Suppose $\boldsymbol{\xi}$ is a k-dimensional random vector with covariance matrix Σ_0. Define the positive definite matrix Σ by the relation

$$
\boldsymbol{\lambda}^T\Sigma_0\boldsymbol{\lambda}=\boldsymbol{\lambda}^T\Sigma\boldsymbol{\lambda}+\int_{R^k}(\boldsymbol{\lambda}^T\boldsymbol{\eta})^2\nu(d\boldsymbol{\eta}),\boldsymbol{\lambda}\in R^k.
\tag{2.14}
$$

Then, for any totally differentiable function $g(\cdot)$ on R^k such that

$$
0<E[\nabla g^T\Sigma\nabla g]+\int_{R^k}E(\nabla_{\boldsymbol{\eta}}g(\boldsymbol{\xi}))^2\nu(d\boldsymbol{\eta})<\infty,
\tag{2.15}
$$

the following inequality holds:

$$
|E[(\boldsymbol{\lambda}^T\boldsymbol{\xi}-E(\boldsymbol{\lambda}^T\boldsymbol{\xi}))\,g(\boldsymbol{\xi})]-UE[\boldsymbol{\lambda}^T\Sigma\nabla g]-UE(\int_{R^k}(\boldsymbol{\lambda}^T\boldsymbol{\eta})(\nabla g(\boldsymbol{\xi}))\nu(d\boldsymbol{\eta}))|^2
$$

$$
\leq U(U-1)\,\boldsymbol{\lambda}^T\Sigma_0\,\boldsymbol{\lambda}\,[E(\nabla g^T\Sigma\nabla g)+\int_{R^k}E(\nabla_{\boldsymbol{\eta}}g(\boldsymbol{\xi}))^2\nu(d\boldsymbol{\eta})]\,.
\tag{2.16}
$$

Suppose $E(\boldsymbol{\xi})=O$ and the covariance matrix of $\boldsymbol{\xi}$ is Σ_0. Choose ν to be null measure. Then $\Sigma=\Sigma_0$ and Theorem 2.1 implies that

$$
|E(\boldsymbol{\lambda}^T\boldsymbol{\xi}\,g(\boldsymbol{\xi}))-UE[\boldsymbol{\lambda}^T\Sigma_0\nabla g]|^2\leq[U(U-1)]\boldsymbol{\lambda}^T\Sigma_0\boldsymbol{\lambda}\,(\nabla g^T\Sigma_0\nabla g)
\tag{2.17}
$$

and hence
$$
|E(\boldsymbol{\lambda}^T\boldsymbol{\xi}\,g(\boldsymbol{\xi}))-E[\boldsymbol{\lambda}^T\Sigma_0\nabla g]|
$$

$$
\begin{aligned}
&\leq [U(U-1)]^{1/2}(\boldsymbol{\lambda}^T\Sigma_0\boldsymbol{\lambda})^{1/2}(E(\nabla g^T\Sigma_0\nabla g))^{1/2} \\
&\quad +(U-1)E(\boldsymbol{\lambda}^T\Sigma_0\nabla g) \\
&\leq [U(U-1)]^{1/2}(\boldsymbol{\lambda}^T\Sigma_0\boldsymbol{\lambda})^{1/2}(E(\nabla g^T\Sigma_0\nabla g))^{1/2} \\
&\quad +(U-1)[E(\boldsymbol{\lambda}^T\Sigma_0\nabla g)^2]^{1/2}
\end{aligned}
\tag{2.18}
$$

$$
\begin{aligned}
&= [U(U-1)]^{1/2}(\lambda^T \Sigma_0 \lambda)^{1/2}(E(\nabla g^T \Sigma_0 \nabla g))^{1/2} \\
&\quad + (U-1)[E(\lambda^T \Sigma_0^{1/2} \Sigma_0^{1/2} \nabla g)^2]^{1/2} \\
&\leq [U(U-1)]^{1/2}(\lambda^T \Sigma_0 \lambda)^{1/2}(E(\nabla g^T \Sigma_0 \nabla g))^{1/2} \\
&\quad + (U-1)[(\lambda^T \Sigma_0 \lambda)E(\nabla g^T \Sigma_0 \nabla g)]^{1/2} \\
&= (U-1)^{1/2}[U^{1/2} + (U-1)^{1/2}][\lambda^T \Sigma_0 \lambda]^{1/2}[E(\nabla g^T \Sigma_0 \nabla g)]^{1/2}
\end{aligned}
$$

$$(2.19)$$

Hence we have the following theorem.

Theorem 2.2 : Suppose ξ is a random vector with mean O and a covariance matrix Σ_0. Define

$$
U \equiv U_\xi(\Sigma_0, 0) = \sup_{g \in \zeta} \frac{\text{Var}[g(\xi)]}{E[\nabla g(\xi)^T \Sigma_0 \nabla g(\xi)]} \tag{2.20}
$$

where ζ is the family of all totally differentiable functions g on R^k such that $0 < E[\nabla g(\xi)^T \Sigma_0 \nabla g(\xi)] < \infty$. Then, for any $\lambda \in R^k$,

$$
|E(\lambda^T \xi\, g(\xi)) - E[\lambda^T \Sigma_0 \nabla g]|
$$

$$
\leq (U-1)^{1/2}[U^{1/2} + (U-1)^{1/2}](\lambda^T \Sigma_0 \lambda)^{1/2}(E[\nabla g^T \Sigma_0 \nabla g])^{1/2} \ . \tag{2.21}
$$

As a special case of Theorem 2.2, Choose $g(\xi) = h(\lambda^T \xi)$ where $\lambda \in R^k$ and $h(\cdot)$ is any real-valued differentiable function. Then $\nabla g = \lambda\, h'(\lambda^T \xi)$ where h' denotes the derivative of h and the relation (2.21) implies that

$$
E(\lambda^T \xi\, h(\lambda^T \xi)) - \lambda^T \Sigma_0 \lambda\, E(h'(\lambda^T \xi))|
$$

$$
\begin{aligned}
&\leq (U-1)^{1/2}[U^{1/2} + (U-1)^{1/2}](\lambda^T \Sigma_0 \lambda)^{1/2}(\lambda^T \Sigma_0 \lambda E[h'(\lambda^T \xi)]^2)^{1/2} \\
&= (U-1)^{1/2}[U^{1/2} + (U-1)^{1/2}](\lambda^T \Sigma_0 \lambda)(E[h'(\lambda^T \xi)]^2)^{1/2}
\end{aligned} \tag{2.22}
$$

for any $\lambda \in R^k$.

3 Probability Bounds

Let

$$
f(\alpha) = 2(\alpha-1)^{1/2}[\alpha^{1/2} + (\alpha-1)^{1/2}], \ \alpha \geq 1 \ . \tag{3.1}
$$

It needs a bit of algebra to show that

$$
f(\alpha) \leq 3(\alpha-1)^{1/2} \quad \text{for} \quad 1 \leq \alpha \leq 9/8 \tag{3.2}
$$

and

$$f(\alpha) \geq 1 \quad \text{for} \quad \alpha \geq 9/8 . \tag{3.3}$$

Let W be any Borel Set on the real line and $\chi_W(t)$ be its indicator function. Define, for $\lambda \in R^k, \lambda \neq 0$,

$$h_W(t) = \exp\{\frac{t^2}{2\lambda^T\Sigma_0\lambda}\} \int_{-\infty}^t (\chi_W(z) - \Phi_{\lambda^T\Sigma_0\lambda}(W)) \exp\{-\frac{z^2}{2\lambda^T\Sigma_0\lambda}\}dz \tag{3.4}$$

where $\Phi_{\tilde{\lambda}_T\Sigma_0\lambda}(W)$ is the probability of the set W under the normal distribution with mean 0 and variance $\lambda^T\Sigma_0\lambda$. It is easy to check that

$$h'_W(t) = (\chi_W(t) - \Phi_{\lambda^T\Sigma_0\lambda}(W)) + \frac{t}{\lambda^T\Sigma_0\lambda}h_W(t) \tag{3.5}$$

and hence

$$\lambda^T\Sigma_0\lambda\, h'_W(t) = \lambda^T\Sigma_0\lambda\, (\chi_W(t) - \Phi_{\lambda^T\Sigma_0\lambda}(W)) + t\, h_W(t) \tag{3.6}$$

or equivalently

$$\lambda^T\Sigma_0\lambda\, h'_W(t) - t\, h_W(t) = \lambda_T\Sigma_0\lambda\, (\chi_W(t) - \Phi_{\lambda^T\Sigma_0\lambda}(W)). \tag{3.7}$$

Now, for $\lambda \in R^k, \lambda \neq 0$ and $W \in \mathcal{B}$, Borel σ-algebra on R, we have
$$|Pr(\lambda^T\xi \in W) - \Phi_{\lambda^T\Sigma_0\lambda}(W)|$$

$$= |E[\chi_W(\lambda^T\xi)] - \Phi_{\lambda^T\Sigma_0\lambda}(W)| \tag{3.8}$$

$$= (\lambda^T\Sigma_0\lambda)^{-1}|\lambda^T\Sigma_0\lambda\, E[h'_W(\lambda^T\xi)] - E[\lambda^T\xi\, h_W(\lambda^T\xi)]|\ \text{(by 3.7)}$$

$$\leq (U-1)^{1/2}[U^{1/2} + (U-1)^{1/2}](E[h'(\lambda^T\xi)]^2)^{1/2}$$

$$\leq 2(U-1)^{1/2}[U^{1/2} + (U-1)^{1/2}] \tag{3.9}$$

since $\sup_t |h'_W(t) \leq 2$. Therefore

$$|Pr(\lambda^T\xi \in W) - Pr(\lambda^T Z \in W)| \leq 2 \min\ [(U-1)^{1/2}\{U^{1/2} + (U-1)^{1/2}\}, 1] \tag{3.10}$$

where Z is $N_k(O, \Sigma_0)$. Relations (3.2), (3.3) and (3.10) prove that

$$|Pr(\lambda^T\xi \in W) - Pr(\lambda^T Z \in W)| \leq 3\, (U-1)^{1/2} \tag{3.11}$$

for all Borel sets W and for all $\lambda \in R^k \neq 0$. Inequality holds trivially for $\lambda = 0$. Hence

$$\sup_{\lambda \in R^k, W \in \mathcal{B}} |Pr(\lambda^T\xi \in W) - Pr(\lambda^T Z \in W)|$$

$$\leq 3\,(U_{\boldsymbol{\xi}} - 1)^{1/2} \qquad (3.12)$$

where $U_{\boldsymbol{\xi}}$ is as defined by (2.20). This result proves the following main theorem of the paper.

Theorem 3.1 : Let $\boldsymbol{\xi}$ be a k-dimensional random vector with mean zero and a positive definite covariance matrix Σ_0. Let $\boldsymbol{Z}$ be a random vector with $N_k(\boldsymbol{O}, \Sigma_0)$ as its probability distribution. Then

$$\sup_{\boldsymbol{\lambda} \in R^k} \sup_{W \in \mathcal{B}} |Pr(\boldsymbol{\lambda}^T \boldsymbol{\xi} \in W) - Pr(\boldsymbol{\lambda}^T \boldsymbol{Z} \in W)| \leq 3\,(U_{\boldsymbol{\xi}} - 1)^{1/2} \qquad (3.13)$$

where $U_{\boldsymbol{\xi}}$ is as defined by (2.20).

4   Applications

(i) As a Corollay to Theorem 3.1, we obtain that $U_{\boldsymbol{\xi}} = 1$ implies that

$$Pr(\boldsymbol{\lambda}^T \boldsymbol{\xi} \in W) = Pr(\boldsymbol{\lambda}^T \boldsymbol{Z} \in W) \qquad (4.1)$$

for all $\boldsymbol{\lambda} \in R^k, W \in \mathcal{B}$. Hence $\boldsymbol{\lambda}^T \boldsymbol{\xi}$ is $N(0, \boldsymbol{\lambda}^T \Sigma_0 \boldsymbol{\lambda})$ for every $\boldsymbol{\lambda} \in R^k$ and therefore $\boldsymbol{\xi}$ is $N_k(\boldsymbol{O}, \Sigma_0)$. It is obvious that $U_{\boldsymbol{\xi}} = 1$ if $\boldsymbol{\xi}$ is $N_k(\boldsymbol{O}, \Sigma_0)$ by the inequality (1.8) and by choosing g to be a linear function in ζ. Hence $U_{\boldsymbol{\xi}} = 1$ iff $\boldsymbol{\xi}$ is $N_k(\boldsymbol{O}, \Sigma_0)$ which was proved earlier by Prakasa Rao and Sreehari (1986) by a different method.

Theorem 4.1 : $U_{\boldsymbol{\xi}} = 1$ iff $\boldsymbol{\xi}$ is $N_k(\boldsymbol{O}, \Sigma_0)$ where $U_{\boldsymbol{\xi}}$ is as defined by (2.20).

(ii) Let us consider another application. Suppose $\boldsymbol{\xi}_n$ is a k-dimensional random vector with mean $\boldsymbol{\mu}_n$ and a positive-definite covariance matrix Σ_n. Define

$$\boldsymbol{\xi}_n^* = \Sigma_n^{-1/2}(\boldsymbol{\xi}_n - \boldsymbol{\mu}_n) \qquad (4.2)$$

and $\boldsymbol{Z}^*$ be a random vector with $N_k(\boldsymbol{O}, I_k)$ as its probability distribution. Applying Theorem 3.1, we obtain that

$$\sup_{\boldsymbol{\lambda} \in R^k, W \in \mathcal{B}} |Pr(\boldsymbol{\lambda}^T \boldsymbol{\xi}_n^* \in W) - Pr(\boldsymbol{\lambda}^T \boldsymbol{Z}^* \in W)|$$

$$\leq 3\,(U(\boldsymbol{\xi}_n^*, I_k) - 1)^{1/2} \qquad (4.3)$$

where we write $U(\boldsymbol{\xi}, \Sigma)$ for $U_{\xi}(\Sigma, 0)$ defined by (2.19). From the definition of $U(\boldsymbol{\xi}_n^*, I_k)$, it is easy to see that

$$U(\boldsymbol{\xi}_n^*, I_k) = U(\boldsymbol{\xi}_n, \Sigma_n).$$

Hence

$$\sup_{\lambda \in R^k, W \in \mathcal{B}} |Pr(\lambda^T \xi_n^* \in W) - Pr(\lambda^T Z^* \in W)|$$

$$\leq 3 \left(U(\xi_n, \Sigma_n) - 1\right)^{1/2}. \tag{4.4}$$

As a corollary, we have the following theorem.

Theorem 4.2 : If $U(\xi_n, \Sigma_n) \to 1$ as $n \to \infty$, then

$$\Sigma_n^{-1/2}(\xi_n - \mu_n) \xrightarrow{\mathcal{L}} N_k(O, I_k). \tag{4.5}$$

Proof : Relation (4.3) implies that $\lambda^T \xi_n^* \xrightarrow{\mathcal{L}} \lambda^T Z^*$ for all $\lambda \in R^k$ and hence $\xi_n^* \xrightarrow{\mathcal{L}} Z^*$.

(iii) Suppose $X_i, i \geq 1$ are independent ℓ-dimensional random vectors with finite covariance matrices $\Sigma_i, i \geq 1$. Let $\{A_i, i \geq 1\}$ be a set of matrices of order $k \times l$. Define $\xi_n = \Sigma_{i=1}^n A_i X_i,$.

Let Γ_n denote the covariance matrix of ξ_n. Then $\Gamma_n = \Sigma_{i=1}^n A_i \Sigma_i A_i^T$. Suppose Γ_n is positive definite for every $n \geq 1$. Define

$$\xi_n^* = \Gamma_n^{-1/2}(\xi_n - E(\xi_n)) .$$

Theorem 4.2 implies that $\xi_n^* \xrightarrow{\mathcal{L}} N_k(O, I_k)$ provided $U(\xi_n, \Gamma_n) \to 1$ as $n \to \infty$.

5 Remarks

Results obtained here extend the work of Utev (1989) from the univariate case to multivariate distributions. Applications of the identity (1.2) to obtain some limit theorems, are given in Prakasa Rao (1979).

Acknowledgement: The author thanks Prof. S.A. Utev for bringing his paper to the author's attention.

References

Borovkov, A.A. and Utev, S.A. (1983) On an inequality and a related characterization of the normal distribution. *Theory of Probability and its Applications* **28**, 219-228.

Chen, L.H.Y. (1982) An inequality for the multivariate normal distribution. *J. Multivariate Anal.* **12**, 306-315.

Chernoff, H.(1981) A note on an inequality involving the normal distribution. *Ann. Probability* **9**, 533-535.

Prakasa Rao, B.L.S. (1979) Characterization of distributions through some identities. *J. Applied Probability* **16**, 903-909.

Prakasa Rao, B.L.S. and Sreehari, M. (1986) Another characterization of multivariate normal distribution. *Statistics and Probability Letters* **4**, 209-210.

Stein, C. (1973) Estimation of the mean of a multivariate normal distribution. Technical Report No.48, Stanford University.

Utev, S.A. (1989) Probability problems connected with a certain integro-differential inequality.*Siberian Mathematical Journal* **30**, 490-493.

Indian Statistical Institute
Delhi Centre
7, S.J.S. Sansanwal Marg
New Delhi 110 016
INDIA

ON THE GAUGE FOR THE
THIRD BOUNDARY VALUE PROBLEM

S.Ramasubramanian

Abstract

For fairly general q, c if

$$E_x \left[\int_0^\infty e_q(s)\hat{e}_c(s)I_A(X(s))d\xi(s) \right] < \infty$$

for some $x \in \bar{D}$, where D is a bounded domain and $A \subset \partial D$ is a nonempty open subset, it is shown that the gauge function for the third boundary value problem is bounded continuous. In the case of the Neumann problem, with the further assumption that q is Holder continuous, it is shown that the gauge is in $C^2(D) \cap C^1(\bar{D})$.

We consider the boundary value problem

$$\frac{1}{2}\Delta u(x) + q(x)u(x) = 0, x \in D$$

$$\frac{\partial u}{\partial n}(x) + c(x)u(x) = -\phi(x), x \in \partial D \qquad (1)$$

where $D \subset \mathbb{R}^d$ is a bounded domain with C^3-boundary and n is the inward normal; here q, c are measurable functions on $\bar{D}$, ∂D respectively so that the measures $q(x)dx$ and $c(x)d\sigma(x)$ belong to the generalized Kato class $GK_d(\bar{D})$ in the sense of Ma and Song [6], pp. 138–139 (or equivalently $q \in K_d(D)$ and $c \in \Sigma_d(\partial D)$ in the sense of Papanicolaou [7], pp. 33–37)

Let $\{P_x : x \in \bar{D}\}$ be the reflecting Brownian motion in $\bar{D}$; let ξ be the local time at the boundary. Note that K_d and Σ_d can be characterised in terms of the reflecting Brownian motion (see Section 2 of [7]). Put

$$e_q(t) = exp(\int_0^t q(X(s))ds)$$

$$\hat{e}_c(t) = exp(\int_0^t c(X(s))d\xi(s)) \qquad (2)$$

where $X(s)$ denotes the s-th coordinate projection on $C([0,\infty) : \mathbb{R}^d)$.

The gauge function for the boundary value problem (1) is given by

$$G(x) = E_x \left[\int_0^\infty e_q(s)\hat{e}_c(s)d\xi(s) \right], x \in \bar{D} \tag{3}$$

One may refer to Ma [5], Ma and Song [6] or Papanicolaou [7] for the role played by the gauge function in solving the third boundary value problem.

In this note we prove the following.

Theorem 1 : *Let A be a nonempty open subset of ∂D. Suppose*

$$E_x \left[\int_0^\infty e_q(s)\hat{e}_c(s)I_A(X(s))d\xi(s) \right] < \infty$$

for some $x \in \bar{D}$. Then the gauge G is a bounded continuous function on $\bar{D}$.

Remark : An analogue of the above result for the Dirichlet problem has been proved by Williams [8], which was a forerunner for the conditional gauge theorem for the Dirichlet problem. (See Falkner [2], p.20).

Lemma 2 : *Suppose the hypothesis of the theorem holds. Let ϕ be a non-negative continuous function on ∂D such that $\{\phi > 0\}$ is a nonempty open subset of A. Put*

$$u(x;\phi) = E_x \left[\int_0^\infty e_q(s)\hat{e}_c(s)\phi(X(s))d\xi(s) \right], x \in \bar{D} \tag{4}$$

Then $u(x;\phi)$ is the continuous stochastic solution to the problem (1). Moreover there exist constants a_1, a_2 such that $0 < a_1 \le u(x;\phi) \le a_2 < \infty$ for all $x \in \bar{D}$.

Proof : It is known that there is a strictly positive integral kernel $\zeta(t,x,z)$ which is continuous on $(0,\infty) \times \bar{D} \times \bar{D}$ such that

$$T_t f(x) \equiv E_x \left[e_q(t)\hat{e}_c(t)f(X(t)) \right] = \int_{\bar{D}} f(z)\zeta(t,x,z)dz, t > 0 \tag{5}$$

See Theorem 5.2 of Ma and Song [6] or Section 3 of Papanicolaou [7] for proof. It is not difficult to see that

$$u(x;\phi) = \frac{1}{2} \int_{\partial D} \int_0^\infty \phi(z)\zeta(s,x,z)dsd\sigma(z) \tag{6}$$

where $d\sigma$ is the surface area measure on ∂D. Using the Chapman- Kolmogorov equation for ζ it is now easily seen that

$$u(x;\phi) = \frac{1}{2} \int_{\partial D} \int_0^t \phi(z)\zeta(s,x,z)dsd\sigma(z) + T_t u(x;\phi) \tag{7}$$

for any $t > 0, x \in \bar{D}$.

By our assumption note that $u(x; \phi) \not\equiv \infty$. Hence, in view of equation (7), and the proofs of Theorem 3.6, Theorem 4.3, Corollary 4.4 of [7], or equivalently the proofs of Theorem 6.1 and Theorem 6.3 of [6] (these in turn are inspired by the corresponding results in Chung and Hsu [1], and Hsu [3]), it follows that u is the bounded continuous stochastic solution to (1). That u is bounded away from zero is clear from (7), continuity and positivity of ζ. This completes the proof. □

We denote by $\{\tilde{P}_x : x \in I\!R^d\}$ the d-dimensional Brownian motion in $I\!R^d$ (that is, without boundary condition). Let $\tilde{P}_{x;z}$ denote the z-conditioned Brownian motion (that is, conditioned to converge to z) starting from x, for $x \in D, z \in \partial D$; see Falkner [2], Zhao [9], [10] or the references given therein for conditioned Brownian motion.

Proof of the theorem : Choose a continuous function ϕ as in the lemma. Put $f(z) = u(z; \phi), z \in \partial D$, where $u(\cdot; \phi)$ is defined by (4).

Since $\{P_x\}$ behaves like $\{\tilde{P}_x\}$ till the time τ of hitting ∂D (provided $x \in D$), in view of the preceding lemma, note that $u(\cdot; \phi)$ is the continuous stochastic solution to the Dirichlet problem

$$\frac{1}{2}\Delta u(x) + q(x)u(x) = 0, \qquad x \in D,$$

$$u(x) = f(x), \quad x \in \partial D \qquad (8)$$

Since $u(\cdot; \phi)$ is bounded away from zero, it now follows (by the gauge theorem for the Dirichlet problem (see Theorem A in Zhao [10])) that the gauge for the Dirichlet problem, viz. $x \to \tilde{E}_x(e_q(\tau))$, is finite. Therefore by the results of Falkner and Zhao (as expressed in Proposition B of Zhao [10]) it follows that for any $x \in D$,

$$u(x; \phi) = \int_{\partial D} f(z)\tilde{E}_{x;z}(e_q(\tau))K_D(x, z)d\sigma(z) \qquad (9)$$

where K_D is the Poisson kernel for the classical Dirichlet problem and $\tilde{E}_{x;z}$ denotes expectation with respect to $\tilde{P}_{x;z}$. Therefore by (6), (9), symmetry of ζ in the space variables we get for any $x \in D$,

$$\infty > u(x; \phi)$$

$$= \frac{1}{2}\int_{\partial D}\left[\int_{\partial D}\int_0^\infty \phi(y)\zeta(s, z, y)dsd\sigma(y)\right]\tilde{E}_{x;z}(e_q(\tau))K_D(x, z)d\sigma(z)$$

$$= \frac{1}{2}\int_{\partial D}\phi(y)\left[\int_{\partial D}\int_0^\infty \tilde{E}_{x;z}(e_q(\tau))K_D(x, z)\zeta(s, y, z)dsd\sigma(z)\right]d\sigma(y)$$

Therefore there exist $y \in \partial D, x \in D$ such that

$$\int_{\partial D} \int_0^\infty \tilde{E}_{x;z}(e_q(\tau)) K_D(x,z) \zeta(s,y,z) ds d\sigma(z) \; < \; \infty \qquad (10)$$

Since the gauge for the Dirichlet problem is finite, by the results of Falkner and Zhao on conditional gauge (as expressed in Corollary 3 Zhao [10]) we have $inf\{\tilde{E}_{x;z}(e_q(\tau)) : x \in D, z \in \partial D\} > 0$. For any $x \in D$, note that $inf\{K_D(x,z) : z \in \partial D\} > 0$. Consequently (10) implies that $G(y) < \infty$ for some $y \in \partial D$. Now by the gauge theorem for the third boundary value problem (Theorem 6.1 of [6] or Theorem 3.6 of [7]) it follows that G is a bounded continuous function on $\bar{D}$. This completes the proof.   □

In the case of the Neumann problem and when q is Holder continuous we can give an alternate proof and get a stronger conclusion (cf. see Williams [8]).

Proposition 3 : *Let* $c \equiv 0$ *and* q *Holder continuous. Let the hypothesis of the theorem hold. Then* $G \in C^2(D) \cap C^1(\bar{D})$.

Proof: Suppose $G \equiv \infty$. Then by the results in Section 5 of Hsu [3], the first eigenvalue λ_0 of the Neumann problem for $\frac{1}{2}\Delta + q$ is nonnegative and there exists a bounded measurable function h on $\bar{D}$ such that

$$h(x) \; \equiv \; \int_D h(z) p(t,x,z) dz + \int_0^t \int_D [(q - \lambda_0)h](z) p(s,x,z) dz ds \, (11)$$

for any $x \in \bar{D}, t > 0$, where p is the transition probability density function of the reflecting Brownian motion $\{P_x\}$. As q, h are bounded, by the properties of p (see pp. 59- 60 of Ito [4]) it follows that the right side of (11) is in $C^2(D) \cap C^1(\bar{D})$ as a function of x. Thus $h \in C^2(D) \cap C^1(\bar{D})$ and solves (in the classical sense)

$$\frac{1}{2}\Delta h(x) + q(x) h(x) = \lambda_0 h(x), \quad x \in D, \qquad \frac{\partial h}{\partial n}(x) = 0, \quad x \in \partial D.$$

Consequently by Proposition 5.9 of Ma and Song [6], $h > 0$ on $\bar{D}$.

Now let ϕ be as in the lemma. Then by the lemma, and the equivalence of continuous weak solution and continuous stochastic solution (Theorem 4.2 of Hsu [3]) we get

$$\lambda_0 \int_D u(x;\phi) h(x) dx \; = \; -\frac{1}{2} \int_{\partial D} \phi(x) h(x) d\sigma(x) \qquad (12)$$

Clearly the left side of (12) ≥ 0, whereas the right side of (12) < 0 as $\phi > 0$ on a set of positive $d\sigma$ measure. This is a contradiction. Thus $G \not\equiv \infty$ and hence by the gauge theorem G is a bounded continuous function.

By (5), (7) note that we may write

$$G(x) = \frac{1}{2} \int_0^t \int_{\partial D} \zeta(s, x, z) d\sigma(z) ds + \int_{\bar{D}} G(z) \zeta(t, x, z) dz \qquad (13)$$

for any $t > 0$. Since q is Holder continuous note that ζ is the fundamental solution for $(-\frac{\partial}{\partial t} + \frac{1}{2}\Delta + q)$ subject to the Neumann boundary condition on ∂D. Hence, as G is bounded, by the properties of fundamental solutions (see Ito [4]) it follows that the two terms on the right side of (13) are in $C^2(D) \cap C^1(\bar{D})$ as functions of x. Hence $G \in C^2(D) \cap C^1(\bar{D})$. □

References

[1] K. L. Chung and P. Hsu : *Gauge theorem for the Neumann problem*, Seminar on Stochastic processes, 1984, pp. 63 – 70. Birkhauser, Boston, 1986.

[2] N. Falkner : *Feynman - Kac functionals and positive solutions of* $\frac{1}{2}\Delta u + qu = 0$, Zeit. Wahr. **65** (1983) 19 – 33.

[3] P. Hsu : *Probabilistic approach to the Neumann problem*, Comm. Pure Appl. Math. **38** (1985) 445 – 472.

[4] S. Ito : *Fundamental solutions of parabolic differential equations and boundary value problems*, Japan J. Math. **27** (1957) 55 – 102.

[5] Z. Ma : *On the probabilistic approach to boundary value problems*, BiBoS **288**, 1987.

[6] Z. Ma and R. Song : *Probabilistic methods in Schrodinger equations*, Seminar on Stochastic processes, 1989, pp. 135 – 164. Birkhauser, Boston, 1990.

[7] V. G. Papanicolaou : *The probabilistic solution of the third boundary value problem for second order elliptic equations*, Prob. Theory Rel. Fields **87** (1990) 27 – 77.

[8] R. J. Williams : *A Feynman - Kac gauge for solvability of the Schrodinger equation*, Adv. Appl. Math. 6 (1985) 1 – 3.

[9] Z. Zhao : *Conditional gauge with unbounded potential*, Zeit. Wahr. **65** (1983) 13 – 18.

[10] Z. Zhao : *Green function for Schrodinger operator and conditioned Feynman - Kac gauge*, J. Math. Anal. Appl. **116** (1986) 309 – 334.

Indian Statistical Institute
8th Mile, Mysore Road, Bangalore 560059, India

A Note on Prediction and an Autoregressive Sequence

MURRAY ROSENBLATT *

Abstract. Prediction for a first order possibly nonGaussian sequence is considered. Remarks are made about prediction with time increasing and with time reversed.

We consider a first order autoregressive stationary nonGaussian sequence with respect to the problem of prediction. Let x_t be the stationary solution of the sequence of equations

$$(1) \qquad x_t - \beta x_{t-1} = v_t , \; t = \ldots, -1, 0, 1, \ldots, 0 < |\beta| < 1 ,$$

where the v_t are independent, identically distributed random variables with mean zero and second moment one. The stationary solution is clearly

$$(2) \qquad x_t = \sum_{j=0}^{\infty} \beta^j v_{t-j} .$$

We also note that the sequence x_t is clearly Markovian because v_j, $j \geq 1$, is independent of x_t, $t \leq 0$. The best predictor of x_t given $x_{t-1}, x_{t-2}, \ldots$ in mean square is linear

$$E(x_t | x_{t-1}, x_{t-2}, \ldots) = E(x_t | x_{t-1}) = \beta x_{t-1}$$

whatever the distribution of v_t with prediction error variance

$$E|x_t - \beta x_{t-1}|^2 = E|v_t|^2 = 1$$
$$< (1 - \beta^2)^{-1} = E|x_t|^2 .$$

Our main interest is to see what happens if we consider the prediction problem for the process x_t with time reversed. Thus our object is to consider the best predictor of x_t given $x_{t+1}, x_{t+2}, \ldots$ in mean square

$$E(x_t | x_{t+1}, x_{t+2}, \ldots) = E(x_t | x_{t+1})$$

which depends only on x_{t+1} since the Markovian property is retained with time reversal. The following proposition is obtained.

* Research supported by ONR contract N00014-81-0003 and NSF grant DMS 83-12106.

Proposition. *The best one-step predictor with time reversed for the process (2)*

$$E(x_t|x_{t+1})$$

is linear if and only if the distribution of v_t is Gaussian.

The argument for this is simple. Let us first note that if the characteristic function of v_t is $\psi(\tau)$ and that of x_t is $\eta(\tau)$, then

(3)
$$\eta(\tau) = \prod_{j=0}^{\infty} \psi(\beta^j \tau) \, .$$

Since

$$\tau_1 x_t + \tau_2 x_{t+1} = \tau_2 v_{t+1} + \sum_{j=0}^{\infty} (\tau_1 \beta^j + \tau_2 \beta^{j+1}) v_{t-j}$$

it follows that the joint characteristic function of x_t, x_{t+1} is

$$\phi(\tau_1, \tau_2) = \psi(\tau_2) \prod_{j=0}^{\infty} \psi(\beta^j \{\beta \tau_2 + \tau_1\})$$
$$= \psi(\tau_2) \eta(\beta \tau_2 + \tau_1) \, .$$

Now

$$\frac{d}{d\tau_1} \phi(\tau_1, \tau_2)_{|\tau_1 = 0} = \phi_{\tau_1}(0, \tau_2)$$
$$= E[ix_t \exp(i\tau_2 x_{t+1})]$$
$$= i \int E[x_t|x_{t+1}] \exp(i\tau_2 x_{t+1}) \, dF_\beta(x_{t+1})$$

where F_β is the distribution function of x_{t+1}. Then

$$\phi_{\tau_1}(0, \tau_2)/\eta(\tau_2) = \sum_{j=0}^{\infty} \beta^j \psi'(\beta^{j+1}\tau_2)/\psi(\beta^{j+1}\tau_2)$$
$$= \frac{1}{\beta} \sum_{j=0}^{\infty} \{\beta^j \psi'(\beta^j \tau_2)/\psi(\beta^j \tau_2)\} - \frac{1}{\beta} \psi'(\tau_2)/\psi(\tau_2)$$
$$= \frac{1}{\beta} \eta'(\tau_2)/\eta(\tau_2) - \frac{1}{\beta} \psi'(\tau_2)/\psi(\tau_2)$$

or

(4)
$$\phi_{\tau_1}(0, \tau_2) = \frac{1}{\beta} \eta'(\tau_2) - \frac{1}{\beta} \frac{\psi'(\tau_2)}{\psi(\tau_2)} \eta(\tau_2) \, .$$

This last relation is valid in the same nontrivial symmetric interval about $\tau_2 = 0$ whether or not the mean of v_t is zero. The last term on the right of (4) is always meaningful since $\eta(\tau)/\psi(\tau)$ is simply the characteristic function of βx_t. If we consider the case of v_t an $N(0,1)$ random variable

$$\psi'(\tau)/\psi(\tau) = -\tau$$

and

$$\frac{-\psi'(\tau_2)}{\psi(\tau_2)}\, \eta(\tau_2) = \tau_2 \eta(\tau_2) = \tau_2 \int e^{i\tau_2 x} f_\beta(x)\, dx$$

$$= i \int e^{i\tau_2 x} f'_\beta(x)\, dx$$

with $f_\beta(x)$ the normal density $N(0, (1 - \beta^2)^{-1})$ and

$$E[x_t | x_{t+1}] = \beta x_{t+1}$$

the standard best linear predictor. Of course, in the normal case this is the best predictor in mean square.

Let us now assume that $E(x_t | x_{t+1})$ is linear and show that one is led to a Gaussian distribution. If $E(x_t | x_{t+1})$ is linear we must have

$$\frac{\psi'(\tau)}{\psi(\tau)}\, \eta(\tau) = c\eta'(\tau)$$

for some constant c. Thus

$$(5) \qquad\qquad \log \psi(\tau) = c \log \eta(\tau)$$

in a symmetrical neighborhood about zero. In fact one can see from (3) and (5) by looking at the second order cumulant that

$$c \sum_{j=0}^{\infty} \beta^{2j} = c(1 - \beta^2)^{-1} = 1, \quad c = (1 - \beta^2)\,.$$

Now

$$\log \psi(\tau) = (1 - \beta^2) \log \eta(\tau)$$

$$= (1 - \beta^2) \log \psi(\tau) + (1 - \beta^2) \sum_{j=1}^{\infty} \log \psi(\beta^j \tau)$$

and so

(6)
$$\beta^2 \log \psi(\tau) = (1 - \beta^2) \sum_{j=1}^{\infty} \log \psi(\beta^j \tau) \,.$$

Let
$$h(\tau) = \log \psi(\tau) + \frac{\tau^2}{2} \,.$$

Then (6) can be rewritten as

$$\beta^2 h(\tau) = (1 - \beta^2) \sum_{j=1}^{\infty} h(\beta^j \tau) \,.$$

However, this implies that

(7)
$$\begin{aligned}
\beta^2 h(\tau) &= (1 - \beta^2) h(\beta \tau) + (1 - \beta^2) \sum_{j=1}^{\infty} h(\beta^{j+1} \tau) \\
&= (1 - \beta^2) h(\beta \tau) + \beta^2 h(\beta \tau) \\
&= h(\beta \tau) \,.
\end{aligned}$$

We know that

(8)
$$h(\tau) = o(\tau^2)$$

as $\tau \to 0$. If $h(\tau) \neq 0$ by (7)

$$h(\beta^j \tau) = \beta^{2j} h(\tau)$$

and thus by (8) we must have $h(\tau) = 0$. It is clear that we must have $\psi(\tau)$ the Gaussian c.f. $\exp(-t^2/2)$.

The proposition clearly implies that the *prediction error variance*

$$E|x_t - E(x_t|x_{t+1})|^2 < 1 - \beta^2$$

if the process (2) *is nonGaussian.* The stationary distribution F_β of x_t clearly satisfies the integral equation

$$\int F_\beta \left(\frac{y - v}{\beta} \right) dF(v) = F_\beta(y)$$

where F is the distribution function of v_t. Except in special cases the distribution F_β cannot be given in elementary terms. If $\beta = 1/2$ and v_t

takes the values ± 1 with probability $1/2$, F_β is a uniform distribution. When $0 < \beta < 1/2$ with the same distribution for v_t, F_β is of Cantorian type (see Davis and Rosenblatt [1] and Garsia [2]). However for $1/2 < \beta < 1$ there are still open questions as to when F_β is singular or absolutely continuous with respect to Lebesgue measure.

If the random variables v_t do not have finite second moment, the measure of prediction has to be changed. Assuming first moments exist and are zero, one could consider mean absolute deviation instead of mean square deviation. The best predictor going forwards in time (predicting x_t given the past) would clearly be the conditional median of the x_t distribution given x_{t-1} assuming that it is well-defined. A similar remark could be made for prediction with time reversed. If v_t were to have a stable distribution with characteristic function $\exp(-|\tau|^\alpha)$, $1 < \alpha < 2$, the best predictor of x_t in absolute mean given the past would be linear

$$\beta x_{t-1}$$

since the distribution with characteristics function $\exp(-|\tau|^\alpha)$ has a density symmetric about zero. The answer for prediction with time reversed is not that transparent.

Related questions for moving average processes are considered in Shepp, et al. [3].

References

[1] R. Davis, and M. Rosenblatt, "Parameter estimates for some time series without contiguity" Statistics & Probability Letters **11** (1991) 515–521.

[2] A. Garsia, "Arithmetic properties of Bernoulli convolutions" Trans. Amer. Math. Soc. **102** (1962) 409–432.

[3] L. Shepp, D. Slepian, and A. Wyner, "On prediction of moving average processes" Bell Systems Tech. J. **59** (1988) 367–415.

Department of Mathematics
University of California, San Diego
La Jolla, California 92093-0112

On generalized stochastic partial differential equations

Yu.A. Rozanov

Abstract

It is shown that most known types of boundary problems for generalized partial differential equations can be set for stochastic equations with stochastic boundary conditions.

We consider a generalized differential equation

$$L\xi = \eta$$

with a linear partial differential operator

$$L = \sum_k a_k \partial^k.$$

We assume that the stochastic source η is a generalized function

$$\eta = (\phi, \eta), \quad \phi \in C_0^\infty(G),$$

in a region $G \subseteq \mathbf{R}^d$. For unexplained notation, we refer the reader to [1]. We are looking for a solution $\xi \in W(G)$ representing a generalized random field

$$\xi = (\phi, \xi), \quad \phi \in C_0^\infty(G)$$

in functional class $W(G)$ associated with the given equation as follows.

Let $W(G)$ be a Banach space of generalized functions (Schwartz distributions) in the region G such that

$$C_0^\infty(G) \subseteq W(G) \subseteq \mathcal{L}_2(G)$$

and

$$\| Lu \|_{\mathcal{L}_2} \le C \| u \|_W, \quad u \in W(G),$$

for the differential operator L considered, where Lu is defined by

$$(\phi, Lu) = (L^*\phi, u), \quad \phi \in C_0^\infty(G).$$

According to the imbedding $W(G) \subseteq \mathcal{L}_2(G)$ with $\| u \|_{L_2} \leq C \| u \|_W$, the dual space

$$X(G) = W(G)^*$$

contains all $x \in C_0^\infty(G)$, for

$$|(x,u)| \leq \| x \|_{\mathcal{L}_2} \| u \|_{\mathcal{L}_2} \leq C \| u \|_W \ .$$

Moreover $C_0^\infty(G)$ is dense in $X(G)$ because if $u \in W(G)$ is such that $(x,u) = 0$ for all $x \in C_0^\infty(G)$, then it follows that $u = 0$. We will denote closure of a subspace by $[\cdot]$ so that $[C_0^\infty(G)] = X(G)$.

Suppose $u = (\phi, u)$ is a generalized random function which is mean square continuous with respect to $\| \cdot \|_X$ on $C_0^\infty(G)$. ($\| \cdot \|_X$ is the norm on $X(G)$). Then we can extend u to all of $X(G)$ and $x \in X(G)$ can be treated as a *test function*. Let us consider the differential equation

(1) $$Lu = f$$

in the region G with boundary conditions of the type

(2) $$(x,u) = (x,u^+), \quad x \in X^+(\Gamma),$$

which are set by means of *some boundary test functions* x, sup $x \subseteq \Gamma$, on a boundary $\Gamma = \partial G$, i.e. such that

$$(x,\phi) \equiv (\phi, x) = 0, \quad \phi \in C_0^\infty(G).$$

(Actually most known PDE boundary problems can be set in this way).

Using the test functions

$$x = L^* \phi = \sum_k (-1)^{|k|} \partial^k(a_k \phi), \quad \phi \in C_0^\infty(G),$$

equation (1) can be recast as

$$(L^* \phi, u) = (\phi, f).$$

Application of these $x = L^* \phi \in X(G)$ to all deterministic $u \in W(G)$ with $Lu \in \mathcal{L}_2(G)$ gives

$$
\begin{aligned}
\| L^* \phi \|_X &= \sup_{\|u\|_W \leq 1} |(\phi, Lu)| \\
&\leq \sup_{\|u\|_W \leq 1} \| \phi \|_{\mathcal{L}_2} \| Lu \|_{\mathcal{L}_2} \\
&\leq C \| \phi \|_{\mathcal{L}_2}, \quad \phi \in C_0^\infty(G).
\end{aligned}
$$

Thus in a case of solvability of the equation (1) for any $f \in L_2(G)$ we have

$$\| L^*\phi \|_X \stackrel{\smile}{\sim} \| \phi \|_{L_2}, \quad \phi \in C_0^\infty(G);$$

indeed, the latter one follows from uniform boundness

$$\| \phi \|_{L_2} \leq C$$

on a set of all $\phi : \| L^*\phi \|_X \leq 1$ where

$$|(\phi, f)| = |(L^*\phi, u)| \leq \| u \|_W$$

for any $f = Lu \in \mathcal{L}_2(G)$.

Let us *assume* that *there is a unique solution* $u \in W(G)$ *of the deterministic boundary problem* (1), (2) *with any* $f \in L_2(G)$ *and the zero–boundary conditions* having in mind that $W(G) = X(G)^*$. Then our test function space $X(G) = [C_0^\infty(G)]$ is of a *direct sum* structure

$$(3) \qquad\qquad X(G) = L^* \mathcal{L}_2(G) + X^+(\Gamma).$$

Note, that this structure holds true if and only if there is the unique linear continuous functional $u = (x, u), \quad x \in X(G)$, which is arbitrarily specified on the subspace $X^-(G) = L^* L_2(G)$ and is zero on the subspace $X^+(\Gamma) \subseteq X(G)$ generated by the boundary test functions in the boundary conditions. And it implies the following result.

Theorem. *There is the unique solution* $u \in \underline{W}(G)$ *of the stochastic boundary problem (1), (2) with any generalized random source* $f = (\phi, f)$ *mean square continuous with respect to* $\| \phi \|_{L_2}, \quad \phi \in C_0^\infty(G)$, *and any random boundary sample* $u^+ \in W(G)$.

Dealing with the probability model (1), (2) one can be interested in the prediction problem or Markov property of the generalized random field $\xi = u \in W(G)$, say.

Let us consider this model as

$$(4) \qquad\qquad\qquad L\xi = \eta$$

in the region $G = G_0$ with the boundary conditions

$$(5) \qquad\qquad (x, \xi) = (x, \xi^+), \quad x \in X^+(\Gamma_0),$$

on the boundary $\Gamma_0 = \partial G_0$.

According to (3) with $G = G_0$ we have for any $G \subseteq G_0$ with its complement in G_0 the corresponding *direct sum* representation

$$X(G_0) = L^* \mathcal{L}_2(G) + L^* \mathcal{L}_2(G^c) + X^+(\Gamma_0),$$

with the non–degenerate operator L^* on $\mathcal{L}_2(G_0) = \mathcal{L}_2(G) + \mathcal{L}_2(G^c)$. Considering the test functions subspace

$$X(G) = [C_0^\infty(G)] \subseteq X(G_0)$$

in the region G we have a *direct sum*

$$X(G) = L^*\mathcal{L}_2(G) + X^+(\Gamma)$$

with $L^*\mathcal{L}_2(G) = [L^*C_0^\infty(G)] \subseteq [C_0^\infty(G)]$,

(6) $$X^+(\Gamma) = X(G) \cap [L^*\mathcal{L}_2(G^c) + X^+(\Gamma_0)].$$

Considering ξ in G as the generalized random field

$$\xi = (x, \xi), \quad x \in X(G) = [C_0^\infty(G)],$$

one can treat $x \in X^+(\Gamma)$ as *boundary test functions*,

$$\sup x \subseteq \Gamma = \partial G,$$

just having in mind

$$(x, \phi) \equiv (\phi, x) = 0, \quad \phi \in C_0^\infty(G_0 \backslash \Gamma).$$

Note in particular that here

$$x = L^*g + x^+, \quad g \in \mathcal{L}_2(G^c), \quad x^+ \in X^+(\Gamma_0),$$

with $(x^+, \phi) \equiv (\phi, x^+) = 0, \quad \phi \in C_0^\infty(G)$. All boundary test functions

$$x \in X(G), \quad \sup x \subseteq \Gamma = \partial G,$$

form a direct sum

$$X(\Gamma) = X^-(\Gamma) + X^+(\Gamma)$$

with

(7) $$X^-(\Gamma) = X(\Gamma) \cup L^*\mathcal{L}_2(G),$$

and all test functions $x \in X(G_0)$ outside G with $\sup x \subseteq G^c$, i.e. such that

$$(x, \phi) = 0, \quad \phi \in C_0^\infty(G),$$

form a direct sum

$$X(G^c) = X^-(\Gamma) + L^*\mathcal{L}_2(G^c) + X^+(\Gamma_0).$$

And this structure of our test functions leads to the following result for the probability model (4), (5) with any generalized stochastic source η with *independent values* in the region G_0 independent of the boundary conditions on the boundary $\Gamma_0 = \partial G_0$.

Theorem. *The generalized random field ξ enjoys the following Markov Property: ξ in any region $G \subseteq G_0$ is conditionally independent of its part outside G conditioned with respect to the boundary values*

$$(x, \xi), \quad x \in X(\Gamma)$$

on $\Gamma = \partial G$ by means of all boundary test functions $x \in X(G_0)$, $\sup x \subseteq \Gamma$.

The forecast $\hat{\xi} = E(\xi | \mathcal{A}(G^c))$ of ξ in any region $G \subseteq G_0$ by means of all data outside G can be given as the unique solution $\hat{\xi} = u \in W(s)$ of the boundary problem (4), (5) with

$$f = E(\eta | \mathcal{A}^-(\Gamma)),$$

and

$$(x, u) = (x, \xi), \quad x \in X^+(\Gamma),$$

associated with the boundary test functions (6), (7). Here,

$$\mathcal{A}^-(\Gamma) = \sigma\{(x, \xi), \quad x \in X^-(\Gamma)\}.$$

Note that here $f = 0$ when there is no $g \in \mathcal{L}_2(G)$, other than $g = 0$ such that

$$L^* g = (\phi, L^* g) = (L\phi, g) = 0, \quad \phi \in C_0^\infty(G).$$

This actually occurs for a variety of differential operators in unbounded regions.

References

[1] Markov Random Fields, Yu.A. Rozanov, (English translation by C.M. Elson), Springer–Verlag, Berlin, 1982.

Steklov Mathematical Institute
Vavilov Street 42
Moscow
117966, GSP-1, Russia

Examples of self-similar stable processes

SHIGEO TAKENAKA

ABSTRACT. Four examples of self-similar symmetric stable processes are introduced. Three of them are constructed by integral geometry and have pure point spectral measures. Using spectral measures we show that these four processes are distinct.

Let $\{X(t); t \in R\}$ be a real symmetric α stable $(S\alpha S)$ process, that is every finite linear combination $\sum_{j=1}^{N} a_j X(t_j)$, $a_j \in R$, has a symmetric stable law:

$$\mathbf{E}[\exp iz \sum_{j=1}^{N} a_j X(t_j)] = \exp\{-c(\{a_j\}, \{t_j\})|z|^\alpha\}.$$

A stochastic process $\{X(t)\}$ is called H-self-similar if for each $k > 0$, the processes $\{k^{-H} X(kt)\}$ and $\{X(t)\}$ have the same finite dimensional marginal laws, which is denoted by $\overset{\mathcal{L}}{\sim}$. Non trivial H-self-similar $S\alpha S$ processes $\{X(t)\}$ with $X(0) = 0$ exist if and only if

$$0 < H \le \frac{1}{\alpha}, \quad 0 < \alpha \le 2 \quad or \quad 0 < H < 1, \quad 0 < \alpha \le 2, \qquad ([7],[8],[14]).$$

For $0 < H < 1$, using stochastic integrals, several examples of H-self-similar $S\alpha S$ processes have been constructed and investigated by various authors (see the references of [5]). The linear fractional stable motion $\{\Delta_H(t)\}$ is a typical example of this type. Here, for $0 < H < 1/\alpha$, we use integral geometry to introduce two processes $\{X_1(t)\}, \{X_2(t)\}$, and by the time change $J(t) = -1/t$, we obtain another example $\{X_1^J(t)\}$. These three processes have pure point spectral measures and have strong determinisms ([11]). For example, all n-dimensional distribution of $(X_1(t_1), \cdots, X_1(t_n))$ are determined by the 2-dimensional marginals $\{(X_1(t_i), X_1(t_j)); 1 \le i, j \le n\}$.

The following table summarizes the results;

	$\Delta_H(t)$	$X_1(t)$	$X_2(t)$	$X_1^J(t)$
H	$0 < H < 1$	\multicolumn{3}{c}{$0 < H < 1/\alpha$}
spectral type	continuous	\multicolumn{3}{c}{discrete}
1-dim distributions	\multicolumn{4}{c}{same}
2-dim distributions	different	\multicolumn{2}{c}{same}	different
3-dim distributions	\multicolumn{4}{c}{different}
determinism		2-dim	4-dim	2-dim
reference	[3],[7]	[10],[14]	[11]	new one

1. Construction of examples.

I. $\{X_1(t); t \in R\}$. Let $E = R \times R_+$ and $d\mu_\beta(x, r) = r^{\beta-2}dxdr, (x, r) \in R \times R_+$, be a measure on the σ-field $\mathcal{B}(E)$. Set

$$(1.1) \qquad\qquad C(t) = \{(x, r) \in E; |x - t| \leq r\},$$
$$(1.2) \qquad\qquad S(t) = C(0)\Delta C(t),$$

where $A\Delta B$ denotes the symmetric difference of two sets A and B.

Note that $\mu_\beta(C(t)) = \infty$ but $\mu_\beta(S(t)) < +\infty$. Let $\mathcal{Y}_\alpha^\beta = \{Y_\alpha^\beta(B); B \in \mathcal{B}(E), \mu_\beta(B) < \infty\}$ be a $S\alpha S$ random measure with control measure μ_β, that is, $\mathcal{Y}_\alpha^\beta$ is a family of $S\alpha S$ random variables which satisfies

(1) $\mathbf{E}[\exp izY_\alpha^\beta(B)] = \exp -\mu_\beta(B)|z|^\alpha$, and
(2) for any disjoint family $\{B_j; j = 1, 2, \cdots\}$, the random variables $\{Y_\alpha^\beta(B_j); j = 1, 2, \cdots\}$ are mutually independent and $Y_\alpha^\beta(\cup_j B_j) = \sum_j Y_\alpha^\beta(B_j)$, a.s.

Define a $S\alpha S$ process by

$$(1.3) \qquad\qquad X_1(t) = Y_\alpha^\beta(S(t)).$$

Then, we have

Theorem 1.1. *The $S\alpha S$ process $\{X_1(t); t \in R\}$ is β/α-self-similar and has stationary increments.*

To prove the theorem, let us introduce a new $S\alpha S$ process. For any $T > 0$ define

$$(1.4) \qquad C^T(t) = \{(x, r) \in R \times R_+; |x - t| \leq r, 0 \leq r \leq T\}.$$

Since $C(t) = C^\infty(t)$, define

$$(1.5) \qquad \tilde{X}_1^T(t) = Y_\alpha^\beta(C^T(t)) - Y_\alpha^\beta(C^T(0)).$$

In the sequel we will see that for each t,

$$(1.6) \qquad \tilde{X}_1(t) = \lim_{T\to\infty} \tilde{X}_1^T(t) \quad a.s.$$

We have,

Lemma 1.2. *The stochastic processes $\{\tilde{X}_1(t)\}$ and $\{X_1(t)\}$ have the same finite dimensional distributions.*

Proof of Lemma 1.2. Let us compare n-dimensional characteristic functions:

$$(1.7) \quad \varphi(z_1, z_2, \cdots, z_n) \equiv \mathbf{E}\left\{\exp i\left(z_1 X_1(t_1) + \cdots + z_n X_1(t_n)\right)\right\}$$

$$= \exp\left\{-\sum_{A\subset\{1,2,\cdots,n\}} \left|\sum_{j\in A} z_j\right|^\alpha \mu_\beta\left(\left(\bigcap_{j\in A} S(t_j)\right)\cap\left(\bigcap_{k\notin A} \mathbf{C}S(t_k)\right)\right)\right\}.$$

Using $\bigcap_k(A_k \triangle B) = (\mathbf{C}B \cap (\bigcap A_k)) \cup (B \cap (\bigcap \mathbf{C}A_k))$ and $\mathbf{C}(A_k \triangle B) = (\mathbf{C}A_k)\triangle B$ we can rewrite the set appearing inside μ_β in (1.7), as
$$(1.8)$$
$$\left(\bigcap_{j\in A} C(t_j) \cap \bigcap_{k\notin A} \mathbf{C}C(t_k) \cap \mathbf{C}C(0)\right) \cup \left(\bigcap_{j\in A} \mathbf{C}C(t_j) \cap \bigcap_{k\notin A} C(t_k) \cap C(0)\right).$$

On the other hand, the corresponding characteristic function of $\{\tilde{X}_1^T\}$ is

$$(1.9) \quad \tilde{\varphi}^T(z_1, z_2, \cdots, z_n) \equiv \mathbf{E}\exp i\left(z_1 \tilde{X}_1^T(t_1) + \cdots + z_n \tilde{X}_1^T(t_n)\right)$$

$$= \mathbf{E}\exp i\left\{\sum_j z_j\left[Y_\alpha^\beta(C^T(t_j)) - Y_\alpha^\beta(C^T(0))\right]\right\}$$

$$= \psi^T\left(-(z_1 + z_2 + \cdots + z_n), z_1, \cdots, z_n\right),$$

where

$$(1.10) \quad \psi^T(z_0, z_1, \cdots, z_n) \equiv \mathbf{E} \exp i \sum_{j=0}^{n} z_j Y_\alpha^\beta(C^T(t_j))$$

$$= \exp - \left(\sum_{\tilde{A} \subset \{0, \cdots, n\}} |\sum_{j \in \tilde{A}} z_j|^\alpha \mu_\beta(\bigcap_{j \in \tilde{A}} C^T(t_j) \cap \bigcap_{k \notin \tilde{A}} \complement C^T(t_k)) \right)$$

$$= \exp - \sum_{A \subset \{1, \cdots, n\}}$$

$$\left\{ \left(|\sum_{j \in A} z_j|^\alpha \mu_\beta(\bigcap_{j \in A} C^T(t_j) \cap \bigcap_{k \notin A} \complement C^T(t_k) \cap \complement C^T(0)) \right) \right.$$

$$\left. + \left(|z_0 + \sum_{j \in A} z_j|^\alpha \mu_\beta(\bigcap_{j \in A} C^T(t_j) \cap \bigcap_{k \notin A} \complement C^T(t_k) \cap C^T(0)) \right) \right\}.$$

It follows that

$$(1.11) \quad \tilde{\varphi}^T(z_1, z_2, \cdots, z_n)$$

$$= \exp - \sum_{A \subset \{1, \cdots, n\}} \left\{ \left(|\sum_{j \in A} z_j|^\alpha \mu_\beta(\bigcap_{j \in A} C^T(t_j) \cap \bigcap_{k \notin A} \complement C^T(t_k) \cap \complement C^T(0)) \right) \right.$$

$$\left. + \left(|-\sum_{k \notin A} z_k|^\alpha \mu_\beta(\bigcap_{j \in A} C^T(t_j) \cap \bigcap_{k \notin A} \complement C^T(t_k) \cap C^T(0)) \right) \right\}$$

$$= \exp - \sum_{A \subset \{1, \cdots, n\}} |\sum_{j \in A} z_j|^\alpha \left\{ \mu_\beta(\bigcap_{j \in A} C^T(t_j) \cap \bigcap_{k \notin A} \complement C^T(t_k) \cap \complement C^T(0)) \right.$$

$$\left. + \mu_\beta(\bigcap_{k \notin A} C^T(t_k) \cap \bigcap_{j \in A} \complement C^T(t_j) \cap C^T(0)) \right\}.$$

In view of

$$\lim_{T \to \infty} \tilde{\varphi}^T(z_1, z_2, \cdots, z_n) = \tilde{\varphi}(z_1, z_2, \cdots, z_n) \equiv \mathbf{E}\{\exp i \sum_{j=1}^{n} z_j \tilde{X}_1(t_j)\}$$

and (1.7),(1.8),(1.11), we have $\varphi(z_1, z_2, \cdots, z_n) = \tilde{\varphi}(z_1, z_2, \cdots, z_n)$. $\square$

Lemma 1.3. $\{\tilde{X}_1(t); t \in R\}$ *is β/α-self-similar and has stationary increments.*

Proof of Lemma 1.3. For the stationarity of the increments, it is easy to see that

$$(1.12) \qquad \tilde{X}_1^T(t+h) - \tilde{X}_1^T(h) = Y_\alpha^\beta(C^T(t+h)) - Y_\alpha^\beta(C^T(h)).$$

Notice that the RHS has the same distribution for each t and take the limit as $T \to \infty$.

For the self-similarity we have

$$(1.13) \quad \tilde{\varphi}^{T,k}(z_1, z_2, \cdots, z_n) \equiv \mathbf{E} \exp i \sum_{j=1}^{n} z_j \tilde{X}_j^T(kt_j)$$

$$= \exp - \sum_{A \subset \{1,\cdots,n\}} |\sum_{j \in A} z_j|^\alpha \left\{ \mu_\beta(\bigcap_{j \in A} C^T(kt_j) \cap \bigcap_{m \notin A} \complement C^T(kt_m) \cap \complement C^T(0)) \right.$$

$$\left. + \mu_\beta(\bigcap_{m \notin A} C(kt_m) \cap \bigcap_{j \in A} \complement C^T(kt_j) \cap C^T(0)) \right\}.$$

$$(1.14) \qquad = \exp - \sum_{A \subset \{1,2,\cdots,n\}} |\sum_{j \in A} z_j|^\alpha$$

$$\left\{ \mu_\beta \left(\bigcap_{j \in A} k \cdot C^{T/k}(t_j) \cap \bigcap_{m \notin A} \complement(k \cdot C^{T/k}(t_m)) \cap \complement(k \cdot C^{T/k}(0)) \right) \right.$$

$$\left. + \mu_\beta \left(k \cdot (\bigcap_{m \notin A} C^{T/k}(t_m) \cap \bigcap_{j \in A} \complement C^{T/k}(t_j) \cap C^{T/k}(0)) \right) \right\}.$$

The measure μ_β satisfies the homogeneity property as $\mu_\beta(k \cdot A) = k^\beta \mu_\beta(A)$. Thus,

$$(1.15) \qquad \tilde{\varphi}^{T,k}(z_1, z_2, \cdots, z_n) = \tilde{\varphi}^{T/k}(k^{\beta/\alpha} z_1, k^{\beta/\alpha} z_2, \cdots, k^{\beta/\alpha} z_n).$$

Finally, letting $T \to \infty$, we have $\{\tilde{X}_1(kt)\} \stackrel{\mathcal{L}}{\sim} \{k^{\beta/\alpha} \cdot \tilde{X}_1(t)\}$. $\square$

Lemma 1.2 and Lemma 1.3 constitute the proof of Theorem 1.1.

II. $\{\Delta_H(t)\}$. If $0 < H < 1$, $H \neq 1/\alpha$, the following H-self-similar $S\alpha S$-process with stationary increments, called linear fractional stable motion, is known ([3])

$$(1.16) \qquad \Delta_H(t) = \int_{-\infty}^{\infty} [|t - u|^{H-\frac{1}{\alpha}} - |u|^{H-\frac{1}{\alpha}}] dZ(u),$$

where $\{Z(u); u \in R\}$ is the $S\alpha S$ motion, i.e. the $S\alpha S$ process with stationary independent increments.

III. $\{X_2(t)\}$.   Define a cone in $E = R \times R_+$ by

(1.17) $$C_2(t) = \{(x, r) \in E; \frac{1}{3}r < |x - t| \le r\}$$

and set

(1.18) $$S_2(t) = C_2(t) \triangle C_2(0).$$

Using the random measure $\mathcal{Y}_\alpha^\beta$ introduced in **I**, define a $S\alpha S$ process,

(1.19) $$X_2(t) = Y_\alpha^\beta(S_2(t)).$$

As in **I**, one can show that $\{X_2(t)\}$ is β/α-self-similar and has stationary increments.

IV. $\{X_1^J(t)\}$.   Let us consider a $S\alpha S$ process which is obtained by the time change , $t \mapsto J(t) = -\frac{1}{t}$, and renormalization:

(1.20) $$X_1^J(t) \equiv |t|^{2\frac{\beta}{\alpha}} X_1(-\frac{1}{t}) = |t|^{2\frac{\beta}{\alpha}} Y_\alpha^\beta(S(-\frac{1}{t})).$$

The 2-dimensional characteristic function of this process is

(1.21) $$\phi_2(z_1, z_2) = \mathbf{E}[\exp i \left(z_1 X_1^J(t_1) + z_2 X_1^J(t_2)\right)]$$

$$= \mathbf{E}[\exp i \left(z_1|t_1|^{2\frac{\beta}{\alpha}} Y_\alpha^\beta(S(-\frac{1}{t_1})) \right) + \left(z_2|t_2|^{2\frac{\beta}{\alpha}} Y_\alpha^\beta(S(-\frac{1}{t_2})) \right)]$$

$$= \exp - \left\{ \left|z_1|t_1|^{2\frac{\beta}{\alpha}} + z_2|t_2|^{2\frac{\beta}{\alpha}}\right|^\alpha \mu_\beta(S(-\frac{1}{t_1}) \cap S(-\frac{1}{t_2})) \right.$$

$$+ \left|z_1|t_1|^{2\frac{\beta}{\alpha}}\right|^\alpha \mu_\beta(S(-\frac{1}{t_1}) \cap \complement S(-\frac{1}{t_2}))$$

$$\left. + \left|z_2|t_2|^{2\frac{\beta}{\alpha}}\right|^\alpha \mu_\beta(\complement S(-\frac{1}{t_1}) \cap S(-\frac{1}{t_2})) \right\}.$$

The 2-dimensional characteristic function of $(X_1^J(kt_1), X_1^J(kt_2))$ is

(1.22) $$\exp - \left\{ \left|z_1|kt_1|^{2\frac{\beta}{\alpha}} + z_2|kt_2|^{2\frac{\beta}{\alpha}}\right|^\alpha \mu_\beta(S(-\frac{1}{kt_1}) \cap S(-\frac{1}{kt_2})) \right.$$

$$+ \left|z_1|kt_1|^{2\frac{\beta}{\alpha}}\right|^\alpha \mu_\beta(S(-\frac{1}{kt_1}) \cap \complement S(-\frac{1}{kt_2}))$$

$$\left. + \left|z_2|kt_2|^{2\frac{\beta}{\alpha}}\right|^\alpha \mu_\beta(\complement S(-\frac{1}{kt_1}) \cap S(-\frac{1}{kt_2})) \right\}$$

$$= \exp - \left\{ \left|z_1|kt_1|^{2\frac{\beta}{\alpha}} + z_2|kt_2|^{2\frac{\beta}{\alpha}}\right|^\alpha k^{-\beta} \mu_\beta(S(-\frac{1}{t_1}) \cap S(-\frac{1}{t_2})) \right.$$

$$+ \left|z_1|kt_1|^{2\frac{\beta}{\alpha}}\right|^\alpha k^{-\beta} \mu_\beta(S(-\frac{1}{t_1}) \cap \complement S(-\frac{1}{t_2}))$$

$$\left. + \left|z_2|kt_2|^{2\frac{\beta}{\alpha}}\right|^\alpha k^{-\beta} \mu_\beta(\complement S(-\frac{1}{t_1}) \cap S(-\frac{1}{t_2})) \right\}$$

$$= \phi_2(k^{\frac{\beta}{\alpha}} z_1, k^{\frac{\beta}{\alpha}} z_2).$$

Generalizing the above argument it is easy to see that this process is β/α-self-similar.

2. Characteristic functions of examples.

Let $\{X(t); t \in R\}$ be a $S\alpha S$ process. If $\alpha \neq 2$, the n-dimensional characteristic function ψ of $\{X(t)\}$ for n fixed time points $t_1, t_2, \cdots, t_n$ is,

$$(2.1) \quad \psi(z_1, \cdots, z_n) \equiv \mathbf{E}[\exp i\{\sum_{j=1}^{n} z_j X(t_j)\}] = \int_{S^{n-1}} |\sum_{j=1}^{n} z_j q_j|^\alpha d\nu(\mathbf{q}),$$

where $\mathbf{q} = (q_1, \cdots q_n) \in S^{n-1}$ and ν is a symmetric measure on S^{n-1}, the unit sphere in R^n. This measure ν characterizes the n-dimensional $S\alpha S$ law, and is called the n-dimensional spectral measure of $\{X(t)\}$ at $t_1, t_2, \cdots, t_n$. In this section, we compare the spectral measures of our examples of self-similar processes.

I. 1-dimensional characteristic functions. Let $\{X(t)\}$ be a self-similar $S\alpha S$ process of index H. The 1-dimensional characteristic function of $X(t)$ is

$$(2.2) \quad \mathbf{E}[e^{izX(t)}] = \mathbf{E}[e^{izt^H X(1)}] = \exp-(|z|^\alpha |t|^{\alpha H} c(\alpha)),$$

where $c(\alpha) = -\log \mathbf{E} \exp(iX(1))$.

Thus all four H-self-similar $S\alpha S$ processes $\{X_1(t)\}$, $\{X_2(t)\}$, $\{\Delta_H(t)\}$ and $\{X_1^J(t)\}$ share the same 1-dimensional distributions up to a constant factor. From now on, let us normalize these processes by multiplicative factors so that they have the same 1-dimensional distribution functions.

II. 2-dimensional characteristic functions . Let us recall the 2-dimensional characteristic function of $\{X_1(t)\}$:

$$(2.3) \quad \varphi_{(X_1(t_1), X_1(t_2))}(z_1, z_2) = \exp -\{\mu_\beta(S(t_1) \cap \mathbf{C}S(t_2))|z_1|^\alpha$$
$$+ \mu_\beta(\mathbf{C}S(t_1) \cap S(t_2))|z_2|^\alpha + \mu_\beta(S(t_1) \cap S(t_2))|z_1 + z_2|^\alpha\}.$$

Thus, the 2-dimensional spectral measure is of pure point type concentrated on the 6 points $\{(\pm 1, 0), (0, \pm 1), \pm(\frac{1}{\sqrt{2}}, \frac{1}{\sqrt{2}})\}$, and their measures are $\mu_\beta(S(t_1) \cap \mathbf{C}S(t_2))$, $\mu_\beta(\mathbf{C}S(t_1) \cap S(t_2))$ and $\mu_\beta(S(t_1) \cap S(t_2))$, respectively.

On the other hand, if $\alpha = 2$, the Gaussian case, we have
(2.4)

$$\varphi_{(X_1(t_1), X_1(t_2))}(z_1, z_2) = \exp -(\frac{1}{2}\sigma(t_1)z_1^2 + \sigma(t_1, t_2)z_1 z_2 + \frac{1}{2}\sigma(t_2)z_2^2),$$

where $\sigma(t_1, t_2)$ is the covariance of $X_1(t_1), X_1(t_2)$ and $\sigma(t)$ is the variance of $X_1(t)$. Thus we have,

$$(2.5) \quad \mu_\beta(S(t_i) \cap \mathbf{C}S(t_j)) = \frac{1}{2}(\sigma(t_i) - \sigma(t_i, t_j)), \quad i, j = 1, 2, \text{ and}$$

$$\mu_\beta(S(t_1) \cap S(t_2)) = \frac{1}{2}\sigma(t_1, t_2) = \frac{1}{4}(\sigma(t_1) + \sigma(t_2) - \sigma(t_1 - t_2)) \qquad .$$

Finaly, we have the following relations between the 1- and 2-dimensional spectral measures of $\{X_1(t)\}$ for any α:

$$(2.6) \quad \mu_\beta(S(t_i) \cap \mathbf{C}S(t_j)) = \frac{1}{4}\{\mu_\beta(S(t_i)) - \mu_\beta(S(t_j)) + \mu_\beta(S(t_i - t_j))\},$$

$$\mu_\beta(S(t_1) \cap S(t_2)) = \frac{1}{4}\{\mu_\beta(S(t_1)) + \mu_\beta(S(t_2)) - \mu_\beta(S(t_1 - t_2))\}.$$

If we take the set $S_2(\cdot)$ instead of $S(\cdot)$, we have the same relations (2.6) for the process $\{X_2(t)\}$. This means that the 2-dimensional characteristic functions of $\{X_1(t)\}$ coincide with those of $\{X_2(t)\}$ at any time points t_1, t_2.

As we saw in (1.21), the spectral measure of the process $\{X_1^J(t)\}$ is also of pure point type, but its support is

$$(2.7) \qquad \{(\pm 1, 0), (0, \pm 1), \pm(\frac{|t_1|^{2\frac{\beta}{\alpha}}}{|t_1|^{2\frac{\beta}{\alpha}} + |t_2|^{2\frac{\beta}{\alpha}}}, \frac{|t_2|^{2\frac{\beta}{\alpha}}}{|t_1|^{2\frac{\beta}{\alpha}} + |t_2|^{2\frac{\beta}{\alpha}}})\}.$$

It is easy to see that the process $\{\Delta_H(t)\}$ has spectral mesure of continuous type.

III. 3-dimensional characteristic functions and determinisms
([10],[11]).   In this sub-section the control measure of white noise need not be μ_β. Instead of μ_β, let us take a measure μ which is equivalent to Lebesgue measure and satisfies $\mu(S(t)) < \infty$ and let us redefine $X_1 \equiv Y_\mu(S(t))$ and so on.

Fix $t_1 < t_2 < t_3$. The spectral measure of $\{X_1(t)\}$ at t_1, t_2, t_3 is of pure point type with weights $\mu_\beta(S^{e_1}(t_1) \cap S^{e_2}(t_2) \cap S^{e_3}(t_3))$, where $S^e = S$ if $e = 1$, $\mathbf{C}S$ if $e = 0$ and $(e_1, e_2, e_3) \in \{0, 1\}^3 \setminus (0, 0, 0)$. It is easy to see that

$$(2.8) \qquad S(t_1) \cap \mathbf{C}S(t_2) \cap S(t_3) = \emptyset, \text{ provided } t_1 > 0.$$

This means that the support of the 3-dimensional spectral measure of $\{X_1(t)\}$ consists of at most 14 points. Using this fact, one can easily show that any finite dimensional distribution of the process $\{X_1(t)\}$ is determined by its 2-dimensional marginals. For instance $S(t_1) \cap S(t_2) \cap S(t_3) = S(t_1) \cap S(t_3)$.

On the other hand, in the case of $\{X_2(t)\}$ one can check:

$$(2.9) \qquad\qquad S_2(1) \cap \mathbf{C}S_2(2) \cap S_2(3) \neq \emptyset.$$

That is, $\{X_1(t)\}$ and $\{X_2(t)\}$ are distinct.

Moreover, one can check that the support of 5-dimensional spectral measures consists of at most 60 points instead of full $62 = 2 \times (2^5 - 1)$ points, for instance,

$$(2.10) \qquad S_2(1) \cap \mathbf{C}S_2(2) \cap S_2(3) \cap \mathbf{C}S_2(4) \cap S_2(5) = \emptyset,$$

and can show that any finite dimensional distribution of $\{X_2(t)\}$ is determined by its 4-dimensional marginals.

References

1. S. Cambanis and M. Maejima, *Two classes of self-similar stable processes with stationary increment*, Stoch. Proc. Appl. **32** (1989), 305–392.

2. N. N. Chentsov, *Lévy's Brownian motion of several parameters and generalized white noise*, Theory Probab. Appl. **2** (1957), 265–266.

3. Y. Kasahara and M. Maejima, *Weighted sums of i.i.d. random variables attracted to integrals of stable processes*, Probab. Th. Rel. Fields **78** (1988), 75–96.

4. K. Kojo and S. Takenaka, *On canonical representations of stable M_t processes*, Prob. and Math. Stat. (to appear).

5. N. Kôno and M. Maejima, *Self-similar stable processes with stationary increments*, Stable Processes and Related Topics, Birkhäuser, 1991.

6. P. Lévy, *Processus Stochastiques et Mouvement Brownien (2nd edition)*, Gauthier-Villars, Paris, 1965.

7. M. Maejima, *On a class of self-similar processes*, Z. Wahrsch. verw. Geb. **62** (1983), 53–72.

8. _____, *A remark on self-similar processes with stationary increments*, Canadian J. Statist **14** (1986), 81-82.

9. Y. Sato, *Structure of Lévy measures of stable random fields of Chentsov type*, preprint (1990).

10. _____, *Distributions of stable random fields of Chentsov type*, Nagoya Math. J. **123** (1991), 119–139.

11. Y. Sato and S. Takenaka, *On determinism of symmetric α stable processes of of generalized Chentsov type*, Gaussian Random Fields, World Scientific, 1991, pp. 332–345.

12. S. Takenaka, *Representations of Euclidean random field*, Nagoya Math. J. **105** (1987), 19–31.

13. _____, *On pathwise projective invariance of Brownian motion I*, Proc. Japan Acad. **64** (1988), 41–44.

14. _____, *Integral geometric construction of self-similar stable processes*, Nagoya Math. J. **123** (1991), 1–12.

DEPT. OF MATH. HIROSHIMA UNIV., 1 KAGAMIYAMA, HIGASHI-HIROSHIMA, 724 JAPAN

E-mail: r0044@math.sci.hiroshima-u.ac.jp

Green Operators of Absorbing Lévy Processes on the Half Line

Hiroshi Tanaka

Abstract

Silverstein's formula ([5]) for Green operators of absorbing Lévy processes on $[0, \infty)$ is proved, without the assumption of the absolute continuity of Green measures, by making use of the Wiener-Hopf factorization for Lévy processes.

1. Introduction.

Let $\{X(t), t \geq 0\}$ be a temporally homogeneous Lévy process in one-dimension with $X(0) = 0$ and assume that all the sample paths are right continuous and have left limits. We consider the Green operator G defined by

$$Gf(x) = E\left\{ \int_0^{\tau(x)} f(x + X(t))dt \right\}, \quad x \geq 0,$$

for any $f \in C_0([0, \infty))$, the space of continuous functions on $[0, \infty)$ with compact supports, where $\tau(x) = \inf\{t > 0 : x + X(t) < 0\}$. The convention $\inf \emptyset = \infty$ is used throughout. For $a > 0$ we put

$$\tau_a^\# = \inf\{t > 0 : X^\#(s) > a \text{ for some } s \in (0, t) \text{ and } X^\#(t) = 0\},$$

$$\mu_a(A) = E\left\{ \int_0^{\tau_a^\#} 1_A(X^\#(t))dt \right\}, \quad A \in \mathcal{B}([0, \infty)),$$

where $X^\#(t) = X(t) - \inf\{X(s) : 0 \leq s \leq t\}$. We also define a measure $\widehat{\mu}_a$ on $[0, \infty)$ for the dual process $\{\widehat{X}(t), t \geq 0\}$ (where $\widehat{X}(t) = -X(t)$) in the same way as we defined μ_a for $\{X(t)\}$. For a random variable Y and an event A $E\{Y; A\}$ denotes the integral of Y over A with respect the underlying probability measure P.

In this paper we prove the following theorems.

Theorem 1. *Assume that $X(t)$ is neither increasing nor decreasing. Then for any $f \in C_0([0, \infty))$*

$$(1.1) \qquad Gf(x) = c(a) \int_{[0,x]} \widehat{\mu}_a(du) \int_{[0,\infty)} \mu_a(dv) f(x - u + v), \quad x \geq 0,$$

where $c(a)$ is a positive constant depending on a.

Theorem 2. *There exist two measures μ and $\widehat{\mu}$ on $[0,\infty)$ determined by*

$$(1.2) \qquad \int_{[0,\infty)} e^{-\theta x}\mu(dx) = \exp\int_0^\infty t^{-1}E\{e^{-\theta X(t)} - e^{-t}; X(t) > 0\}dt,$$

$$(1.3) \qquad \int_{[0,\infty)} e^{-\theta x}\widehat{\mu}(dx) = \exp\int_0^\infty t^{-1}E\{e^{\theta X(t)} - e^{-t}; X(t) < 0\}dt,$$

θ *being positive, and if $X(t)$ is not identically zero then for any $f \in C_0([0,\infty))$ we have*

$$(1.4) \qquad Gf(x) = c\int_{[0,x]} \widehat{\mu}(du) \int_{[0,\infty)} \mu(dv)f(x - u + v),$$

where

$$c = \exp\int_0^\infty t^{-1}(1 - e^{-t})P\{X(t) = 0\}dt < \infty.$$

The above theorems are essentially due to Silverstein[5], Theorem 6 of page 543, in which the absolute continuity of the Green measure was assumed. Under this condition μ and $\widehat{\mu}$ have densities ψ and $\widehat{\psi}$, respectively, relative to the Lebesgue measure and the density version of (1.4) takes the form

$$(1.5) \qquad g(x,y) = \int_0^{\min\{x,y\}} \widehat{\psi}(x - u)\psi(y - u)du.$$

A similar representation of the Green function of an absorbing random walk on $\mathbf{Z}^+$ was obtained by Spitzer[6], page 209.

Silverstein([5]) argued from a potential theoretic viewpoint. Our proof is based on the Wiener-Hopf factorization for Lévy processes (Pecherskii and Rogozin[3], see also Greenwood and Pitman[2] and Sato[4]). So far as one admits this factorization our proof is quite elementary. Since μ_a and $\widehat{\mu}_a$ are constant multiples of μ and $\widehat{\mu}$, respectively, Theorem 1 is a consequence of Theorem 2. However, by our method we must prove first Theorem 1 and then Theorem 2.

2. Proof of Theorem 1.

We put

$$N(t) = \inf\{X(s) : 0 \le s \le t\}, \quad \widehat{N}(t) = \inf\{\widehat{X}(s) : 0 \le s \le t\},$$
$$X^{\#}(t) = X(t) - N(t), \qquad \widehat{X}^{\#}(t) = \widehat{X}(t) - \widehat{N}(t).$$

The proof of Theorem 1 is based on the following fact:

Wiener-Hopf factorization (Pecherskii and Rogozin[3], see also Greenwood and Pitman[2] and Sato[4]). Let $\lambda > 0$ and let T be a random time which is independent of the process $\{X(t)\}$ and is distributed according to the exponential distribution with mean $1/\lambda$. Then for $\theta, \zeta \geq 0$

$$(2.1) \qquad E \exp\{\zeta N(T) - \theta X^{\#}(T)\} = \varphi_\lambda^+(\theta)\varphi_\lambda^-(\zeta),$$

where

$$(2.2) \qquad \varphi_\lambda^+(\theta) = \exp \int_0^\infty t^{-1} e^{-\lambda t} E\{e^{-\theta X(t)} - 1; X(t) > 0\}dt,$$

$$(2.3) \qquad \varphi_\lambda^-(\zeta) = \exp \int_0^\infty t^{-1} e^{-\lambda t} E\{e^{\zeta X(t)} - 1; X(t) < 0\}dt.$$

In particular, $N(T)$ and $X^{\#}(T)$ are independent.

The following lemma is well-known; it can be proved by considering $Y(s) = X(t-) - X((t-s)-), 0 \leq s \leq t$, which is equivalent in law to the process $X(s), 0 \leq s \leq t$.

Lemma 1. *For each fixed $t > 0$ $- N(t)$ and $\widehat{X}^{\#}(t)$ have the same law.*

To proceed to the proof of Theorem 1, let $\lambda > 0$ and put

$$G_\lambda f(x) = E\left\{\int_0^{\tau(x)} e^{-\lambda t} f(x + X(t))dt\right\}, \qquad x \geq 0.$$

Since $\{\tau(x) > t\} \subset \{x + N(t) \geq 0\}$ and the difference of the two events is contained in $\{\tau(x) = t\}$ which has probability zero except for (at most) countable number of t, we have for $f \in C_0([0, \infty))$

$$(2.4) \qquad G_\lambda f(x) = \int_0^\infty e^{-\lambda t} E\{f(x + X(t)); \tau(x) > t\}dt$$

$$= \int_0^\infty e^{-\lambda t} E\{f(x + X(t)); -N(t) \leq x\}dt$$

$$= \lambda^{-1} E\{f(x + X(T)); -N(T) \leq x\}$$

$$= \lambda^{-1} E\{f(x + N(T) + X^{\#}(T)); -N(T) \leq x\}$$

$$= \lambda^{-1} \int_{[0,x]} \widehat{\mu}^\lambda(du) \int_{[0,\infty)} \mu^\lambda(dv) f(x - u + v),$$

where

$$\widehat{\mu}^\lambda(du) = P\{-N(T) \in du\} = P\{\widehat{X}^{\#}(T) \in du\} \quad \text{(by Lemma 1)},$$

$$\mu^\lambda(dv) = P\{X^{\#}(T) \in dv\}.$$

Next we put

$$\mu_a^\lambda(A) = E\left\{ \int_0^{\tau_a^\#} e^{-\lambda t} 1_A(X^\#(t))dt \right\}, \quad A \in \mathcal{B}([0,\infty)).$$

Then

(2.5) μ_a^λ increases to μ_a as $\lambda \downarrow 0$.

Moreover, using the strong Markov property of $X^\#(t)$ we obtain

$$\lambda^{-1}\mu^\lambda(A) = E\left\{ \int_0^{\tau_a^\#} e^{-\lambda t} 1_A(X^\#(t))dt \right\} + E\left\{ \int_{\tau_a^\#}^\infty e^{-\lambda t} 1_A(X^\#(t))dt \right\}$$
$$= \mu_a^\lambda(A) + E\{e^{-\lambda \tau_a^\#}\}\lambda^{-1}\mu^\lambda(A),$$

and hence

(2.6) $\mu^\lambda = c_\lambda \mu_a^\lambda, \quad c_\lambda = \lambda\{E(1 - e^{-\lambda \tau_a^\#})\}^{-1}.$

We define $\widehat{\mu}_a^\lambda$ for $\widehat{X}^\#(t)$ in the same way as we defined μ_a^λ for $X^\#(t)$. Then we have statements for $\widehat{\mu}_a^\lambda$ similar to (2.5) and (2.6). Therefore we have

(2.7) $G_\lambda f(x) = \lambda^{-1} c_\lambda \widehat{c}_\lambda \int_{[0,x]} \widehat{\mu}_a^\lambda(du) \int_{[0,\infty)} \mu_a^\lambda(dv) f(x - u + v).$

We now let $\lambda \downarrow 0$ in (2.7) and use (2.5). Note that under the assumption of Theorem 1 μ_a and $\widehat{\mu}_a$ are finite for any compact subset of $[0,\infty)$. We then obtain (1.1) for any $x > 0$ which is a point of continuity of $\widehat{\mu}_a$, as well as the existence of $\lim_{\lambda \downarrow 0} \lambda^{-1} c_\lambda \widehat{c}_\lambda = c(a)$. Since both sides of (1.1) are right continuous in $x > 0$ and the set of points of continuity of $\widehat{\mu}_a$ is dense in $[0,\infty)$, (1.1) holds for all $x \geq 0$. This proves Theorem 1.

Remark 1. If we put

$$G_0 f(x) = E\left\{ \int_0^{\sigma(x)} f(x + X(t))dt \right\}, \quad x > 0,$$

where $\sigma(x) = \inf\{t > 0 : x + X(t) \leq 0\}$, then

$$G_0 f(x) = c(a) \int_{[0,x)} \widehat{\mu}_a(du) \int_{[0,\infty)} \mu_a(dv) f(x - u + v), \quad f \in C_0([0,\infty)).$$

Remark 2. If $X(t)$ enters $(-\infty, -\lambda)$ a.s. for any $\lambda > 0$ in addition to the assumption in Theorem 1, then μ_a is an invariant measure of the Markov process $\{X^{\#}(t)\}$ which is unique up to a multiplicative constant.

3. Proof of Theorem 2.

The idea for proving Theorem 2 is simply to let λ tend to 0 in (2.4) but we must use the fact that the t-integrals in the right hand sides of (1.2) and (1.3) converge absolutely, which is a part of Theorem 9.1 of Fristedt[1]. This absolute convergence will be verified in the course of our proof. To avoid trivial complications we assume that $X(t)$ is neither increasing nor decreasing. For a measure m on $[0, \infty)$ we denote by $L(\theta, m)$ the Laplace transform $\int_{[0,\infty)} e^{-\theta x} m(dx)$, $\theta > 0$.

Lemma 2. *Under the assumption of Theorem 1* $L(\theta, \mu_a) < \infty$ *and* $L(\theta, \hat{\mu}_a)$ $< \infty$ *for all* $\theta > 0$.

Proof. We consider the case where $X(t)$ is recurrent in the sense that $X(t)$ hits any half line $([x, \infty), (-\infty, y), y < 0 < x)$ with probability 1. For $x > 0$ denote by $h_a(x)$ the probability that $X(t)$ enters $(-\infty, -x)$ before it enters (a, ∞). Since

$$1 \geq h_a(x + r)/h_a(x)$$
$$\geq P\{-x + X(t) \text{ enters } (-\infty, -x - r) \text{ before it enters } (a, \infty)\}$$
$$\to 1 \quad \text{as } x \to \infty$$

for any $r > 0$, the function $h_a(\log t)$ (and also its reciprocal) is slowly varying at infinity. Next for $y \geq 0$ denote by $G(y, A)$ the expected amount of time $y + X(t)$ spends in a Borel subset A of $[0, \infty)$ up to $\tau(y)$. Let $\sigma_0 = 0, \tau_k$ be the infimum of $t > \sigma_k$ such that $y + X(t) < x - a$ $(k \geq 0)$ and σ_k be the infimum of $t > \tau_k$ such that $y + X(t) > x$ $(k \geq 1)$. Then $0 = \sigma_0 < \tau_0 < \sigma_1 < \tau_1 < \cdots$ and for $a < x < y$ we can write

$$G(y, (x, x + a]) = \sum_{k=0}^{\infty} E\left\{\int_{\sigma_k}^{\tau_k} 1_{(x, x+a]}(y + X(t))dt; \sigma_k < \tau(y)\right\}.$$

It is easy to see that the k-th term in the above is dominated by $\rho^k e_a$ where ρ is the probability that $x - a + X(t)$ enters (x, ∞) before it enters $(-\infty, 0)$ and e_a is the expected exit time of $X(t)$ from $[-a, a]$. Therefore

$$G(y, (x, x + a]) \leq e_a(1 - \rho)^{-1} = e_a/h_a(x - a)$$

for any $y > x$ and hence $\mu_a((x, x+a])$ is also dominated by $e_a/h_a(x - a)$ for $x > a$. Since $1/h_a(\log t)$ is slowly varying at infinity, $\mu_a((x, x+a]) = o(e^{\varepsilon x})$

as $x \to \infty$ for any $\varepsilon > 0$ and this implies $L(\theta, \mu_a) < \infty$ for all $\theta > 0$. In the transient case, as is easily seen $\mu_a((x, x+a])$ is dominated by the expected amount of time $X(t)$ spends in $[-a, a]$ if $x > a$, from which it follows that $\mu_a((x, x+a])$ is bounded in x and hence $L(\theta, \mu_a) < \infty$ for all θ.

Lemma 3. $\int_0^\infty t^{-1} |E\{e^{-\theta|X(t)|} - e^{-t}\}| dt < \infty$ for all $\theta > 0$.

Proof. Using (2.1), (2.2) and (2.3) we can compute $\lambda^{-1}L(\theta, \mu^\lambda)L(\theta, \widehat{\mu}^\lambda)$ easily. The result is

$$\lambda^{-1}L(\theta, \mu^\lambda)L(\theta, \widehat{\mu}^\lambda) = \exp\left\{\int_0^\infty g(t, \lambda)dt - \log(1+\lambda)\right\},$$

$$g(t, \lambda) = t^{-1}e^{-\lambda t}E\{e^{-\theta|X(t)|} - e^{-t}\}.$$

Since the limit of $\lambda^{-1}c_\lambda \widehat{c}_\lambda$ as $\lambda \downarrow 0$ exists, the product measure $\lambda^{-1}\mu^\lambda \otimes \widehat{\mu}^\lambda = \lambda^{-1}c_\lambda \widehat{c}_\lambda \mu_a^\lambda \otimes \widehat{\mu}_a^\lambda$ is vaguely convergent by virtue of (2.5), and hence by Lemma 2 $\lambda^{-1}L(\theta, \mu^\lambda)L(\theta, \widehat{\mu}^\lambda)$ must tend to a nonzero limit as $\lambda \downarrow 0$. Therefore

(3.1) $\displaystyle\lim_{\lambda \downarrow 0} \int_0^\infty g(t, \lambda)dt$ exists.

On the other hand the validity of the formulas (2.2) and (2.3) entails the absolutely integrability of $g(t, 0)$ over the interval $(0, 1)$. This combined with (3.1) implies the absolute integrability of $g(t, 0)$ over $(0, \infty)$ as was to be proved.

Now Theorem 2 can be proved as follows. We put $\nu^\lambda = c_\lambda^+ \mu^\lambda$ and $\widehat{\nu}^\lambda = c_\lambda^- \widehat{\mu}^\lambda$ where

$$c_\lambda^\pm = \exp\int_0^\infty t^{-1}e^{-\lambda t}(1 - e^{-t})P\{\pm X(t) > 0\}dt.$$

Making use of (2.2) and (2.3) we can compute $L(\theta, \nu^\lambda)$ and $L(\theta, \widehat{\nu}^\lambda)$. As a result we have

$$L(\theta, \nu^\lambda) = \exp\int_0^\infty t^{-1}e^{-\lambda t}E\{e^{-\theta X(t)} - e^{-t}; X(t) > 0\}dt$$

which tends to the right hand side of (1.2) as $\lambda \downarrow 0$ by Lemma 3. Similarly $L(\theta, \widehat{\nu}^\lambda)$ tends to the right hand side of (1.3) as $\lambda \downarrow 0$. Therefore ν^λ and $\widehat{\nu}^\lambda$ converge vaguely to μ and $\widehat{\mu}$ as $\lambda \downarrow 0$, respectively. Since $\nu^\lambda \otimes \widehat{\nu}^\lambda = \lambda^{-1}(1+\lambda)\overline{c}_\lambda \mu^\lambda \otimes \widehat{\mu}^\lambda$ where $\overline{c}_\lambda = \exp\{-\int_0^\infty t^{-1}e^{-\lambda t}(1-e^{-t})P(X(t) = 0)dt\}$, we obtain (1.4) together with the finiteness of c by letting λ tend to 0 in (2.4).

Acknowledgment. The author wishes to thank the referee for valuable comments through which the content of the paper was much improved; in particular, a simple derivation of Theorem 2 was found in the course of making a revision.

REFERENCES

1 Fristedt, B., *Sample functions of stochastic processes with stationary independent increments*, Adv. Probab. **3** (1973), 241-396.
2 Greenwood, P. and Pitman, J., *Fluctuation identities for Lévy processes and splitting at the maximum*, Adv. Appl. Probab. **12** (1980), 893-902.
3 Pecherskii, E. A. and Rogozin, B. A., *On joint distributions of random variables associated with fluctuations of a process with independent increments*, Theor. Probab. Appl. **14** (1969), 410-423.
4 Sato, K., "Processes with Independent Increments (in Japanese)," Kinokuniya, Tokyo, 1990.
5 Silverstein, M. L., *Classification of coharmonic and coinvariant functions for a Lévy process*, Ann. Probab. **8** (1980), 539-575.
6 Spitzer, F., "Principles of Random Walk," Van Nostrand, New York, 1964.

Department of Mathematics
Faculty of Science and Technology
Keio University
Yokohama 223, Japan

Moments of Sums
of Independent Random Variables

K. Urbanik

Abstract
For any positive real number p a p-equivalence of random variables is defined in terms of moments of order p. For p being not an integer it is shown that p-equivalent nonnegative random variables are identically distributed.

Throughout this paper all random variables under consideration will tacitly be assumed to be nonnegative. Two random variables X and Y are said to be equivalent, in symbols $X \sim Y$, if they are identically distributed. Given a positive real number p by M_p we denote the set of all random variables X with finite p-th moment EX^p. The random variables X and Y from M_p are said to be *p-equivalent*, in symbols $X \sim_p Y$, if for every positive integer n the equality

$$E \left(\sum_{j=1}^n X_j \right)^p = E \left(\sum_{j=1}^n Y_j \right)^p$$

holds, where the two n-tuples $X_1, X_2, ..., X_n$ and $Y_1, Y_2, ..., Y_n$ consist of independent random variables fulfilling the conditions $X_j \sim X$ and $Y_j \sim Y$ $(j = 1, 2, ..., n)$, respectively. It is evident that

$$X \sim_p 0 \quad \text{yields} \quad X \sim 0. \tag{1}$$

Moreover, it is easy to show that for p being positive integer number the relation $X \sim_p Y$ is equivalent to the conditions $X, Y \in M_p$ and $EX^k = EY^k$ for $k = 1, 2, ..., p$. For the remaining values of p we shall prove a rather surprising result.

Theorem. *Let p be not an integer. Then $X \sim_p Y$ if and only if $X, Y \in M_p$ and $X \sim Y$.*

Let us begin with some notation. In the sequel C will denote the space of all real–valued functions continuous on the compactified half–line $[0, \infty]$

with the norm $\| f \| = \max\{|f(t)| : t \in [0, \infty]\}$. Given a positive number p, $L_{2,p}$ will stand for the space of all real–valued Borel functions g defined on the half–line $[0, \infty)$ and with finite norm

$$\| g \|_{2,p} = \left(\int_0^\infty | g(t) |^2 t^{-1-p} dt \right)^{\frac{1}{2}}.$$

Put $C_p = C \cap L_{2,p}$. The space C_p is equipped with the norm

$$\| \|_p = \| \| + \| \|_{2,p}.$$

Observe that C_p is a Banach algebra under pointwise multiplication and

$$\| fg \|_p \leq \| f \| \ \| g \|_p \tag{2}$$

for $f, g \in C_p$. It is clear that the set C_p is dense in $L_{2,p}$ in the $\| \|_{2,p}$–topology and

$$f(0) = 0 \qquad \text{for} \ \ f \in C_p. \tag{3}$$

By a subalgebra of C_p we mean a subset of C_p closed under linear combinations and multiplication. We say that the subset A of C_p separates points, if for every pair of distinct points $a, b \in [0, \infty]$ there exists a function $f \in A$ such that $f(a) \neq f(b)$. In the sequel lin A will denote the linear span of A. A sequence $\{e_n\}$ of functions of C_p is called an approximate unit if $\lim_{n \to \infty} \| f - f e_n \|_p = 0$ for every $f \in C_p$.

Lemma 1 *Let h be a decreasing nonnegative function from C fulfilling the condition* $(1 - h)^s \in L_{2,p}$ *for a certain positive integer s. Then the set*

$$A_s(h) = lin \left\{ \prod_{j=1}^s (1 - h^{n_j}) : \ n_j = 1, 2, ...; \ j = 1, 2, ..., s \right\}$$

is a subalgebra of C_p *which separates points and contains an approximate unit.*

Proof. Observe that, by (3), $h(0) = 1$, which yields $\| h \| = 1$ and

$$\prod_{j=1}^s (1 - h^{n_j}) \leq \sum_{j=1}^s n_j^s (1 - h)^s,$$

for $n_j = 1, 2, ...; j = 1, 2, ..., s$. Hence we get the inclusion $A_s(h) \subset C_p$. Further from the equality

$$\prod_{j=1}^s (1 - h^{n_j}) \prod_{j=1}^s (1 - h^{m_j}) = \sum \prod_{j \in I} (1 - h^{n_j}) \prod_{j \in J} (1 - h^{m_j}) \prod_{j \in K} (h^{n_j + m_j} - 1),$$

$$\tag{4}$$

where the summation runs over all partitions of the index set $\{1, 2, ..., s\}$ into disjoint subsets I, J and K it follows that the set $A_s(h)$ is closed under multiplication and, consequently, is a subalgebra of C_p. Since the function $(1-h)^s$ is increasing and belongs to $A_s(h)$ we conclude that the set $A_s(h)$ separates points. Finally, setting $e_n = (1 - h^n)^s$ we have $e_n \in A_s(h)$,

$$\| e_n \| \leq 1 \quad (n = 1, 2, ...) \tag{5}$$

and

$$\lim_{n \to \infty} e_n(u) = 1, \tag{6}$$

for every $u \in (0, \infty)$. Since the function $1 - e_n$ is decreasing, we have the inequality

$$\| f - f e_n \| \leq \| 1 - e_n \| \max \{| f(t) |:\ t \in [0, u]\} + \| f \| (1 - e_n(u)), \tag{7}$$

for $f \in C_p$ and $u \in (0, \infty)$. Observe that, by (3), $\lim_{u \to 0} f(u) = 0$. Combining this with (5) and (6) we get from (7) the formula

$$\lim_{n \to \infty} \| f - f e_n \| = 0.$$

On the other hand by the bounded convergence theorem we derive from (5) and (6)

$$\| f - f e_n \|_{2,p}^2 = \int_0^\infty | 1 - e_n(t) |^2 | f(t) |^2 t^{-1-p} dt \to 0,$$

as $n \to \infty$ which shows that the sequence $\{e_n\}$ is an approximate unit. The lemma is thus proved. $\square$

Lemma 2 *Let A be a subalgebra of C_p separating points and containing an approximate unit. Then A is dense in C_p in the $\| \ \|_p$-topology.*

Proof. Let $\{e_n\}$ be an approximate unit belonging to A. Put

$$B = \{f e_n : f \in C_p,\ n = 1, 2, ...\}.$$

It is clear that the set B is dense in C_p in the $\| \ \|_p$-topology. Since the set A separates points and condition (3) holds we conclude, by Stone–Weierstrass Theorem ([1], Theorem 4E), that the set A is dense in C_p in the $\| \ \|$-topology. Consequently, for every $f \in C_p$ we can find a sequence $f_n \in A$ $(n = 1, 2, ...)$ such that $\| f - f_n \| \to 0$ as $n \to \infty$. Since for every index k, $f_n e_k \in A$ and, by (2), $\| f e_k - f_n e_k \|_p \leq \| f - f_n \| \| e_k \|_p$ $(n = 1, 2, ...)$ we infer that the set A is dense in B in $\| \ \|_p$-topology, which completes the proof. $\square$

Lemma 3 *Let A be a subalgebra of C_p separating points and containing an approximate unit. Let U_o be a linear and multiplicative $\| \; \|_{2,p}$-isometry from A into C_p. Then U_o can be extended to a linear $\| \; \|_{2,p}$-isometry U from C_p into $L_{2,p}$ and $(Uf)(t) = f(v(t))$ for every $f \in C_p$ and almost every $t \in [0, \infty]$, where v is a nonnegative Borel function defined on $[0, \infty]$.*

Proof. From the inequality $\| \; \|_{2,p} \leq \| \; \|_p$ it follows that the mapping U_o from the set A equipped with the $\| \; \|_p$-topology into the set $L_{2,p}$ with the $\| \; \|_{2,p}$-topology is continuous. Since, by Lemma 2, the set A is dense in the $\| \; \|_p$-topology, the mapping U_o can be extended to a linear and multiplicative $\| \; \|_{2,p}$-isometry U from C_p into $L_{2,p}$. In particullar we have

$$U(fg) = U(f)\, U(g) \qquad \text{for } f, g \in C_p. \tag{8}$$

Denote by k the positive integer fulfilling the condition $k - 1 < p \leq k$. It is clear that the function $h(t) = \exp(-t^k)$ and $s = 1$ fulfill the conditions of Lemma 1. Thus setting

$$g_n(t) = 1 - \exp(-nt^k) \qquad (n = 1, 2, ...) \tag{9}$$

we conclude that lin $\{g_n : n = 1, 2, ...\}$ is a dense subset of C_p in the $\| \; \|_p$-topology. Introduce the notation

$$Ug_n = 1 - w_n \qquad (n = 1, 2, ...). \tag{10}$$

By (8) we have the inequality

$$U(g_n g_m) = (1 - w_n)(1 - w_m) \qquad (n, m = 1, 2, ...). \tag{11}$$

On the other hand using the formula $g_n g_m = g_n + g_m - g_{n+m}$ we have

$$U(g_n g_m) = 1 - w_n - w_m + w_{n+m} \qquad (n, m = 1, 2, ...).$$

Combining this with (11) we get equality $w_{n+m} = w_n w_m$ for $n, m = 1, 2, ...$, which yields

$$w_n = w_1^n \qquad (n = 1, 2, ...). \tag{12}$$

Again by (8) $U(g_n^r) = (1 - w_n)^r$ $(n = 1, 2, ...)$. Since U is a $\| \; \|_{2,p}$-isometry, we have

$$\int_0^\infty \left(1 - w_n(t)\right)^{2r} t^{-1-p} \, dt = \int_0^\infty \left(1 - \exp(-nt^k)\right)^{2r} t^{-1-p} \, dt.$$

Observe that for every n the right–hand side of the above equality tends to 0 as $r \to \infty$. This yields the inequality $|\, 1 - w_n(t)\,| < 1$ for every n and almost every $t \in [0, \infty)$. Using (12) we conclude that $0 < w_1(t) \leq 1$ for almost every $t \in [0, \infty)$. Put $v(t) = (-\log w_1(t))^{1/k}$. Changing if necessary the function v on a set of the Lebesgue measure 0 we may assume without loss of generality that it is a non–negative Borel function on the half–line $[0, \infty)$ and, by (9), (10) and (12),

$$U(1 - \exp(-nt^k)) = 1 - \exp(-nv(t)^k) \quad (n = 1, 2, ...)$$

or, equivalently,

$$(Ug)(t) = g(v(t)) \quad \text{for} \ g \in \lim\{g_n : \ n = 1, 2, ...\}.$$

Suppose now that $f \in C_p$. As it was mentioned before, the set $\lim \{g_n : n = 1, 2, ...\}$ is dense in C_p in the $\| \ \|_p$–topology. Consequently, there exists a sequence $f_m \in \lim \{g_n : \ n = 1, 2, ...\}$ $(m = 1, 2, ...)$ such, that $\| f - f_m \|_p \to 0$ as $m \to \infty$. Hence it follows that $f(v(t)) - f_m(v(t)) \to 0$ almost everywhere and $\| Uf - Uf_m \|_{2,p} \to 0$ as $m \to \infty$. Noting that $(Uf_m)(t) = f_m(v(t))$ $(m = 1, 2, ...)$ we get the equality $(Uf)(t) = f(v(t))$ almost everywhere, which completes the proof. $\quad\square$

Given a random variable Z, the function $\varphi(t) = Ee^{-tZ}$ for $t \in [0, \infty)$ will be called *the Laplace transform of Z*.

Lemma 4 *Let k be a positive integer and $k-1 < p < k$. Let $Z_1, Z_2, ..., Z_k$ be a sequence of independent random variables from M_p with the Laplace transforms $\varphi_1, \varphi_2, ..., \varphi_k$, respectively. Then*

$$\int_0^\infty \prod_{j=1}^k (1 - \varphi_j(t)) \ t^{-1-p} \ dt = \Gamma(-p) \sum_{r=1}^k (-1)^r \sum_{j_1, ..., j_r} E \ (Z_{j_1} + ... + Z_{j_r})^p,$$

where the summation $\sum_{j_1, ..., j_r}$ runs over all r–element subsets $\{j_1, ..., j_r\}$ of the set of indices $\{1, 2, ..., k\}$.

Proof. For $k = 1$ our statement is a consequence of Fubini's Theorem. In fact we have

$$\int_0^\infty (1 - \varphi_1(t)) \ t^{-1-p} \ dt = E \int_0^\infty (1 - \exp(-tZ_1)) \ t^{-1-p} \ dt = -\Gamma(-p) \ EZ_1^p.$$

Suppose now that $k \geq 2$. Then $EZ_j < \infty$ $(j = 1, 2, ..., k)$ and, consequently, the functions $t^{-1}(1 - \varphi_j(t))$ $(j = 1, 2, ..., k)$ are bounded on the half–line $[0, \infty)$. Hence it follows that

$$\int_0^\infty \prod_{j=1}^k (1 - \varphi_j(t)) \ t^{-1-p} \ dt < \infty. \tag{13}$$

Setting $v_k(t) = e^{-t} - \sum_{j=1}^{k-1} \frac{(-t)^j}{j!}$ we have $(-1)^k v_k(t) \geq 0$ for $t \in [0, \infty)$ and

$$\int_0^\infty E\, v_k(tZ)\, t^{-1-p}\, dt = \Gamma(-p)\, EZ^p \tag{14}$$

for every random variable $Z \in M_p$. Since

$$\prod_{j=1}^k (1 - \varphi_j(t)) = 1 + \sum_{r=1}^k (-1)^r \sum_{j_1,\ldots,j_r} E \exp(-t(Z_{j_1} + \ldots + Z_{j_r})),$$

we have the formula

$$\prod_{j=1}^k (1 - \varphi_j(t)) - \sum_{r=1}^k (-1)^r \sum_{j_1,\ldots,j_r} E\, v_k(t(Z_{j_1} + \ldots + Z_{j_r})) = \sum_{m=0}^{k-1} c_m t^m \tag{15}$$

with some constans $c_0, c_1, \ldots, c_{k-1}$. By (13) and (14) the left–hand side of the above equality is integrable on the half–line $[0, \infty)$ with respect to the measure $t^{-1-p}\, dt$. Hence it follows $c_0 = c_1 = \ldots = c_{k-1} = 0$. Now integrating the left–hand side of (15) with respect to the measure $t^{-1-p}\, dt$ and using formula (14) we get the assertion of the Lemma. $\quad\square$

As an immediate consequence of Lemma 4 we get the following statement.

Lemma 5 *Let k be a positive integer and $k - 1 < p < k$. Let X and Y be random variables with the Laplace transforms φ and ψ, respectively. If $X \sim_p Y$, then*

$$\int_0^\infty \prod_{j=1}^k (1 - \varphi^{n_j}(t))\, t^{-1-p}\, dt = \int_0^\infty \prod_{j=1}^k (1 - \psi^{n_j}(t))\, t^{-1-p}\, dt$$

for every k–tuple $n_1, n_2, \ldots, n_k$ of positive integers. Moreover, the above integrals are finite.

We are now in a position to prove the Theorem.

Proof of the Theorem. Denote by k the positive integer fulfilling the condition $k - 1 < p < k$. Suppose that $X \sim_p Y$ and denote by φ and ψ the Laplace transforms of X and Y, respectively. By (1) it suffices to consider the case of random variables X and Y, which are nondegenerate at the origin. Then we have $\varphi(\infty) < 1$ and $\psi(\infty) < 1$. Moreover without loss of generality we may assume that

$$\varphi(\infty) \leq \psi(\infty). \tag{16}$$

From formula (4) and Lemma 5 it follows that $(1 - \varphi)^k \in L_{2,p}$. Consequently, the function $h = \varphi$ and $s = k$ fulfill the conditions of Lemma 1. Applying Lemma 2 we conclude that the set $A_k(\varphi)$ is dense in C_p in the $\| \ \|_p$–topology. From formula (4) and Lemma 5 it follows that the mapping U_o defined by the formula

$$U_o \left(\prod_{j=1}^{k} (1 - \varphi^{n_j}) \right) = \prod_{j=1}^{k} (1 - \psi^{n_j}) \tag{17}$$

for any k–tuple $n_1, n_2, ..., n_k$ of positive integers can be extended to a linear and multiplicative $\| \ \|_{2,p}$–isometry U_o from $A_k(\varphi)$ onto $A_k(\psi)$. Now applying Lemma 3 we conclude that the mapping U_o has an extension to a linear $\| \ \|_{2,p}$–isometry U from C_p into $L_{2,p}$ of the form

$$(Uf)(t) = f(v(t)) \tag{18}$$

for almost all $t \in [0, \infty)$ and $f \in C_p$, where v is a nonnegative Borel function defined on the half–line $[0, \infty)$. In particular we have

$$U \left((1 - \varphi)^k \right)(t) = (1 - \varphi(v(t)))^k$$

almost everywhere, which, by (17), yields the equality

$$\psi(t) = \varphi(v(t)) \tag{19}$$

for almost all $t \in [0, \infty)$. Let ϑ be the inverse function of φ. The function ϑ maps the interval $(\varphi(\infty), 1)$ onto the half–line $[0, \infty)$ and is infinitely differentiable on the open interval $(\varphi(\infty), 1)$. By inequality (16) the superposition $\vartheta(\psi(t))$ is well defined for $t \in [0, \infty)$ and, by (19), $\vartheta(\psi(t)) = v(t)$ almost everywhere. Changing if necessary the function v on a set of the Lebesgue measure 0 we may assume without loss of generality that the above equality holds for all $t \in [0, \infty)$. Hence it follows that the function v is continuously differentiable on $(0, \infty)$ and its derivative v' fulfils the inequality $v'(t) = \psi'(t)/\varphi'(v(t)) > 0$ for $t \in (0, \infty)$. Consequently, the function v is increasing. Moreover, $v(0) = \vartheta(1) = 0$. Let w be the inverse function of v. It is clear that the function w maps the interval $[0, v(\infty))$ onto the half–line $[0, \infty)$, is increasing and continuously differentiable on $(0, v(\infty))$. Since U is a $\| \ \|_{2,p}$–isometry we have, by (18),

$$\int_0^\infty f^2(t) \, t^{-1-p} \, dt = \int_0^\infty f(v(t))^2 \, t^{-1-p} \, dt$$

for every $f \in C_p$. Substituting $t = w(s)$ for $s \in [0, v(\infty))$ we get the formula

$$\int_0^\infty f^2(t) \, t^{-1-p} \, dt = \int_0^{v(\infty)} f^2(s) \, w'(s) \, w(s)^{-1-p} \, ds,$$

which yields $v(\infty) = \infty$ and $w'(t)\,w(t)^{-1-p} = t^{-1-p}$ for $t \in [0, \infty)$. Solving the above equation under the condition $w(\infty) = \infty$ we get the formula $w(t) = t$ or, equivalently, $v(t) = t$ for $t \in [0, \infty)$. Taking into account (19) we have $\varphi = \psi$ and, consequently, $X \sim Y$. The converse implication is obvious. That completes the proof. $\square$

References

[1] L.H. Loomis, (1953) *An Introduction to Abstract Harmonic Analysis*, D.Van Nostrand, Toronto New York London.

Institute of Mathematics
University of Wrocław
pl. Grunwaldzki 2–4
50–384 Wrocław, Poland.

Relative Entropy and Hydrodynamic Limits

S. R. S. Varadhan*

Abstract

For the non-gradient version of the Ginzburg-Landau type model for which we established a hydrodynamic scaling limit earlier, we now show that if we start from an initial distribution that is locally Gibbsian then at any later time the distibutions remain close to locally Gibbsian distributions.

In [1] we considered the following model. N lattice sites are arranged periodically in one dimension with a lattice spacing of $1/N$. We have spin variables x_j attached to each site j/N, the sites being viewed as equally spaced points on the circle of unit circumference. The spins x_j vary in time in such a manner that they undergo a diffusion on R^N denoted by $\{x_1(t), \ldots, x_N(t)\}$. The diffusion process is described by

$$dx_i(t) = dz_{i-1,i}(t) - dz_{i,i+1}(t)$$

$$dz_{i,i+1}(t) = \frac{N^2}{2} [\phi'(x_i(t)) - \phi'(x_{i+1}(t))] \, dt + N \, d\beta_{i,i+1}(t)$$

Here $\phi'(x) = d\phi/dx$ for a suitable function $\phi(x)$. One can think of $\phi(x)$ as somewhat like $x^2/2$. If ϕ where $x^2/2$ we would have a linear model and our model should be thought of, as a nonlinear version. The random process $\beta_{i,i+1}(t)$ are N, independent Brownian motions. Equivalently one could write down the generator of the process

$$L_N = \frac{N^2}{2} \sum \left(\frac{\partial}{\partial x_i} - \frac{\partial}{\partial x_{i+1}} \right)^2 - \frac{N^2}{2} \sum [\phi'(x_i) - \phi'(x_{i+1})] \left[\frac{\partial}{\partial x_i} - \frac{\partial}{\partial x_{i+1}} \right]$$

One can also describe the process through its Dirichlet form

$$\frac{N^2}{2} \int \sum \left(\frac{\partial u}{\partial x_i} - \frac{\partial u}{\partial x_{i+1}} \right)^2 e^{-\Sigma \phi(x_i)} \prod dx_i$$

We normalize ϕ so that $\int \exp[-\phi(x)] \, dx = 1$. We assume

$$M(\theta) = \int \exp [\theta x - \phi(x)] \, dx$$

---

* Research partially supported by U. S. National Science Foundation grants DMS 89001682 and DMS 9100383.

is finite for all $\theta \in R$. Then

$$h(x) = \sup_\theta [\theta x - \log M(\theta)]$$

is strictly convex and

$$\lambda = h'(x)$$

$$x = \frac{M'(\lambda)}{M(\lambda)} = [\log M(\lambda)]'$$

are inverse functions. The measure $\prod e^{-\phi(x_i)} dx_i$ is invariant for the diffusion. It is however not ergodic. $x_1 + \cdots + x_N$ is a conserved quantity. The ergodic pieces are restrictions of $\prod e^{-\phi(x_i)} dx_i$ to various hyperplanes $\frac{1}{N} \sum x_i = a$. The factor N^2 corresponds to speeding up of time. Since space has already been scaled by a factor of N, we are clearly anticipating a diffusive limit in our choice of parabolic scaling of space and time. We would like to think of

$$\nu_N(t) = \frac{1}{N} \sum x_j(t) \, \delta_{j/N}$$

as a randomly changing signed measure on the circle S. At time 0, we have a random initial configuration with density

$$f_N^0(x_1, \ldots, x_N) \exp \left[- \sum \phi(x_i) \right] \prod dx_i$$

on R^N. The distribution of the configuration at time t will be given by a density

$$f_N^t(x_1, \ldots, x_N) \exp \left[- \sum \phi(x_i) \right] \prod dx_i$$

where f_N^t satisfies the Kolmogorov equation

$$\frac{\partial f_N^t}{\partial t} = L_N f_N^t$$

$$f_N^t\big|_{t=0} = f_N^0$$

Although in princple we should use L_N^*, because L_N is symmetric with respect to the weight $\exp[- \sum \phi(x_i)]$ we do not have to take the adjoint. Suppose at time 0 for some smooth deterministic profile $m_0(\theta)$

$$\nu_N(0) \sim m_0(\theta) \, d\theta \ ,$$

or more precisely

$$\frac{1}{N} \sum J\left(\frac{i}{N}\right) x_i \to \int J(\theta) \, m_0(\theta) \, d\theta$$

in probability with respect to $f_N^0(x_1, \ldots, x_N) \exp[-\sum \phi(x_i)] \prod dx_i$ for every nice test function $J(\theta)$ on S. Under suitable regularity conditions of a simple nature on f_N^0 we want to conclude that for some profile $m(\theta, t)$ we have

$$\frac{1}{N} \sum J(\frac{i}{N}) x_i \to \int J(\theta) m(\theta, t) d\theta$$

in probability with respect to $f_N^t(x_1, \ldots, x_N) \exp[-\sum \phi(x_i)] \prod dx_i$. In addition we should be able to describe how to compute $m(\theta, t)$ from the initial profile $m_0(\theta)$. Hopefully $m(\theta, t)$ does not depend on f_N^0 in any manner other than through the initial profile $m_0(\theta)$ determined by it. The main step is to compute

$$d \frac{1}{N} \sum J(\frac{i}{N}) x_i(t) = -\frac{N^2}{2} \cdot \frac{1}{N} \sum [J(\frac{i+1}{N}) - J(\frac{i}{N})]$$
$$\cdot [\phi'(x_{i+1}(t)) - \phi'(x_i(t))] \, dt$$
$$+ N \cdot \frac{1}{N} \sum [J(\frac{i+1}{N}) - J(\frac{i}{N})] \, d\beta_{i,i+1}(t)$$
$$= \frac{1}{2N} \cdot N^2 \sum [J(\frac{i+1}{N}) - 2J(\frac{i}{N}) + J(\frac{i-1}{N})] \, \phi'(x_i(t))$$
$$+ d(\text{"noise"})$$

Approximately "noise" $= \frac{1}{N} \sum J'(\frac{i}{N}) \, d\beta_{i,i+1}(t)$ and goes to zero by the law of large numbers. We therefore have roughly

$$d(\frac{1}{N} \sum J(\frac{i}{N}) x_i(t)) \simeq \frac{1}{2N} \sum J''(\frac{i}{N}) \, \phi'(x_i(t)) \, dt$$

The problem is that the measure

$$\frac{1}{2N} \sum \phi'(x_i) \delta_{i/N}$$

is not so easily expressed in terms of $\frac{1}{N} \sum x_i \delta_{i/N}$. We need to know how the spins arrange themselves so that a knowledge of

$$\frac{1}{N\epsilon} \sum_B x_j = m$$

determines

$$\frac{1}{N\epsilon} \sum_B \phi'(x_j)$$

over a block B of size $N\epsilon$ (ϵ is a small positive number).

On the infinite lattice the invariant measures for the diffusion are

$$\prod \frac{1}{M(\lambda)} e^{\lambda x_i - \phi(x_i)} dx_i$$

for various values of λ. These are the "Gibbs" measures in this context.

$$\text{The average spin} = \int x \cdot \frac{1}{M(\lambda)} e^{\lambda x - \phi(x)} dx$$
$$= \frac{M'(\lambda)}{M(\lambda)} = m .$$

Therefore we know that λ has to be chosen so that

$$\lambda = h'(m) .$$

Moreover

$$\int \phi'(x) \frac{1}{M(\lambda)} e^{\lambda x - \phi(x)} dx = \lambda$$

so that, if the average spin is m and the spins are organized according to a Gibbs ensemble then the average of $\phi'(x)$ is $h'(m)$.

If we assume that locally the system is close to Gibbs ensembles then we get

$$\frac{\partial}{\partial t} \int J(\theta) m(\theta, t) d\theta = \frac{1}{2} \int J''(\theta) h'(m(\theta, t) d\theta$$

or equivalently the nonlinear heat equation

$$\frac{\partial m(\theta, t)}{\partial t} = \frac{1}{2} [h'(m(\theta, t))]_{\theta\theta}$$

with initial condition

$$m(\theta, 0) = m_0(\theta)$$

This is the hydrodynamic equation. The problem is to prove that the density

$$f_N^t(x_1, \ldots, x_N) \exp \left[- \sum \phi(x_i) \right] \prod dx_i$$ has the correct profile. This was done in [1]. In [3], H. T. Yau used a method of relative entropy to study this problem.

Assume that $m(\theta, t)$ is given as a smooth solution of

$$\frac{\partial m}{\partial t} = \frac{1}{2} (h'(m(\theta, t))_{\theta\theta}$$

We go from $m(\theta, t)$ to $\lambda(\theta, t)$ where

$$\lambda(\theta, t) = h'(m(\theta, t))$$

λ is also a smooth function. We expect f_N^t to be approximately of the form

$$(1) \qquad g_N^t(x_1, \ldots, x_N) = \prod_{j=1}^{N} \frac{1}{M(\lambda(j/N, t))} \, e^{\lambda(j/N,t)x_j}$$

Let us asume $f_N^0 = g_N^0$, a special locally Gibbsian ensemble. We want to see how close f_N^t is to g_N^t for $t > 0$. As a measure of closeness we take relative entropy

$$\overline{H}_N(t) = \int \log \frac{f_N^t}{g_N^t} \cdot f_N^t \, e^{-\sum \phi(x_i)} \prod dx_i$$

$$\overline{H}_N(0) = 0$$

It is shown in [3] that

$$\lim_{N \to \infty} \frac{1}{N} \overline{H}_N(t) = 0 \quad \text{for} \quad t > 0$$

This proves hydrodynamic scaling. The profile is correct for g_N^t with exponentially small error probabilities using the large deviation theory. The entropy inequality

$$Q(A) \le \frac{H(Q, P) + 1}{\log 1/P(A)}$$

takes care of the rest.

We can use ideas from large deviation theory to handle the following modification of the model that we just considered. We change the Dirichlet form to

$$\frac{N^2}{2} \int \sum a(x_i, x_{i+1}) \left(\frac{\partial u}{\partial x_i} - \frac{\partial u}{\partial x_{i+1}} \right)^2 e^{-\sum \phi(x_i)} \prod dx_i$$

The corresponding operator is

$$(2) \qquad \begin{aligned} L_N = {}& \frac{N^2}{2} \sum a(x_i, x_{i+1}) \left(\frac{\partial}{\partial x_i} - \frac{\partial}{\partial x_{i+1}} \right)^2 \\ & - \frac{N^2}{2} \sum W(x_i, x_{i+1}) \left(\frac{\partial}{\partial x_i} - \frac{\partial}{\partial x_{i+1}} \right) \end{aligned}$$

where

$$W(x, y) = a(x, y) \left[\phi'(x) - \phi'(y) \right] - \left[a_x(x, y) - a_y(x, y) \right]$$

If we compute

$$d\left(\frac{1}{N}\sum J\left(\frac{i}{N}\right)x_i(t)\right)$$

and ignore the noise terms that are negligible, we now get

$$d\left(\frac{1}{N}\sum J\left(\frac{i}{N}\right)x_i(t)\right) \simeq \sum J'\left(\frac{i}{N}\right)W(x_i(t),x_{i+1}(t))\,dt$$

There is a factor of $1/N$ missing because we could do summation by parts only once. On the other hand $E\{W(x_i,x_{i+1})\} = 0$ under every Gibbs measure so we need to do more analysis. By using ideas from large deviations one can establish hydrodynamical limits to an equation

$$(3) \qquad \frac{\partial m}{\partial t} = \frac{1}{2}\frac{\partial}{\partial \theta}\left(\hat{a}(m(\theta,t))\,(h'(m(\theta,t))_\theta)\right)$$

where $\hat{a}(m)$ is the Green-Kubo diffusion coefficient. It turns out one can actually give a variational characterization of $\hat{a}(m)$. This was carried out in [2].

A question that arises naturally is the validity of the analog of Yau's result [3] in this more general context. A direct proof along the lines of [3] seems impossible at the moment. The purpose of the note is to demonstrate the tools and the results in [2] are capable of yielding Yau's result on relative entropy in this context. We define for any solution $m(t,\theta)$ of (2) the local Gibbs density g_N^t by (1) and f_N^t is defined as the solution of Kolmogorov's equation

$$\frac{\partial f_N^t}{\partial t} = L_N f_N^t$$

$$f_N^t\big|_{t=0} = g_N^0$$

where L_N is now given by (3). We define relative entropy as before by

$$H_N(t) = \int \log \frac{f_N^t}{g_N^t} \cdot f_N^t\, e^{-\Sigma\phi(x_i)}\,\Pi\,dx_i$$

We also define the entropy

$$H_N^0(t) = \int f_N^t \log f_N^t\, e^{-\Sigma\phi(x_i)}\,\Pi\,dx_i$$

relative to the equilibrium and the entropy production

$$I_N(t) = \frac{N^2}{2}\int \frac{1}{f_N^t}\sum\left(\frac{\partial f_N^t}{\partial x_i} - \frac{\partial f_N^t}{\partial x_{i+1}}\right)^2 e^{-\Sigma\phi(x_i)}\,\Pi\,dx_i$$

$$= -\frac{d}{dt}H_N^0(t)\,.$$

We have the following lemma:

Lemma. *For every $t > 0$*

$$\liminf_{N\to\infty} \frac{1}{N} I_N(t) \geq \frac{1}{2} \int \hat{a}(m(t,\theta))\, (\lambda_\theta(t,\theta))^2 \, d\theta$$

Proof: The proof is based on two observations:

$$\lim_{N\to\infty} \int \sum J(\frac{i}{N})(\phi'(x_i) - \phi'(x_{i+1}))\, f_N^t \, e^{-\Sigma\phi(x_i)} \, \Pi \, dx_i$$

$$= \lim_{N\to\infty} \int \frac{1}{N} \sum J'(\frac{i}{N})\, \phi'(x_i)\, f_N^t \, e^{-\Sigma\phi(x_i)} \, \Pi \, dx_i$$

$$= \int \lambda(t,\theta)\, J'(\theta) \, d\theta \ .$$

which is contained in [2] and used repeatedly. On the other hand it is shown in [2] that

$$\lim_{\epsilon\to 0} \limsup_{N\to\infty} \Big[\int \Big\{ \sum J(\frac{i}{N})(\phi'(x_i)-\phi'(x_{i+1})) - \frac{2}{N} \sum J^2(\frac{i}{N}) \frac{1}{\hat{a}(\bar{x}_{i,N\epsilon})} \Big\}$$

$$f_N^t \, e^{-\Sigma\phi(x_i)} \, \Pi \, dx_i - \frac{1}{4N} I_N^t \Big] \leq 0$$

Here $\bar{x}_{i,N\epsilon}$ denotes $\frac{1}{2N\epsilon} \sum_{|j-i|\leq N\epsilon} x_j$. It follows now that

$$\int \lambda(t,\theta)\, J'(\theta)\, d\theta - 2 \int J^2(\theta) \frac{d\theta}{\hat{a}(m(t,\theta))} \leq \liminf_{N\to\infty} \frac{I_N(t)}{4N}$$

Since J is arbitrary we get the lemma.

Theorem. *For any $t \geq 0$*

$$\lim_{N\to\infty} \frac{1}{N} H_N^0(t) = \int h(m(t,\theta))\, d\theta \ .$$

Proof: By the choice of g_N^0

$$\lim_{N\to\infty} \frac{1}{N} H_N^0(0) = \int h(m(0,\theta))\, d\theta$$

and

$$H_N^0(t) = H_N^0(0) - \int_0^t I_N(s)\, ds$$

By Fatou's lemma and our earlier lemma

$$\limsup_{N \to \infty} \frac{1}{N} H_N^0(t) \leq \int h(m(0, \theta)) - \frac{1}{2} \int_0^t \int \hat{a}(m(t, \theta))(\lambda_\theta(t, \theta))^2 \, d\theta$$

$$= \int h(m(0, \theta)) + \int \int_0^t [\frac{d}{ds} h(m(s, \theta))] \, ds \, d\theta$$

$$= \int h(m(t, \theta)) \, d\theta \ .$$

By general large deivation theory

$$\liminf_{N \to \infty} \frac{1}{N} H_N^0(t) \geq \int h(m(t, \theta)) \, d\theta \ ,$$

and we are done.

Corollary.

$$\lim_{N \to \infty} \frac{1}{N} H_N(t) = 0.$$

Proof: By direct computation

$$\frac{1}{N} H_N(t) = \frac{1}{N} H_N^0(t) - \frac{1}{N} \int \sum \lambda(t, \frac{i}{N}) x_i f_N^t \, e^{-\Sigma \phi(x_i)} \, \pi \, dx_i$$

$$- \frac{1}{N} \sum \log M(\lambda(t, \frac{i}{N}))$$

Taking limits

$$\lim_{N \to \infty} \frac{1}{N} H_N(t) = \int h(m(t, \theta)) \, d\theta - \int \lambda(t, \theta) \, m(t, \theta) \, d\theta$$

$$+ \int \log M(\lambda(t, \theta)) \, d\theta$$

$$= 0$$

by properties of Legendre transform.

References
[1] Guo, M. Z., Papanicolaou, G. C., Varadhan, S. R. S., Comm. Math. Phys. 118, 31 (1988).
[2] Varadhan, S. R. S., to appear.
[3] Yau, H. T., Lett. math. Phys. 22, 63 (1991).

Courant Institute of mathematical sciences,
New York University, New York.

Donsker's δ-function and its applications in the theory of white noise analysis

HISAO WATANABE

Abstract. We dicuss the renormalization of the second order moment of the local time of self-intersections for Brownian motion.

Let $B(t) = (B_1(t), B_2(t), \cdots, B_d(t))$ be a d-dimentional Brownian motion with $B(0) = 0$ on a probability space $(\Omega, \mathcal{F}, P)$ and let $\delta(\cdot)$ be the Dirac δ-function. We shall formally consider $\delta(B(t))$, which will be called Donsker's δ-function. It was firstly considered by Kuo [2] in the framework of white noise analysis for $d = 1$. After that, Kallianpur and Kuo [1] dicussed its regurality properties. For $d \geq 2$, a point is a polar set for Brownian motion. Therefore, in almost sure sense, $\delta(B(t))$ might be zero. But, in the stochastic calculus (also, in the Malliavin calculus), it is meaningful. In fact, Donsker's δ-function $\delta(B(t) - a), a \in R^d$, is a generaliged Brownian functional which is characterized by its S-transformation , i.e.,

$$(1) \qquad S(\delta(B(t) - a)(\xi) = p(t, \int_0^t \xi(u)du - a),$$

where $\xi = (\xi_1, \cdots, \xi_d), \xi_i$ is an element of the space of rapidly decreasing smooth functions on R and $p(t, x) = (2\pi t)^{-d/2} exp(-|x|^2/(2t))$. As it can be seen easily, we have the following equality:

$$(2) \qquad p(t, \int_0^t \xi(u)du - a) = p(t, a)$$
$$+ \sum_{i=1}^d \int_0^t \frac{\partial}{\partial x_i} p(t, \int_0^r \xi(u)du - a)\xi_i(r)dr.$$

In terms of Brownian functionals, we can rewrite the above relation (2) as follows:

$$(3) \qquad \delta(B(t) - a) = p(t, a) + \sum_{i=1}^d \int_0^t \frac{\partial}{\partial x_i} p(t - r, B(r) - a)dB_i(r).$$

However, we have to keep in mind that equation (3) has no meaning in the ordinary sense, but it is meaningful in the theory of white noise analysis of Hida.

In H.Watanabe [5], we discussed the renormalization of local time of self-intersections of Brownian motions which play an important role in polymer physics. In fact, for $d = 2$ *or* 3, if we put $\{\delta(B(u) - B(s))\} = \delta(B(u) - B(s)) - p(s - t, 0)$, we can show that

$$(4) \qquad \int_0^1 \int_0^1 \{\delta(B(u) - B(s))\}duds$$

is a generalized Brownian functional in the sense of Hida. This result is analogous to Varadhan's renormalization for $d = 2$ (see [4]).

In the following, we shall discuss the 2nd moment of (4). It is written as follows,

$$(5) \qquad \int_0^1 \int_0^1 \int_0^1 \int_0^1 \{X(u, v, s, t)\}dudvdsdt,$$

where $\{X(u, v, s, t)\} = \{\delta(B(v) - B(u))\}\{\delta(B(t) - B(s))\}$. We have to calculate the S-transformation of $\{X(u, v, s, t)\}$. Assume that $0 \leq t_1 \leq t_2 \leq t_3 \leq t_4 \leq 1$. Then, the S-transformation of $\{X(t_1, t_2, t_3, t_4)\}$ is easy to calculate as in H.Watanabe[5]. We have

$$V_1(\xi) = \iiiint_{0 \leq t_1 \leq t_2 \leq t_3 \leq t_4 \leq 1} S(\{X(t_1, t_2, t_3, t_4)\})(\xi)dt_1 dt_2 dt_3 dt_4 =$$

$$= \iiiint_{0 \leq t_1 \leq t_2 \leq t_3 \leq t_4 \leq 1} (p(t_2 - t_1, \int_{t_1}^{t_2} \xi du) - p(t_2 - t_1, 0))$$

$$\times (p(t_4 - t_3, \int_{t_3}^{t_4} \xi du) - p(t_4 - t_3, 0))dt_1 dt_2 dt_3 dt_4,$$

which converges. Also, it turns out that V_1 satisfies the properties required in the characterization theorem of Hida distributions by Potthoff and Streit[3].

However, for $X(t_1, t_3, t_2, t_4) = \delta(B(t_3) - B(t_1))\delta(B(t_4) - B(t_2))$, the above renormalization is not applicable. We propose another normalization. Now, we consider the following formal expression:

$$X(t_1, t_3, t_2, t_4)$$

$$= \int_{R^d} \delta(B(t_2) - B(t_1) - a)\delta(B(t_3) - B(t_2) + a)\delta(B(t_4) - B(t_3) - a)da.$$

As suggested from this identity, we define the renormalization of X as follows:

$$\{X(t_1, t_3, t_2, t_4)\} =$$

$$= \int_{R^d} \{\delta(B(t_2) - B(t_1) - a)\}\{\delta(B(t_3) - B(t_2) + a)\}\{\delta(B(t_4) - B(t_3) - a)\}da.$$

Then, we can calculate its S-transformation, i.e.

$$V_2(\xi) = \iiiint\limits_{0 \le t_1 \le t_2 \le t_3 \le t_4 \le 1} S(\{X(t_1, t_3, t_2, t_4)\})(\xi) dt_1 dt_2 dt_3 dt_4$$

$$= \iiiint\limits_{0 \le t_1 \le t_2 \le t_3 \le t_4 \le 1} \int_{R^d} \left(p(t_2 - t_1, \int_{t_1}^{t_2} \xi du - a) - p(t_2 - t_1, a) \right)$$

$$\times \left(p(t_3 - t_2, \int_{t_2}^{t_3} \xi du + a) - p(t_3 - t_2, a) \right)$$

$$\times \left(p(t_4 - t_3, \int_{t_3}^{t_4} \xi du - a) - p(t_4 - t_3, a) \right) da\, dt_1 dt_2 dt_3 dt_4.$$

After some calculation, we can show that $V_2(\xi)$ converges and also enjoys the properties required in the characterization theorem of the Hida distributions. We can apply the same procedure to $X(t_1, t_4, t_2, t_3)$.

REFERENCES

1. G. Kalianpur and H.H. Kuo, *Regularity property of Donsker's delta function*, Appl. Math. Optim. **12** (1984), 89–95.
2. H.H. Kuo, *Donsker's delta function as a generalized Brownian functional and its application*, in "Lecture Notes in Control and Inforfation Science vol 49," Springer-Verlag, New York, 1983, pp. 167–178.
3. J. Potthoff and L. Streit, *A characterization of Hida distribution*, J. Funct. Anal. (to appear).
4. S. R. S. Varadhan, *Appendix to Euclidean Quantum Field Theory by K. Symanzik*, in "Local Quantum Theory (edited by R. Jost)," Academic Press, New York, 1969.
5. H. Watanabe, *The local time of self-intersections of Brownian motions as generalized Brownian functionals*, Lett. Math. Phys. **23** (1991), 1–9.

Department of Applied Mathematics
Okayama University of Science
Ridaicho 1-1 Okayama 700 Japan

A Fractional Calculus on Wiener Space

SHINZO WATANABE

Abstract. The regularity in the sense of Hölder continuity of a class of conditional expectation is obtained on a Wiener space by using the notion of Donsker's delta functions, typical generalized Wiener functionals belonging to fractional order Sobolev spaces with negative differentiability indices.

1. Introduction

One of the remarkable features in the recent development of Malliavin calculus is that it has introduced a *smooth* or *differentiable* structure on the Wiener space so we can develop thereon a *smooth* probability theory, whereas standard theories of probability have been developed so far only on a measurable structure. As a typical merit, we can discuss thereby a smooth theory of conditional expectations for given smooth and nondegenerate finite dimensional Wiener maps, in other words, we can discuss a smooth disintegration theory on a foliation of hypersurfaces imbedded in the Wiener space. Here, an important role is played by what we call Donsker's delta functions, an important class of generalized Wiener functionals, first introduced by H.-H. Kuo ([3], [7]).

Our idea is as follows. Suppose we are given a d-dimensional Wiener map

$$F : B \to \mathbf{R}^d$$

defined on an abstract Wiener space B which is smooth and non-degenerate in the sense of Malliavin. Then for each $x \in \mathbf{R}^d$ which is in the strict support of the laws of F, a hypersurface S_x in B is defined by

$$S_x = F^{-1}(x) = \{w \in B|\ F(w) = x\}$$

so that we have a foliation $\{S_x\}$ of hypersurfaces imbedded in B. Arguing quite formally and heuristically, consider the composite

$$\delta_x \circ F = \delta_x(F(w)),$$

where δ_x is the Dirac delta function at $x \in \mathbf{R}^d$. Then, a natural measure on S_x induced from the Wiener measure should be the measure having $\delta_x \circ F$

as its density with respect to the Wiener measure. It is needless to say that $\delta_x \circ F$ is no longer a Wiener functional in the sense of random variable on the Wiener space and the induced measure on S_x is singular to Wiener measure. The idea is, however, to define $\delta_x \circ F$ as a *generalized Wiener functional* or a *Schwartz distribution on the Wiener space*. Once this notion can be properly defined, then its generalized expectation

$$p_F(x) = E[\delta_x \circ F]$$

should be the density at x of the law of F and, for a smooth Wiener functional G, we should have the following identity concerning the conditional expectation given $F = x$:

$$E[G \cdot \delta_x \circ F] = E[G|F = x]p_F(x).$$

In the Malliavin calculus, a family of Sobolev-type Banach spaces formed of Wiener and generalized Wiener functionals has been introduced and therefore it is quite natural to define this composite $\delta_x \circ F$, called a *Donsker's delta function*, to be an element in some Sobolev space with negative differentiability index. We can then discuss the smoothness of the Banach space valued map

$$x \mapsto \delta_x \circ F$$

and thereby the smoothness of conditional expectation given $F = x$. Also, by a general result obtained by Sugita ([6]) and Airault-Malliavin ([1]), among others, we can always associate a finite Borel measure to any positive generalized Wiener functional and hence the induced measure on S_x can be obtained as the one associated to Donsker's delta function. We should remark that a somewhat different approach has been given recently by Feyel and De la Pradelle ([2]) in which they first constructed a Hausdorff measure of finite codimension on the Wiener space, a quite general and intrinsic notion, so that it induces a finite measure on each S_x by restriction, and then, established a disintegration theory by obtaining a co-area formula of Federer.

An approach via Malliavin calculus, so far, was based essentially on integration by parts on Wiener space and chain rules for derivatives of composite functions so that, in general statements of results, the differentiability indices of Sobolev spaces were usually restricted to integers. Main aim in the present exposition is to *fractionalize* these results to general nonintegral indices with help of fractional calculus on Sobolev spaces based essentially on complex interpolation and L^p-boundedness of a class of operators. We can thereby sharpen the results obtained so far and widen their applicability.

2. Donsker's delta functions in the Malliavin calculus

Let (B, H, μ) be an abstract Wiener space in the usual standard notation; in particular, H is the Cameron-Martin space. The Malliavin calculus has been developed as a differential-integral calculus or a Schwartz distribution theory on B. Main ingredients are notions of differential operators like the gradient operator (or Shigekawa's H-derivative) D, its dual divergence operator (or the Skorohod operator) D^*, the Ornstein-Uhlenbeck operator $L = -D^*D$ and notions of Sobolev space $\mathcal{D}_p^s$, $1 < p < \infty$, $s \in \mathbf{R}$, of real Wiener and generalized Wiener functionals. Roughly

$$(1) \qquad \mathcal{D}_p^s = (I - L)^{-s/2}(\mathcal{L}_p)$$

with the norm

$$(2) \qquad \|F\|_{p,s} = \|(I - L)^{s/2}F\|_{\mathcal{L}_p}$$

where $\mathcal{L}_p$ is the usual L_p-space. In particular,

$$(3) \quad \mathcal{D}_p^0 = \mathcal{L}_p, \quad \mathcal{D}_p^s \hookrightarrow \mathcal{D}_{p'}^{s'} \text{ if } p' < p \text{ and } s' < s, \quad (\mathcal{D}_p^s)' = \mathcal{D}_{p/(p-1)}^{-s}.$$

So, roughly, $F \in \mathcal{D}_p^s$ if and only if the derivatives of F up to the s-th order belong to $\mathcal{L}_p$ and therefore, p and s are called the *integrability index* and the *differentiability index* of the Sobolev space, respectively. Similarly as in the Schwartz theory of distributions, we define the space of test Wiener functionals by

$$\mathcal{D}_{\infty-}^\infty = \bigcap_{s>0} \bigcap_{1<p<\infty} \mathcal{D}_p^s$$

and then its dual, the space of generalized Wiener functionals, is given by

$$\mathcal{D}_{1+}^{-\infty} = \bigcup_{s>0} \bigcup_{1<p<\infty} \mathcal{D}_p^{-s}.$$

For every $\alpha \geq 0$ and $1 < p, q, r < \infty$ such that $p^{-1} + q^{-1} = r^{-1}$, the multiplication

$$(F, G) \in \mathcal{D}_p^\alpha \times \mathcal{D}_q^\alpha \to F \cdot G \in \mathcal{D}_r^\alpha$$

is continuous. This is seen easily by the Leibniz rule when α is an integer and by a similar interpolation argument as in [9], in general (cf. also [4]). Hence, $\mathcal{D}_{\infty-}^\alpha = \bigcap_{1<p<\infty} \mathcal{D}_p^\alpha$ is an algebra. In particular, $\mathcal{D}_{\infty-}^\infty$ is an algebra and $\mathcal{D}_{1+}^{-\infty}$ is an $\mathcal{D}_{\infty-}^\infty$-module. Furthermore, the *generalized expectation* $E(\Phi)$ of $\Phi \in \mathcal{D}_{1+}^{-\infty}$ is, by definition, the natural coupling of Φ with 1, the test Wiener functional identically equal to 1. If we consider generally E-valued Wiener functionals, E being a real separable Hilbert space, the corresponding spaces are denoted by $\mathcal{L}_p(E)$, $\mathcal{D}_p^s(E)$, etc. We refer to, e.g., [8] for details of the above notions and their basic facts like Meyer's equivalence of norms.

Let $S'(\mathbf{R}^d)$ be the real space of Schwartz tempered distributions on $\mathbf{R}^d$ and $S(\mathbf{R}^d)$ be its subspace of rapidly decreasing C^∞-functions. Choose a system of increasing norms $\| \ \|_n$, $n \in \mathbf{Z}$, such that $S(\mathbf{R}^d) = \bigcap_{n=1}^\infty S_n$ and $S'(\mathbf{R}^d) = \bigcup_{n=1}^\infty S_{-n}$, S_n being the closure of $S(\mathbf{R}^d)$ w.r.t. $\| \ \|_n$, (cf. [8] for an example of such norms; a particular choice is irrelevant, however, in future discussions).

Let F be a d-dimensional Wiener functional or map, i.e., a μ-measurable map

$$F: B \to \mathbf{R}^d.$$

For given $T \in S'(\mathbf{R}^d)$, we want to define the composite $T \circ F = T(F(w))$ as an element in $\mathcal{D}_{1+}^{-\infty}$. If $T = \varphi \in S(\mathbf{R}^d)$, then $\varphi \circ F$ is obviously a Wiener functional such that $\varphi \circ F \in \bigcap_{1<p<\infty} \mathcal{D}_p^0 \ (= \bigcap_{1<p<\infty} \mathcal{L}_p)$.

DEFINITION 1. We say that $T \circ F$ is defined in $\mathcal{D}_{1+}^{-\infty}$ and $T \circ F = \Phi$ if $\Phi \in \mathcal{D}_{1+}^{-\infty}$ exists and also, $1 < p < \infty$, $\alpha \geq 0$ and $n \in \mathbf{Z}^+$ exist such that $T \in S_{-n}$, $\Phi \in \mathcal{D}_p^{-\alpha}$ and the following holds: For any sequence $\{\varphi_k\}$ in $S(\mathbf{R}^d)$ such that $\|\varphi_k - T\|_{-n} \to 0$ as $k \to \infty$, it holds that $\|\varphi_k \circ F - \Phi\|_{p,-\alpha} \to 0$ as $k \to \infty$.

It is obvious that Φ is uniquely determined whenever $T \circ F$ is defined in the sense of Definition 1. In the following, we are particularly interested in the case of $T = (1 - \Delta)^{\alpha/2}\delta_x \in S'(\mathbf{R}^d)$, $\alpha \geq 0$, where Δ is the Laplace operator. As stated in the introduction, $\delta_x \circ F$, when it is defined, is called Donsker's delta function.

DEFINITION 2. A d-dimensional Wiener map F is called *regular in the sense of Malliavin* if (i) $F \in \mathcal{D}_{\infty-}^\infty(\mathbf{R}^d)$, i.e. $F = (F^1, \ldots, F^d)$ with $F^i \in \mathcal{D}_{\infty-}^\infty$ and (ii) the Malliavin covariance σ_F of F (defined by $\sigma_F^{ij}(w) = \langle DF^i(w), DF^j(w) \rangle_H$) is nondegenerate in the sense that

$$(\det \sigma_F)^{-1} \in \bigcap_{1<p<\infty} \mathcal{L}_p =: \mathcal{L}_{\infty-}.$$

It is well-known ([8]) that, if F is regular in the sense of Malliavin then $T \circ F$ is defined in $\mathcal{D}_{1+}^{-\infty}$ for every $T \in S'(\mathbf{R}^d)$ and $T \circ F \in \bigcup_{\alpha>0} \bigcap_{1<p<\infty} \mathcal{D}_p^{-\alpha}$. In particular, for every $m = 0, 1, \cdots$ and $x \in \mathbf{R}^d$,

$$(4) \qquad [(1 - \Delta)^m \delta_x] \circ F \in \bigcap_{1<p<\infty} \mathcal{D}_p^{-2k} \quad \text{provided } k > m + d/2.$$

From this, we can deduce that the generalized expectation

$$p_F(x) = E[\delta_x \circ F],$$

which coincides with the density at x of the law of F, is C^∞ in x, more precisely, $p_F \in S(\mathbf{R}^d)$. Define the *strict support of the law of F* by

$$SSP(F) = \{x; p_F(x) > 0\}.$$

Then we can also deduce from (4) that, if $G \in \bigcup_{1<p<\infty} \mathcal{D}_p^{2k}$ and $k > m + d/2$, then

$$x \in SSP(F) \mapsto E[G|F = x] = E[G \cdot \delta_x \circ F]/p_F(x)$$

is C^{2m} and hence, it is C^∞ provided $G \in \bigcap_{k=1}^{\infty} \bigcup_{1<p<\infty} \mathcal{D}_p^{2k}$. These results can be made sharp by fractionalizing the differentiability indices as follows.

THEOREM 1. *Supose $F : B \to \mathbf{R}^d$ be regular in the Malliavin sense. Then, for $\alpha > \beta \geq 0$ and $1 < p < \infty$ satisfying*

$$(5) \qquad d - p\{d - (\alpha - \beta)\} > 0 \iff 0 \leq \beta < \alpha - d/q, \ p^{-1} + q^{-1} = 1$$
$$\iff 1 < p < d/\{d - (\alpha - \beta)\}_+ \ (d/0 = \infty \quad by \ convention),$$

we have
$$(6) \qquad \{(1 - \Delta)^{\beta/2}\delta_x\} \circ F \in \mathcal{D}_p^{-\alpha} \ \textit{for every } x \in \mathbf{R}^d$$

and furthermore, the map

$$x \in \mathbf{R}^d \mapsto \{(1 - \Delta)^{\beta/2}\delta_x\} \circ F \in \mathcal{D}_p^{-\alpha}$$

is bounded and continuous.

Let $C^\nu(\mathbf{R}^d)$, $\nu \geq 0$, be the Banach space of $[\nu]$-times continuously differentiable functions whose $[\nu]$-th derivatives are uniformly Hölder continuous with exponent $\nu - [\nu]$ with the norm

$$\|f\|_{C^\nu} = \sum_{k;|k|\leq[\nu]} \|D^k f\|_\infty + \sum_{k;|k|=[\nu]} \sup_{x\neq y} \frac{|D^k f(x) - D^k f(y)|}{|x - y|^{\nu-[\nu]}}.$$

Since $(1 - \Delta)^{-\beta/2} (C^0(\mathbf{R}^d)) \subset C^\beta(\mathbf{R}^d)$ (cf. [5]), we have the following

COROLLARY 2. *Let $\alpha > 0$ and $q > 1 \vee (d/\alpha)$. Take $G \in \mathcal{D}_q^\alpha$. Then, $u(x) := E[G \cdot \delta_x \circ F] \in C^\beta(\mathbf{R}^d)$ for every $0 < \beta < \alpha - d/q$.*

Remark. The above result is sharp in the sense that if (5) is not satisfied by α, β, and p, then there is a regular d-dimensional Wiener map such that $\{(1 - \Delta)^{\beta/2}\delta_0\} \circ F \notin \mathcal{D}_p^{-\alpha}$, cf. [9].

A similar result under a more relaxed condition for d-dimensional Wiener map F is given in the following

THEOREM 3. *Let* $F : B \to \mathbf{R}^d$ *be a Wiener map such that*

(i) $F \in \mathcal{D}_{\infty-}^{1+\delta}(\mathbf{R}^d) := \bigcap_{1<p<\infty} \mathcal{D}_p^{1+\delta}(\mathbf{R}^d)$ *for some* $\delta > 0$,

(ii) $(\det \sigma_F)^{-1} \in \mathcal{L}_{\infty-}$, *and*

(iii) *the density* p_F *of the law of* F *is bounded.*

Then, for $0 \le \beta < \delta \wedge \alpha$ *and* $1 < p < \infty$ *satisfying*

$$
(7) \qquad d - p\{d - (\alpha \wedge \delta - \beta)\} > 0 \Longleftrightarrow 0 \le \beta < \alpha \wedge \delta - d/q
$$
$$
\Longleftrightarrow 1 < p < d/\{d - (\alpha \wedge \delta - \beta)\}_+,
$$

we have

$$
(8) \quad \{(1 - \Delta)^{\beta/2} \delta_x\} \circ F \in \mathcal{D}_p^{-\alpha} \quad \text{boundedly and continuously in } x .
$$

3. Proofs

For Theorem 1, a proof in the case $\beta = 0$ was given in [9] and it can be applied in the same way to the general case. The continuous and bounded dependence in x of $\{(1 - \Delta)^{\beta/2} \delta_x\} \circ F$ is not stated explicitly there but this can be obtained in the same way if we notice the following fact: For $\alpha > \beta \ge 0$ and $1 < p < \infty$ satisfying (5), we can find $\tilde{p} > p$ such that the map

$$
x \in \mathbf{R}^d \mapsto (1 - \Delta)^{-(\alpha-\beta)/2} \delta_x \in L_{\tilde{p}}(\mathbf{R}^d)
$$

is bounded and continuous. Here, we denote by $L_p(\mathbf{R}^d)$ the usual L_p-space on $\mathbf{R}^d$ with respect to the Lebesgue measure and by $|\ |_p$ its norm .

For Theorem 3, suppose $F : B \to \mathbf{R}^d$ satisfies the conditions (i),(ii), (iii) of Theorem 3.

LEMMA 1. *For every* $0 \le \rho \le 1$ *and* $1 < p < \tilde{p}$, *there exists a positive constant* $C = C(\rho, p, \tilde{p}; F)$ *such that*

$$
(9) \qquad \|(I - L)^{\rho/2} g \circ F\|_p \le C |(1 - \Delta)^{\rho/2} g|_{\tilde{p}}
$$

for every $g \in \mathcal{S}(\mathbf{R}^d)$. *Here* $\|\ \|_p = \|\ \|_{p,0}$ *is the norm of* $\mathcal{L}_p$.

Proof. By the Meyer's equivalence of norms (cf. [8]), we have (writing $\partial_i = \partial/\partial x_i$ and omitting the summation sign for repeated indices),

$$
\begin{aligned}
\|(I - L)^{1/2} g \circ F\|_p &\le c(\|g \circ F\|_p + \|D(g \circ F)\|_p) \\
&= c(\|g \circ F\|_p + \|\partial_i g \circ F \cdot DF^i\|_p) \\
&\le c'(\|g \circ F\|_{\tilde{p}} + \|\partial_i g \circ F\|_{\tilde{p}})
\end{aligned}
$$

because $DF^i \in \mathcal{D}_{\infty-}^\delta \subset \mathcal{L}_{\infty-}$. Noting (iii), this is further dominated by $c''|(1 - \Delta)^{1/2} g|_{\tilde{p}}$. Now appealing to the same interpolation technique as in [9], we obtain the lemma.

Now we turn to the proof of Theorem 3. Let $0 \leq \rho < \delta \wedge 1$. If $g \in \mathcal{S}(\mathbf{R}^d)$, we have by the chain rule ([8]) that

$$(\partial_i g) \circ F = \langle D(g \circ F), DF^j \rangle_H \cdot \gamma^{ij}$$

where $\gamma = (\gamma^{ij}) = \sigma_F^{-1}$. Then, if $p^{-1} + q^{-1} = 1$,

$$\|(\partial_i g) \circ F\|_{p,-\rho}$$
$$= \sup\{E[\langle D(g \circ F), DF^j \rangle_H \cdot \gamma^{ij} \cdot G] \mid G \in \mathcal{D}_{\infty-}^\infty, \|G\|_{q,\rho} \leq 1\}$$
$$= \sup\{E[(I - L)^{1/2}(g \circ F) \cdot [(I - L)^{-1/2}D^*](\gamma^{ij} \cdot G \cdot DF^j)]$$
$$\mid G \in \mathcal{D}_{\infty-}^\infty, \|G\|_{q,\rho} \leq 1\}.$$

It is proved in [10] that $\gamma^{ij} \in \mathcal{D}_{\infty-}^\rho$ and hence we deduce by the continuity of multiplication: $\mathcal{D}_{\infty-}^\rho \times \mathcal{D}_q^\rho \subset \mathcal{D}_{q'}^\rho$, $1 < q' < q < \infty$, that

$$\|[(I - L)^{-1/2}D^*(\gamma^{ij} \cdot G \cdot DF^j)\|_{q',\rho} \leq c\|G\|_{q,\rho}$$

Therefore, if $p' + q' = 1$ then $1 < p < p'$ and

$$\|(\partial_i g) \circ F\|_{p,-\rho} \leq c\|(I - L)^{1/2}(g \circ F)\|_{p',-\rho} = c\|(I - L)^{(1-\rho)/2}(g \circ F)\|_{p'}.$$

Combining this with the above lemma, we can conclude the following: If $1 < p < \tilde{p}$ then $c' = c'(p, \tilde{p}; F) > 0$ exists such that

$$\|(\partial_i g) \circ F\|_{p,-\rho} \leq c'|(1 - \Delta)^{(1-\rho)/2}g|_{\tilde{p}}$$

for every $g \in \mathcal{S}(\mathbf{R}^d)$. Changing g suitably, we deduce easily, from this and L_p-boundedness of the operator $(1 - \Delta)^{-1/2}\partial_i$, that

$$\|g \circ F\|_{p,-\rho} \leq c''|(1 - \Delta)^{-\rho/2}g|_{\tilde{p}} \quad \text{for every } g \in \mathcal{S}(\mathbf{R}^d).$$

This implies that

$$\|A_\rho\|_{L_{\tilde{p}}(\mathbf{R}^d) \to \mathcal{L}_p} < \infty$$

where $A_\rho : \mathcal{S}(\mathbf{R}^d) \to \mathcal{D}_{1+}^{-\infty}$ is defined by

$$A_\rho g = (I - L)^{-\rho/2}\{[(1 - \Delta)^{\rho/2}g] \circ F\}.$$

Now, Theorem 3 can be concluded in the same way as in Theorem 1 for, at least, $0 \leq \alpha \wedge \delta \leq 1$. The general case can also be obtained in the same way.

References

[1] H. Airault and P. Malliavin: Intégration géométrique sur l'espace de Wiener, Bull. Sc. math., 2e série, **112** (1988), 3–52.

[2] D. Feyel and A. de La Pradelle: Mesures de Hausdorff de codimension finie sur l'espace de Wiener, CR. Acad. Sci. Paris, **310**, série I, (1990), 153-154.

[3] H.-H. Kuo: Donsker's delta function as a generalized Brownian functionals and its application, in *Theory and Applications of Random Fields, Proc. IFIP Conf. Bangalore 1982, (ed. G. Kallianpur)*, **LNCI 49**, Springer (1983), 167–178.

[4] S. Kusuoka: On the foundation of Wiener-Riemannian manifolds, in *Stochastic analysis, path integration and dynamics, (ed. K.D. Elworthy and J-C. Zambrini)*, Longman (1989), 130-164.

[5] E. M. Stein: *Singular Integrals and Differentiable Properties of Functions*, Princeton Univ. Press, 1970.

[6] H. Sugita: Positive generalized Wiener functions and potential theory over abstract Wiener spaces, Osaka J. Math., **25** (1988), 665–696.

[7] S. Watanabe: Malliavin's calculus in terms of generalized Wiener functionals, in *Theory and Applications of Random Fields, Proc. IFIP Conf. Bangalore 1982, (ed. G. Kallianpur)*, **LNCI 49**, Springer (1983), 284–290.

[8] S. Watanabe: *Lectures on Stochastic Differential Equations and Malliavin Calculus*, Tata Institute of Fundamental Research/Springer, 1984.

[9] S. Watanabe: Donsker's δ-functions in the Malliavin calculus, in *Stochastic Analysis, Libre Amicorum for Moshe Zakai, (ed. E. Mayer-Wolf, E. Merzbach and A. Shwartz)*, Academic Press (1991), 495-502.

[10] S. Watanabe: Fractional order Sobolev spaces on Wiener spaces, preprint.

Department of Mathematics
Kyoto University
Kyoto, 606, Japan

Inequalities for Products of White Noise Functionals[*]

J.A. YAN

Abstract. Some sharp inequalities for the Wiener and Wick products of white noise functionals in the new settings of white noise analysis are established.

1. Introduction

Let $S'(I\!R)$ denote the Schwartz space of tempered distributions, the dual of $S(I\!R)$. Let A denote the self-adjoint operator $-\frac{d^2}{dx^2} + 1 + x^2$ on $L^2(I\!R)$. For each $p \geq 0$ we put $S_p(I\!R) = Dom(A^p)$ and denote by $S_{-p}(I\!R)$ the dual of $S_p(I\!R)$. Then we have

$$S(I\!R) = pr \lim S_p(I\!R) = \bigcap_{p \geq 0} S_p(I\!R), \ S'(I\!R) = \bigcup_{p \geq 0} S_{-p}(I\!R).$$

Here "prlim" stands for the projective limit. Let $S_p(I\!R^n)$ denote the L^2-domain of $(A^{\otimes n})^p$. The norm of $S_p(I\!R^n)$ is defined by

$$|f|_{2,p} = |(A^{\otimes n})^p f|_2 = |(A^p)^{\otimes n} f|_2,$$

where $|\cdot|_2$ is the norm of $L^2(I\!R^n)$. Let μ be the white noise measure on $S'(I\!R)$. We denote by (L^2) the space $L^2(S'(I\!R), \mu)$. We can construct Sobolev spaces over $(S'(I\!R), \mu)$ as follows. Let $\Gamma(A^p)$ denote the second quantization of A^p. For $p \geq 0$, we set $(S)_p = Dom(\Gamma(A^p))$ and denote by $(S)_{-p}$ the dual of $(S)_p$. Then the well-known Hida spaces of test functionals and distributions are given by

$$(S) = pr \lim_{p \geq 0}(S)_p = \bigcap_{p \in I\!R_+} (S)_p, \ (S)^* = \bigcup_{p \in I\!R_+} (S)_{-p}.$$

Recently, Kondratiev and Streit introduced a new setting of the white noise analysis as follows. For $\beta \in [0, 1)$ and $p \geq 0$ let $(S)_p^\beta$ denote the subspace of (L^2) which is the collection of those elements φ of (L^2) such that the following norm $\|\varphi\|_{p,\beta}$ of φ is finite:

$$\|\varphi\|_{p,\beta}^2 = \sum_{n=0}^{\infty} (n!)^{1+\beta} |f_n|_{2,p}^2, \tag{1.1}$$

[*] Research supported by the National Science Foundation of China.

AMS Subject classification: 60H99, 46F10, 46F25.

Key Words: White noise functional, Wiener product, Wick procduct.

where $(f_n, n \in I\!N_0)$ is the sequence of kernels corresponding to φ by the Wiener-Itô decomposition:

$$\varphi = \sum_n I_n(f_n). \tag{1.2}$$

We put

$$(S)^\beta = pr \lim_{p \geq 0} (S)^\beta_p = \bigcap_{p \geq 0} (S)^\beta_p, \tag{1.3}$$

$$(S)^{-\beta} = \bigcup_{p \geq 0} (S)^{-\beta}_{-p}$$

where $(S)^{-\beta}_{-p}$ is the dual of $(S)^\beta_p$ with the norm $\| \cdot \|_{-p,-\beta}$:

$$\|\varphi\|^2_{-p,-\beta} = \sum_{n=0}^\infty (n!)^{1-\beta} |f_n|^2_{2,-p}. \tag{1.4}$$

It is easy to see that $(S)^0 = (S), (S)^{-0} = (S)^*$ and for $1 > \beta > 0$ we have $(S) \supset (S)^\beta, (S)^* \subset (S)^{-\beta}$. Since for $\beta_1 < \beta_2$ we have $(S)^{\beta_2} \subset (S)^{\beta_1}$, we can put

$$(S)^+ = pr \lim_{\beta \in [0,1)} (S)^\beta = pr \lim_{\substack{p \in I\!R_+ \\ \beta \in [0,1)}} (S)^\beta_p. \tag{1.5}$$

and consider $(S)^+$ as a space of test functionals of white noise. Its dual $(S)^-$ is then given by

$$(S)^- = \bigcup_{\beta \in [0,1)} (S)^{-\beta} = \bigcup_{\substack{\beta \in [0,1) \\ p \geq 0}} (S)^{-\beta}_{-p}. \tag{1.6}$$

It had been introduced by Meyer-Yan [2] another space of test functionals of white noise which seems to be the smallest reasonable space of test functionals containing all exponential vectors $\mathcal{E}(\xi)$, $\xi \in S(I\!R)$. This space can be described as follows (see Kondratiev-Streit [1]). For $p \geq 0$, $q > 0$, let $M_{p,q}$ be the subspace of (L^2) which is the collection of those elements φ of (L^2) such that the following norm $\|\varphi\|_{M_{p,q}}$ is finite:

$$\|\varphi\|_{M_{p,q}} = \sum_{n=0}^\infty (n!)^2 2^{-qn} |f_n|^2_{2,p}. \tag{1.7}$$

For fixed $p \geq 0$ we have $M_{p,q_1} \subset M_{p,q_2}$ for $q_1 < q_2$. We put $M_p = \bigcup_{q>0} M_{p,q}$ and introduce on M_p the inductive limit topology. It is easy to see that $M_{p_1} \subset M_{p_2}$ for $p_1 > p_2$. So we can put

$$M = \bigcap_{p \geq 0} M_p = \bigcap_{p \geq 0} \bigcup_{q > 0} M_{pq}. \tag{1.8}$$

Then $\mathcal{M}$ is a candidate for spaces of test functionals of white noise. Its dual is given by

$$M^* = \bigcup_{p \geq 0} M_{-p} = \bigcup_{p \geq 0} \bigcap_{q > 0} M_{-p,-q}, \tag{1.9}$$

where $M_{-p,-q}$ is the dual of $M_{p,q}$ with the norm $\| \cdot \|_{M_{-p,-q}}$:

$$\|\varphi\|^2_{M_{-p,-q}} = \sum_{n=0}^{\infty} 2^{qn} |f_n|^2_{2,-p}. \tag{1.10}$$

Obviously, we have $M^* \supset (S)^- \supset (S)^{-\beta} \supset (S)^* \supset (L^2) \supset (S) \supset (S)^{\beta} \supset (S)^- \supset \mathcal{M}$. We refer the reader to the remarkable work [1] by Kondratiev and Streit for a complete study in the subject.

2. Inequalities for Wiener Product

Lemma 2.1. Let $\beta \in [0, 1)$. Then for any $a > 0$ we have

$$\sum_{k=0}^{m \wedge n} \left[\binom{m}{k} \binom{n}{k} \right]^{\frac{1-\beta}{2}} \binom{m+n-2k}{n-k}^{\frac{1+\beta}{2}}$$

$$\leq \left(\frac{1 + a^2 \wedge a^{-2}}{1 + a \wedge a^{-1}} \right)^{\frac{(1+\beta)(m+n)}{2}} (1+a)^{m(1+\beta)} (1+a^{-1})^{n(1+\beta)}, \tag{2.1}$$

$$\sum_{k=0}^{m \wedge n} \left[\binom{m}{k} \binom{n}{k} \right]^{\frac{1-\beta}{2}} \binom{m+n-2k}{n-k}^{\frac{1+\beta}{2}} \leq (\sqrt{3})^{(m+n)(1+\beta)}. \tag{2.2}$$

Proof. By the symmetry between a and a^{-1} in (2.1) we may assume $0 < a \leq 1$, so that $a^2 \wedge a^{-2} = a^2$, $a \wedge a^{-1} = a$. Since

$$\binom{m}{k} \binom{n}{k} \leq \binom{m+n}{2k}, \quad \binom{m+n-2k}{n-k} \leq \binom{m+n}{n-k}$$

and $\frac{1-\beta}{2} + \frac{1+\beta}{2} = 1$, we have

$$\sum_{k=0}^{m \wedge n} \left[\binom{m}{k} \binom{n}{k} \right]^{\frac{1-\beta}{2}} \binom{m+n-2k}{n-k}^{\frac{1+\beta}{2}} \leq \sum_{k=0}^{m \wedge n} \binom{m+n}{2k}^{\frac{1-\beta}{2}} \binom{m+n}{n-k}^{\frac{1+\beta}{2}}$$

$$=a^{-n(1+\beta)}\sum_{k=0}^{m\wedge n}\left[\binom{m+n}{2k}a^{\frac{1+\beta}{1-\beta}2k}\right]^{\frac{1-\beta}{2}}\left[\binom{m+n}{n-k}a^{2(n-k)}\right]^{\frac{1+\beta}{2}}$$

$$\leq a^{-n(1+\beta)}\left[\sum_{k=0}^{m\wedge n}\binom{m+n}{2k}a^{\frac{1+\beta}{1-\beta}2k}\right]^{\frac{1-\beta}{2}}\left[\sum_{k=0}^{m\wedge n}\binom{m+n}{n-k}a^{2(n-k)}\right]^{\frac{1+\beta}{2}}$$

$$\leq a^{-n(1+\beta)}\left[\sum_{j=0}^{m+n}\binom{m+n}{j}a^{\frac{1+\beta}{1-\beta}j}\right]^{\frac{1-\beta}{2}}\left[\sum_{j=0}^{m+n}\binom{m+n}{j}a^{2j}\right]^{\frac{1+\beta}{2}}$$

$$=a^{-n(1+\beta)}\left(1+a^{\frac{1+\beta}{1-\beta}}\right)^{\frac{(m+n)(1-\beta)}{2}}\left(1+a^2\right)^{\frac{(m+n)(1+\beta)}{2}}$$

$$\leq a^{-n(1+\beta)}\left(1+a\right)^{\frac{(m+n)(1+\beta)}{2}}\left(1+a^2\right)^{\frac{(m+n)(1+\beta)}{2}},$$

from which follows (2.1). Here we have used the fact that $(1+a^\alpha)^{\frac{1}{\alpha}}\leq 1+a$.
Now we turn to prove (2.2). We have

$$\sum_{k=0}^{m\wedge n}\left[\binom{m}{k}\binom{n}{k}\right]^{\frac{1-\beta}{2}}\binom{m+n-2k}{n-k}^{\frac{1+\beta}{2}}$$

$$\leq\sum_{k=0}^{m\wedge n}\left[\binom{m}{k}\binom{n}{k}\right]^{\frac{1-\beta}{2}}2^{(m+n-2k)\frac{1+\beta}{2}}$$

$$=2^{(m+n)\frac{1+\beta}{2}}\sum_{k=0}^{m\wedge n}\left[\binom{m}{k}\binom{n}{k}2^{-\frac{1+\beta}{1-\beta}k}\right]^{\frac{1-\beta}{2}}$$

$$\leq2^{(m+n)\frac{1+\beta}{2}}\left\{\sum_{k=0}^{m\wedge n}\left[\binom{m}{k}\binom{n}{k}2^{-\frac{1+\beta}{1-\beta}k}\right]^{\frac{1}{2}}\right\}^{1-\beta}$$

$$\leq2^{(m+n)\frac{1+\beta}{2}}\left\{\left(\sum_{k=0}^{m\wedge n}\binom{m}{k}2^{-\frac{1+\beta}{1-\beta}k}\right)^{\frac{1}{2}}\left(\sum_{k=0}^{m\wedge n}\binom{n}{k}2^{-\frac{1+\beta}{1-\beta}k}\right)^{\frac{1}{2}}\right\}^{1-\beta}$$

$$\leq2^{(m+n)\frac{1+\beta}{2}}\left(1+2^{-\frac{1+\beta}{1-\beta}}\right)^{\frac{m(1-\beta)}{2}}\left(1+2^{-\frac{1+\beta}{1-\beta}}\right)^{\frac{n(1-\beta)}{2}}$$

$$\leq2^{(m+n)\frac{1+\beta}{2}}\left(1+\frac{1}{2}\right)^{\frac{m(1+\beta)}{2}}\left(1+\frac{1}{2}\right)^{\frac{n(1+\beta)}{2}}$$

$$=(\sqrt{3})^{(m+n)(1+\beta)}.$$

Here we have also used the fact that $(1+x^\alpha)^{\frac{1}{\alpha}}\leq 1+x$ for $x\geq 0$, $\alpha\geq 0$.
The lemma is proved.

Remark. For the case $\beta=0$ (2.1) and (2.2) were proved by Yan [4].
Now let $f_m\in\widehat{L^2}(\mathbb{R}^m)$ and $g_n\in\widehat{L^2}(\mathbb{R}^n)$. It is well known that

$$I_n(f_m)I_n(g_n)=\sum_{k=0}^{m\wedge n}k!\binom{m}{k}\binom{n}{k}I_{m+n-2k}(f_m\hat{\otimes}_k g_n),\qquad(2.3)$$

where $f_m \hat{\otimes}_k g_n$ stands for the symmetrization of $f_m \otimes_k g_n$. The latter is defined by

$$f_m \otimes_k g_n = \int_{R^k} f_m(u_1, \cdots, u_k, \cdots) g_n(u_1, \cdots, u_k, \cdots) \, du_1 \cdots du_k.$$

Theorem 2.2. Let $p \geq 0$ and $0 \leq \beta < 1$. Let $\alpha, \gamma > 0$ be such that $2^{-\alpha} + 2^{-\gamma} = 1$ and $\alpha \neq \gamma$. Then we have

$$\|\varphi\psi\|_{p,\beta} \leq C_{\alpha,\beta} \|\varphi\|_{p+\alpha(1+\beta),\beta} \|\psi\|_{p+\gamma(1+\beta),\beta} \tag{2.4}$$

For the case $\alpha = \gamma = 1$ we have for any $\delta > \frac{1+\beta}{2} \log_2 3 \triangleq \delta_0$

$$\|\varphi\psi\|_{p,\beta} \leq D_{\delta,\beta} \|\varphi\|_{p+\delta,\beta} \|\psi\|_{p+\delta,\beta} \tag{2.5}$$

Here $C_{\alpha,\beta} = [1 - (2^{\alpha\wedge\gamma} - 2 + 2^{1-\alpha\wedge\gamma})^{1+\beta}]^{-1} < \infty$, $D_{\delta,\beta} = [1 - 2^{-2(\delta-\delta_0)}]^{-1}$.

Proof. Assume $\varphi = \sum_m I_m(f_m)$ and $\psi = \sum_n I_n(g_n)$. Let $a > 0$. We put

$$\lambda(a) = \frac{1 + a^2 \wedge a^{-2}}{1 + a \wedge a^{-1}}.$$ By (2.3) and (2.1) we have

$$\|\varphi\psi\|_{p,\beta} \leq \sum_{m,n=0}^{\infty} \sum_{k=0}^{m\wedge n} k! \binom{m}{k}\binom{n}{k} \|I_{m+n-2k}(f_m \hat{\otimes}_k g_n)\|_{p,\beta}$$

$$\leq \sum_{m,n=0}^{\infty} \sum_{k=0}^{m\wedge n} k! \binom{m}{k}\binom{n}{k} [(m+n-2k)!]^{\frac{1+\beta}{2}} |f_m|_{2,p} |g_n|_{2,p}$$

$$= \sum_{m,n=0}^{\infty} \sum_{k=0}^{m\wedge n} \frac{1}{(k!)^\beta} \left[\binom{m}{k}\binom{n}{k}\right]^{\frac{1-\beta}{2}} \binom{m+n-2k}{n-k}^{\frac{1+\beta}{2}} (m!n!)^{\frac{1+\beta}{2}} |f_m|_{2,p} |g_n|_{2,p}$$

$$\leq \sum_{m,n=0}^{\infty} \lambda(a)^{\frac{(1+\beta)(m+n)}{2}} (1+a)^{m(1+\beta)} (m!\, n!)^{\frac{1+\beta}{2}} |f_m|_{2,p} (1+a^{-1})^{n(1+\beta)} |g_n|_{2,p}.$$

Set $a = 2^\alpha - 1$. Then $1 + a^{-1} = 2^\gamma$. Since $\alpha \neq \gamma$, $2^{-\alpha} + 2^{-\gamma} = 1$ we obtain that $\frac{1+a^2\wedge a^{-2}}{1+a\wedge a^{-1}} = 2^{\alpha\wedge\gamma} - 2 + 2^{1-\alpha\wedge\gamma} < 1$. Thus we have

$$\|\varphi\psi\|_{p,\beta}$$
$$\leq \sum_{m,n=0}^{\infty} \lambda(a)^{\frac{(1+\beta)(m+n)}{2}} (m!)^{\frac{1+\beta}{2}} |f_m|_{2,p+\alpha(1+\beta)} (n!)^{\frac{1+\beta}{2}} |g_n|_{2,p+\gamma(1+\beta)}$$
$$\leq C_{\alpha,\beta} \|\varphi\|_{p+\alpha(1+\beta),\beta} \|\psi\|_{p+\gamma(1+\beta),\beta}.$$

(2.4) is proved. Similarly we can prove (2.5).

Remark. For the case $\beta = 0$, (2.4) and (2.5) were proved by Yan [4].

Lemma 2.3. For any $a > 0$ we have

$$\sum_{k=0}^{n\wedge m} \binom{m+n-2k}{n-k} \leq (1+a)^m (1+a^{-1})^n. \tag{2.6}$$

Proof. We have

$$
\begin{aligned}
\sum_{k=0}^{m\wedge n} \binom{m+n-2k}{n-k} &= \sum_{k=0}^{m\wedge n} \binom{m+n-2k}{m-k}^{\frac{1}{2}} \binom{m+n-2k}{n-k}^{\frac{1}{2}} \\
&\le \sum_{k=0}^{m\wedge n} \binom{m+n}{m-k}^{\frac{1}{2}} \binom{m+n}{n-k}^{\frac{1}{2}} \\
&= a^{\frac{m-n}{2}} \sum_{k=0}^{m\wedge n} \left[\binom{m+n}{m-k} a^{k-m}\right]^{\frac{1}{2}} \left[\binom{m+n}{n-k} a^{n-k}\right]^{\frac{1}{2}} \\
&\le a^{\frac{m-n}{2}} (1+a^{-1})^{\frac{m+n}{2}} (1+a)^{\frac{m+n}{2}} \\
&= (1+a)^m (1+a^{-1})^n.
\end{aligned}
$$

Theorem 2.4. Let $p \ge 0$, $q_1, q_2 > 0$. Assume $\varphi \in M_{p,q_1}$, $\psi \in M_{p,q_2}$. Then for any $\varepsilon > 0$ we have

$$
\|\varphi\psi\|_{M_{p,q+\epsilon}} \le \exp\{2^{q+\epsilon}\}(1-2^{-\epsilon})^{-1} \|\varphi\|_{M_{p,q_1}} \|\psi\|_{M_{p,q_2}}, \qquad (2.7)
$$

where $q = 2\log_2[2^{\frac{q_1}{2}} + 2^{\frac{q_2}{2}}]$.

Proof. Assume $\varphi = \sum_m I_m(f_m)$ and $\psi = \sum_n I_n(g_n)$. By (2.3) and (2.6)

$$
\begin{aligned}
\|\varphi\psi\|_{M_{p,q+\epsilon}} &\le \sum_{m,n=0}^{\infty} \sum_{k=0}^{m\wedge n} k! \binom{m}{k}\binom{n}{k} \|I_{m+n-2k}(f_m \hat{\otimes}_k g_n)\|_{M_{p,q+\epsilon}} \\
&\le \sum_{m,n=0}^{\infty} \sum_{k=0}^{m\wedge n} k! \binom{m}{k}\binom{n}{k} \frac{(m+n-2k)!}{m!n!} 2^{-\frac{q+\epsilon}{2}(m+n-2k)} m!n! |f_m|_{2,p} |g_n|_{2,p} \\
&= \sum_{m,n=0}^{\infty} 2^{-\frac{\epsilon}{2}(m+n)} \sum_{k=0}^{m\wedge n} \frac{2^{(q+\epsilon)k}}{k!} \binom{m+n-2k}{n-k} 2^{-\frac{q}{2}(m+n)} m!n! |f_m|_{2,p} |g_n|_{2,p} \\
&\le \exp\{2^{q+\epsilon}\} \sum_{m,n=0}^{\infty} 2^{-\frac{q+\epsilon}{2}(m+n)} (1+a)^m (1+a^{-1})^n m!n! |f_m|_{2,p} |g_n|_{2,p}.
\end{aligned}
$$

Letting $a = 2^{\frac{q-q_1}{2}} - 1$ gives $1+a^{-1} = 2^{\frac{q-q_2}{2}}$, so by Schwarz inequality we get (2.7).

As an immediate consequence of Theorem 2.2 and 2.4 we obtain the following

Theorem 2.5. Let $p \ge 0$ and $\beta \in [0,1)$.

(1) $(S)^\beta$, M_p, $(S)^+$ and M are algebras under Wiener product.

(2) The mapping $(\varphi, \psi) \mapsto \varphi\psi$ from $M_p \times M_p$ to M_p is continuous, where $p \ge 0$. The mapping $(\varphi, \psi) \mapsto \varphi\psi$ from $(S)_p^\beta \times (S)^\beta$ to $(S)_{p-}^\beta$ is continuous, where $p \ge 0$ and $(S)_{p-}^\beta = \bigcup_{0<q<p}(S)_q^\beta$ is provided with the inductive limit topology.

(3) Let $\varphi \in M_p$ and $G \in M_{-p}$. The Wiener product φG is well defined via the following relation:

$$\langle \varphi G, \psi \rangle = \langle \varphi \psi, G \rangle, \ \forall \psi \in M_p.$$

Then $\varphi G \in M_{-p}$ and the mapping $(\varphi, G) \mapsto \varphi G$ from $M_p \times M_{-p}$ to M_{-p} is continuous.

(4) Assume $p > 0$. Let $\varphi \in (S)_p^\beta$ and $G \in (S)_{-q}^{-\beta}, 0 < q < p$. The Wiener product φG is well defined via the following relation:

$$\langle \varphi G, \psi \rangle = \langle \varphi \psi, G \rangle, \ \forall \psi \in (S)^\beta.$$

Then $\varphi G \in (S)^{-\beta}$ and the mapping $(\varphi, G) \mapsto \varphi G$ from $(S)_p^\beta \times (S)_{-q}^{-\beta}$ to $(S)^{-\beta}$ is continuous.

Remark. The fact that $(S)^\beta$ and M are algebras was proved in Kondratiev-Streit [1] by a topological argument.

3. Inequalities for Wick Product, Application to Transforms of White Noise Functionals

In this section some sharp inequalities for Wick product of two elements of $(S)^{-\beta}$ or M^* will be established. An application of these inequalities to several transforms of white noise functionals is also given.

Let $\varphi = \sum_n I_n(f_n)$ and $\psi = \sum_n I_n(g_n)$. Put

$$h_n = \sum_{k+j=n} f_n \hat{\otimes} g_j \qquad (3.1)$$

If the sequence (h_n) corresponds to a distribution we denote by $\varphi : \psi$ this distribution and call it the Wick product of φ and ψ. This Wick product is characterized by its S-transform as follows:

$$S(\varphi : \psi)(\xi) = S\varphi(\xi)S\psi(\xi), \qquad (3.2)$$

where

$$S\varphi(\xi) = \langle \mathcal{E}(\xi), \varphi \rangle$$

and $\mathcal{E}(\xi) = \exp\{\langle \cdot, \xi \rangle - \frac{1}{2}|\xi|_2^2\}$.

Theorem 3.1. Let $p \in \mathbb{R}$ and $\beta \in (-1, 1)$. Let $\alpha, \gamma > 0$ be such that $2^{-2\alpha} + 2^{-2\gamma} = 1$. Then we have

$$\|\varphi : \psi\|_{p,\beta} \leq \|\varphi\|_{p+\alpha(1+\beta),\beta} \|\psi\|_{p+\gamma(1+\beta),\beta} \qquad (3.3)$$

Proof. We have by (3.1)

$$(n!)^{\frac{1+\beta}{2}} |h_n|_{2,p} \leq (n!)^{\frac{1+\beta}{2}} \sum_{k+j=n} |f_k|_{2,p} |g_j|_{2,p}$$

$$= \sum_{k+j} \binom{n}{k}^{\frac{1+\beta}{2}} (k!)^{\frac{1+\beta}{2}} |f_k|_{2,p} (j!)^{\frac{1+\beta}{2}} |g_j|_{2,p} \qquad (3.4)$$

Let $a = 2^{2\gamma} - 1$. Then $1 + a^{-1} = 2^{2\alpha}$. By (3.4) we have

$$(n!)^{(1+\beta)}|h_n|_{2,p}^2$$

$$\leq \left(\sum_{k=0}^n \binom{n}{k}^{1+\beta} a^{k(1+\beta)}\right)\left(\sum_{k+j=n}(k!)^{1+\beta}|f_n|_{2,p}^2(j!)^{1+\beta}|g_j|_{2,p}^2 a^{-k(1+\beta)}\right)$$

$$\leq \left(\sum_{k=0}^n \binom{n}{k} a^k\right)^{1+\beta}\sum_{k+j=n}(k!)^{1+\beta}|f_n|_{2,p}^2(j!)^{1+\beta}|g_j|_{2,p}^2 a^{-k(1+\beta)}$$

$$= \sum_{k+j=n}(1+a^{-1})^{k(1+\beta)}(1+a)^{j(1+\beta)}(k!)^{1+\beta}|f_k|_{2,p}^2(j!)^{1+\beta}|g_j|_{2,p}^2$$

$$\leq \sum_{k+j=n}(k!)^{1+\beta}|f_k|_{2,p+\alpha(1+\beta)}^2(j!)^{1+\beta}|g_j|_{2,p+\gamma(1+\beta)}^2,$$

from which follows (3.3).

Remark. For the case $\beta = 0$. (3.3) was proved by Yan [4].

Theorem 3.2. Let $p \geq 0$ and $q_1, q_2 > 0$. Assume $\varphi \in M_{p,q}$, $\psi \in M_{p,q}$. Then we have

$$\|\varphi : \psi\|_{M_{p,q}} \leq \|\varphi\|_{M_{p,q_1}}\|\psi\|_{M_{p,q_2}} \tag{3.5}$$

where $q = 2\log_2[2^{\frac{q_1}{2}} + 2^{\frac{q_2}{2}}]$.

Proof. Assume $\varphi = \sum_m I_m(f_m)$, $\psi = \sum_n I_n(g_n)$. By (3.1) we have for $0 < a < \infty$

$$n!2^{-\frac{q}{2}n}|h_n|_{2,p} \leq 2^{-\frac{q}{2}n}\sum_{k+j=n}\binom{n}{k}k!|f_k|_{2,p}j!|g_j|_{2,p}$$

$$= 2^{-\frac{q}{2}n}\sum_{k+j=n}\binom{n}{k}a^k[k!|f_n|_{2,p}j!|g_j|_{2,p}a^{-k}]$$

$$\leq 2^{-\frac{q}{2}n}\left(\sum_{k=0}^n\binom{n}{k}^2 a^{2k}\right)^{\frac{1}{2}}\left[\sum_{k+j=n}(k!|f_k|_{2,p}j!|g_j|_{2,p}a^{-k})^2\right]^{\frac{1}{2}}$$

$$\leq 2^{-\frac{q}{2}n}\sum_{k=0}^n\binom{n}{k}a^k[\cdots]^{\frac{1}{2}} = 2^{-\frac{q}{2}n}(1+a)^n[\cdots]^{\frac{1}{2}}$$

$$= \left(\sum_{k+j=n}(1+a^{-1})^{2k}2^{-qk}(k!)^2|f_n|_{2,p}^2(1+a)^{2j}2^{-qj}(j!)^2|g_j|_{2,p}^2\right)^{\frac{1}{2}}. \tag{3.6}$$

If we take $a = 2^{\frac{q-q_2}{2}} - 1$, then $1 + a^{-1} = 2^{\frac{q-q_1}{2}}$. Thus from (3.6) we get immediately (3.5).

Theorem 3.3. Let $p \geq 0$ and $q > 0$. Then we have

$$\|\varphi : \psi\|_{M_{-p,-q}} \leq \|\varphi\|_{M_{-p,-q}}\|\psi\|_{M_{-p,-q}} \tag{3.7}$$

Proof. Immediately follows from (3.1).

As a consequence of Theorems 3.1-3.3 we obtain the following

Theorem 3.4. Let $p \geq 0$ and $\beta \in [0, 1)$. All the spaces $(S)^\beta$, $(S)^{-\beta}$, M_p, M_{-p}, $(S)^+$, $(S)^-$, M and M^* are stable under the Wick product, and the corresponding mappings $(\varphi, \psi) \mapsto \varphi : \psi$ are continuous.

Now we turn to apply Wick product inequalities to the study of transforms of white noise functionals. To begin with , let us recall some results about the Hida spaces case (i.e. $\beta = 0$) in this direction. Let $\varphi \in (S), \lambda \in \mathbb{R}, y \in S'(\mathbb{R})$. It was shown in Potthoff-Yan [3] that the following mappings, called scaling, translation and Gâteaux differentiation respectively, are well defined on (S) and continuous from (S) into itself:

$$\sigma_\lambda \varphi(x) = \tilde{\varphi}(\lambda x), \quad x \in S'(\mathbb{R}) \tag{3.8}$$

$$\tau_y \varphi(x) = \tilde{\varphi}(x + y), \quad x \in S'(\mathbb{R}) \tag{3.9}$$

$$D_y \varphi = \lim_{t \downarrow 0} \frac{\tau_{ty} \varphi - \varphi}{t} \tag{3.10}$$

where $\tilde{\varphi}$ is the continuous version of φ. Moreover we have

$$\langle \sigma_\lambda \varphi, G \rangle = \langle \varphi, (\Gamma(\lambda)G) : \frac{d\mu^{(\lambda)}}{d\mu} \rangle \tag{3.11}$$

$$\langle \tau_y \varphi, G \rangle = \langle \varphi, G : \mathcal{E}(y) \rangle \tag{3.12}$$

$$\langle D_y \varphi, G \rangle = \langle \varphi, G : I_1(y) \rangle \tag{3.13}$$

where $G \in (S)^*$ and $\Gamma(\lambda)$ is the second quantization of the multiplication by λ; $\frac{d\mu^{(\lambda)}}{d\mu}$ is the generalized Radon-Nikodym derivative w.r.t.μ of the gaussian measure $\mu^{(\lambda)}$ with the variance parameter λ^2, whose S-transform is $\exp\{-\frac{1-\lambda^2}{2}|\xi|_2^2\}$, $\xi \in S(\mathbb{R})$; $\mathcal{E}(y)$ is the exponential vector at y, whose S-transform is $\exp\langle y, \xi \rangle$, $\xi \in S(\mathbb{R})$. In [4], we have used (3.11)–(3.13) to extend the mappings σ_λ, τ_y and D_y to Sobolev spaces $(S)_p$. Now we shall do the same thing for new Sobolev spaces $(S)_p^\beta$ and M_p. To this end, we need the following lemma, the proof of which is very easy and will be omitted.

Lemma 3.5. (1) For $0 < \beta < 1$ and $p > \frac{1}{4}$ we have $\frac{d\mu^{(\lambda)}}{d\mu} \in (S)_{-p}^{-\beta}$ (so $\frac{d\mu^{(\lambda)}}{d\mu} \in M_{-p}$).

(2) Let $p \geq 0$, $y \in (S)_{-p}(\mathbb{R})$ and $0 \leq \beta < 1$. Then $\mathcal{E}(y), I_1(y) \in (S)_{-p}^{-\beta} \subset M_{-p}$. If $y \in S_p(\mathbb{R})$ then $\mathcal{E}(y), I_1(y) \in M_p \subset (S)_p^\beta$.

(3) Let $p \in \mathbb{R}$ and $\beta \in (-1, 1)$. Then for any $y \in S_p(\mathbb{R})$ we have

$$\|G : \mathcal{E}(y)\|_{p,\beta} \leq \|G\|_{p+\alpha(1+\beta),\beta} \exp\{2^{2\gamma(1+\beta)-1}|y|_{2,p}^2\}, \tag{3.14}$$

where $\alpha, \gamma > 0$ are such that $2^{-2\alpha} + 2^{-2\gamma} = 1$.

(4) Let $p \in \mathbb{R}$ and $y \in S_p(\mathbb{R})$. Then for any $\beta \in (-1, 1)$ and $G \in (S)_p^\beta$, we have $G : I_1(y) \in (S)_{p-}^\beta = \bigcup_{q<p} (S)_q^\beta$.

By using (3.3)-(3.7), (3.11)-(3.13) and by Lemma 3.5 we can obtain easily the following theorem. We leave the proof to the reader.

Theorem 3.6. (1) The restrictions to $(S)^\beta$, $\beta \in [0,1)$ or to M of the mappings σ_λ, τ_y and D_y are continuous into itself.

(2) Let $p > \frac{1}{4}$ and $0 < \beta < 1$. The scaling mapping σ_λ can be extended to a continuous mapping from M_p to M_p or from $(S)_p^\beta$ to $(S)_q^\beta$ for some $q < p$, q depending on p and λ.

(3) Let $p \in I\!R$ and $y \in S_p(I\!R)$. The mappings τ_y and D_y can be extended to continuous mapping from M_{-p} to M_{-p}. Moreover, for any fixed $\varphi \in M_{-p}$, the mappings $y \mapsto \tau_y \varphi$ and $y \mapsto D_y \varphi$ are continuous from $S_p(I\!R)$ to M_{-p}.

(4) Let $p > 0$ and $0 < q < p$. Let $y \in S_{-p}(I\!R)$. The mappings τ_y and D_y can be extended to continuous mappings from $(S)_p^\beta$ to $(S)_q^\beta$. Moreover, for any fixed $\varphi \in (S)_p^\beta$, the mappings $y \mapsto \tau_y \varphi$ and $y \mapsto D_y \varphi$ are continuous from $S_{-p}(I\!R)$ to $(S)_q^\beta$.

(5) Let $p \geq 0$ and $q > p$. Let $y \in S_p(I\!R)$. The mappings τ_y and D_y can be extended to continuous mappings from $(S)_{-p}^{-\beta}$ to $(S)_{-q}^{-\beta}$. Moreover, for any fixed $\varphi \in (S)_{-p}^{-\beta}$, the mappings $y \mapsto \tau_y \varphi$ and $y \mapsto D_y \varphi$ are continuous from $S_p(I\!R)$ to $(S)_{-q}^{-\beta}$.

References

[1] Kondratiev, Ju.G., Streit, L.: Spaces of White Noise Distributions: Constructions, Discriptions, Applications, I. Preprint (1991).

[2] Meyer, P.A., Yan, J.A.: Les "fonctions caractéristiques" des distributions sur l'espace de Wiener, Sém.Prob. XXV, LN in Math. 1485 (1991), 61-78.

[3] Potthoff, J., Yan, J.A.: Some results about test and generalized functionals of white noise, BiBoS preprint (1989), to appear in: Proc. Singapore Probab. Conf. (1989), L.H.Y. Chen (ed.).

[4] Yan, J.A.: Products and transforms of white noise functionals, Preprint (1990).

J.A. Yan
Institute of Applied Mathematics
Academia Sinica, Beijing 100080, China

A Note on the Consistency of M–Estimates in Linear Models *

L.C. Zhao, C. Radhakrishna Rao and X.R. Chen

Abstract

Weak consistency of M–estimates of regression parameters in a general linear model is established under the condition $(X_n' X_n)^{-1} \to 0$ as $n \to \infty$, where X_n is the design matrix for the first n observations. The M– estimate is obtained by minimizing the sum of $\rho(\varepsilon_i)$, $i = 1, \ldots, n$, where ρ is a convex function satisfying some minimal regularity conditions, and ε_i is the i–th residual.

1 Introduction

We consider the linear model

$$(1.1) \qquad y_i = x_i'\beta + e_i, \quad i = 1, \ldots, n$$

where x_i is a p–vector, β is a p–vector of unknown parameters and $e_1, \ldots, e_n$ are i.i.d. random errors with a common distribution function F. Several authors studied the asymptotic properties of $\hat{\beta}$, an M–estimate of β, obtained by minimizing

$$(1.2) \qquad \sum_{i=1}^{n} \rho(y_i - x_i'\beta)$$

for a suitable function ρ, or by solving an estimating equation of the type

$$(1.3) \qquad \sum_{i=1}^{n} \psi(y_i - x_i'\beta)x_i = 0$$

*Research supported by the Air Force Office of Scientific Research under Grant AFOSR-89-0279. The Research work of L.C. Zhao and X.R. Chen is also partially supported by the National Science Foundation of China.

for a suitable function ψ. For detailed references on this subject starting with the seminal work of Huber (1973), the reader is referred to papers by Rao (1988) and Bai, Rao and Wu (1989).

In this paper, we consider the method of estimation mentioned in (1.2) with some mild restrictions on ρ and prove the weak consistency of an M–estimate $\hat{\beta}$, when p is fixed, under the condition

$$(1.4) \qquad S_n^{-1} \to 0 \text{ as } n \to \infty, \text{ with } \quad S_n = \sum_{i=1}^{n} x_i x_i'.$$

It may be noted that in most of the earlier papers, the strong condition

$$(1.5) \qquad \max_{1 \le i \le n} x_i' X_n^{-1} x_i \to 0 \text{ as } n \to \infty$$

was used. Yohai and Maronna (1979) proved the consistency of an M–estimate $\hat{\beta}$ under the condition (1.4) when $\hat{\beta}$ is obtained from the estimating equation (1.3), but the additional conditions imposed on the estimating equation are somewhat severe and they exclude the important case of least absolute deviations (LAD) estimate. Pollard (1991) studied LAD estimators and his basic convexity lemma extends the findings to a wider class of convex $\rho(\cdot)$. The case of nonconvex ρ, with possibly discontinuous derivative, has been considered by Jurečková (1989). Our proof of the sufficiency of the condition (1.4) covers several important special cases. It may be noted that, when $\hat{\beta}$ is the least squares estimate of β in a Gauss–Markoff model, the condition (1.4) is also necessary for the consistency of $\hat{\beta}$, although in our generality of $\rho(\cdot)$, the problem of necessity of (1.4) is open.

2   The Main Result

We make the following assumptions on ρ:

(A1) ρ is a convex function on R^1 with right and left derivatives $\psi_+(\cdot)$ and $\psi_-(\cdot)$. Choose $\psi(\cdot)$ such that $\psi_-(u) \le \psi(u) \le \psi_+(u)$ for all $u \in R^1$.

(A2) $E\psi(e_1) = 0$ and there exist positive constants, c_0, c_1 and Δ such that
$$(2.1) \qquad |E\psi(e_1 + u)| \ge c_0|u| \quad \text{for} \quad |u| < \Delta$$
and
$$(2.2) \qquad E\psi^2(e_1 + u) \le c_1 < \infty \quad \text{for} \quad |u| < \Delta.$$

(A3) $\qquad\qquad\qquad S_n^{-1} \to 0 \quad \text{as} \quad n \to \infty.$

Theorem 1. *Assume that (A_1) – (A_3) are satisfied and let $\hat{\beta}$ be an M–estimate of β obtained by minimizing $\sum \rho(y_i - x_i'\beta)$ as in (1.2). Then $\hat{\beta}$ is weakly consistent for β.*

In the special case of LAD estimate, treated by Pollard (1991) and others, the weak consistency of $\hat{\beta}$ follows as a corollary to Theorem 1, where we may not need the continuity of F at 0, but the median of F is assumed to be 0.

3   Proof of the Main Result

We start with some notations. Assume that $S_{n_0} > 0$ for some integer n_0 and that $n \geq n_0$. Write

$$(3.1) \qquad \beta_n = S_n^{1/2}\beta, \quad x_{in} = S_n^{-1/2}x_i, \quad i = 1, \ldots, n.$$

The model (1.1) can be rewritten as

$$(3.2) \qquad Y_i = x_{in}'\beta_n + e_i, \quad i = 1, \ldots, n,$$

with

$$(3.3) \qquad \sum_{i=1}^{n} x_{in}x_{in}' = I_p$$

where I_p is the $p \times p$ identity matrix. It is easy to see that $\hat{\beta}_n = S_n^{1/2}\hat{\beta}$ is an M–estimate of β_n in (3.2). Without loss of generality, we assume that the true parameter $\beta = 0$ in model (1.1), i.e., $\beta_n = 0$ in (3.2). In order to prove Theorem 2.1, by virtue of $(A3)$, we need only to show that

$$(3.4) \qquad \hat{\beta}_n = O_p(1).$$

Denote by U the unit sphere $\{\beta \in R^p, \|\beta\| = 1\}$ in R^p. Define

$$
\begin{aligned}
D_n(\beta) &= \sum_{i=1}^{n} \{\rho(e_i - x_{in}'\beta) - \rho(e_i)\} \\
(3.5) \qquad &= \sum_{i=1}^{n} \int_0^{-x_{in}'\beta} \psi(e_i + t)\,dt, \quad \beta \in R^p,
\end{aligned}
$$

$$(3.6)\ \ D(\gamma, L) = D_n(L\gamma) = \sum_{i=1}^{n} \int_0^{-Lx_{in}'\gamma} \psi(e_i + t)\,dt, \quad L > 0, \quad \gamma \in U.$$

By the convexity of $D_n(\beta)$ it is easily seen that for $L > 0$,

(3.7) $$P(\| \hat{\beta}_n \| \geq L) \leq P\{\inf_{\gamma \in U} D(\gamma, L) \leq 0\},$$

where $\| \cdot \|$ stands for the Euclidean norm.

If for some $L_1 > 0$ and $L_2 > 0$, $\Lambda \subseteq U$, $\Lambda^c = U - \Lambda$,

$$\inf_{\gamma \in \Lambda} D(\gamma, L_1) > 0 \quad \text{and} \quad \inf_{\gamma \in \Lambda^c} D(\gamma, L_2) > 0,$$

then for $L \geq \max(L_1, L_2)$, we have $\inf_{\gamma \in U} D(\gamma, L) > 0$ by the convexity of $D_n(\beta)$. In other words, we have

(3.8) $\{\inf_{\gamma \in U} D(\gamma, L) \leq 0\} \subset \{\inf_{\gamma \in \Lambda} D(\gamma, L_1) \leq 0\} \cap \{\inf_{\gamma \in \Lambda^c} D(\gamma, L_2) \leq 0\}.$

Λ is chosen below in (3.11). For $\delta > 0$, $\varepsilon_1 > 0$ and $\eta \in (0, 1)$, define

(3.9) $\quad J = \{j : 1 \leq j \leq n, \ \| x_{jn} \| > \delta\}, \quad J^c = \{1, \ldots, n\} - J,$

(3.10) $\quad S(\delta) = \sum_{i \in J^c} x_{in} x'_{in} = \sum_i x_{in} x'_{in} I(\| x_{in} \| \leq \delta),$

(3.11) $\quad \Lambda = \{\gamma \in U, \ \gamma' S(\delta) \gamma \geq \varepsilon_1\}, \quad \Lambda^c = U - \Lambda,$

(3.12) $\quad I_\gamma = \{i : 1 \leq i \leq n, \ |x'_{in} \gamma| \geq \eta\}, \quad I_\gamma^c = \{1, \ldots, n\} - I_\gamma.$

By the property of ψ, there exits a constant $K > 0$ such that $\psi(-\infty) < -K < K < \psi(\infty)$. Given $\varepsilon > 0$, take

(3.13) $$0 < \varepsilon_2 < 2^{-8}(c_1 p)^{-1/2} \varepsilon^{1/2} K,$$

and divide U into M parts, $\tilde{U}_1, \ldots, \tilde{U}_M$, such that the diameter of each part is less than ε_2. Then take

$$0 < \varepsilon_1 < \min(1/2, 2^{-16}(M c_1)^{-1} K^2 \varepsilon),$$

(3.14) $$0 < \varepsilon_3 < \min(1, \varepsilon_1/(60)),$$

and divide U into N parts, $\tilde{V}, \ldots, \tilde{V}_N$, such that the diameter of each part is less than $\varepsilon_3/2$. Let

(3.15) $\quad L_1 = 2^6 (c_1 N P)^{1/2}(c_0 \varepsilon_1 \varepsilon^{1/2})^{-1}, \quad 0 < \delta < \Delta/L_1,$

(3.16) $\quad m = [p\delta^{-2}], \text{ the integer part of } p\delta^{-2}.$

Further take $\alpha > 0$ such that

$$P\{\psi(e_1 + \alpha) \leq K\} < \varepsilon/(16m),$$

(3.17) $$P\{\psi(e_1 - \alpha) \geq -K\} \leq \varepsilon/(16m).$$

Finally take

$$(3.18) \qquad 0 < \eta < \min\{(2m^{1/2})^{-1}, \ (2^8 mc_1^{1/2})^{-1} K\varepsilon\},$$

and

$$(3.19) \qquad L_2 \geq \min\{2\alpha\eta^{-1}, \ 2^7\alpha mc_1^{1/2}(K\varepsilon)^{-1}\}.$$

At first we consider the case of $\gamma \in \Lambda$. By (3.6) and the monotonicity of ψ,

$$
D(\gamma, L_1) = \sum_{i=1}^{n} \int_0^{-L_1 x_{in}'\gamma} (\psi(e_i + t) - \psi(e_i))\, dt - L_1 \sum_{i=1}^{n} x_{in}'\gamma \psi(e_i)
$$

$$
(3.20) \qquad \geq \sum_{i \in J^c} \int_0^{-L_1 x_{in}'\gamma} (\psi(e_i + t) - \psi(e_i))\, dt - L_1 \ \| \sum_{i=1}^{n} x_{in}\psi(e_i) \ \| .
$$

Let the non–empty sets among $\tilde{V}_1 \cap \Lambda, \ldots, \tilde{V}_N \cap \Lambda$ be $V_1, \ldots V_{N_1}$. Since the diameter of $L_1 V_j = \{L_1\gamma : \gamma \in V_j\}$ is less than $L_1\varepsilon/2$, it can be covered by a p–dimensional closed hyperrectangle T_j with a diameter less than $L_1\varepsilon_3$, $j = 1, \ldots, N_1$. Take a point $\gamma_j \in V_n$, $j = 1, \ldots N_1$. There are three cases for a fixed T_j. 1°. $(-x_{in}'\beta) \geq 0$ for each $\beta \in T_j$, then there exists a $\beta_{ij} \in T_j$ such that $(-x_{in}'\beta_{ij}) = \inf\{-x_{in}'\beta : \beta \in T_j\}$. 2°. $(-x_{in}'\beta) \leq 0$ for each $\beta \in T_j$, then there exists a $\beta_{ij} \in T_j$ such that $(-x_{in}'\beta_{ij}) = \sup\{-x_{in}'\beta : \beta \in T_j\}$. 3°. $(-x_{in}'\beta) > 0$ for some $\beta \in T_j$ and $(-x_{in}'\beta) < 0$ for some $\beta \in T_j$, then there exists a $\beta_{ij} \in T_j$ such that $(-x_{in}'\beta_{ij}) = 0$. Write

$$(3.21) \qquad \Psi(e_i, t) = \psi(e_i + t) - \psi(e_i) - G(t),$$

where

$$(3.22) \qquad G(t) = E\psi(e_i + t), \quad t \in R^1.$$

By the monotonicity of ψ and the selection of β_{ij},

$$
(3.23) \quad \inf_{\gamma \in \Lambda} \sum_{i \in J^c} \int_0^{-L_1 x_{in}'r} \{\psi(e_1 + t) - \psi(e_i)\}\, dt
$$

$$
\geq \inf_{1 \leq j \leq N_1} \sum_{i \in J^c} \int_0^{-x_{in}'\beta_{ij}} \{\psi(e_i + t) - \psi(e_i)\}\, dt
$$

$$
\geq \inf_{1 \leq j \leq N_1} \sum_{i \in J^c} \int_0^{-x_{in}'\beta_{ij}} G(t)\, dt - \sup_{1 \leq j \leq N_1} | \sum_{i \in J^c} \int_0^{-x_{in}'\beta_{ij}} \Psi(e_i, t)\, dt|.
$$

Since $\| \beta_{ij} - L_1\gamma_j \| < L_1\varepsilon_3$, by (3.14) we have

$$
| \sum_{i \in J^c} (x_{in}'\beta_{ij})^2 - \sum_{i \in J^c} (L_1 x_{in}'\gamma_j)^2 | \leq L_1^2(2 + \varepsilon_3)\varepsilon_3 \sum_i \| x_{in} \|^2
$$

$$
(3.24) \qquad\qquad\qquad\qquad < 3L_1^2\varepsilon_3 p < L_1^2\varepsilon_1/2.
$$

By the selection of β_{ij}, (3.9) and (3.15),

(3.25) $\quad |-x'_{in}\beta_{ij}| \le |L_1 x'_{in}\gamma_j| < \Delta \quad$ for $\quad i \in J^c, \quad j = 1, \ldots, N_1.$

By (2.1), (3.10), (3.11) and (3.24), and noting that $\gamma_j \in \Lambda$, we have

$$\inf_j \sum_{i \in J^c} \int_0^{-x'_{in}\beta_{ij}} G(t)\, dt \ge 2^{-1}c_0 \inf_j \sum_{i \in J^c}(x'_{in}\beta_{ij})^2$$

(3.26) $\qquad \ge \quad 2^{-1}c_0 L_1^2 \inf_j(\sum_{i \in J^c}(x'_{in}\gamma_j)^2 - \varepsilon_1/2) \quad \ge \quad c_0 L_1^2 \varepsilon_1/4.$

By (2.2), (3.15), (3.25), and the fact that $\| \beta_{ij} \| < L_1(1 + \varepsilon_3) < 2L_1$, we get

(3.27) $\quad P\{\sup_j | \sum_{i \in J^c} \int_0^{-x'_{in}\beta_{ij}} \Psi(e_i, t)\, dt| \ge c_0 L_1^2 \varepsilon_1/8\}$

$$\le \quad \sum_j P\left\{| \sum_{i \in J^c} \int_0^{-x'_{in}\beta_{ij}} \Psi(e_i, t)\, dt| \ge c_0 L_1^2 \varepsilon_1/8\right\}$$

$$\le \quad \sum_j 64(c_0 L_1^2 \varepsilon_1)^{-2} \sum_{i \in J^c} E\left(\int_0^{-x'_{in}\beta_{ij}} \Psi(e_i, t)\, dt\right)^2$$

$$\le \quad \sum_j 2^6(c_0 L_1^2 \varepsilon_1)^{-2} \sum_{i \in J^c} |x'_{in}\beta_{ij}| \, |\int_0^{-x'_{in}\beta_{ij}} E|\psi(e_i + t) - \psi(e_i)|^2\, dt|$$

$$\le \quad \sum_j 2^8 c_1 (c_0 L_1^2 \varepsilon_1)^{-2} \sum_{i \in J^c}(x'_{in}\beta_{ij})^2$$

$$< \quad 2^{10} N c_1 p(c_0 L_1 \varepsilon_1)^{-2} = \varepsilon/4.$$

By (2.2) and (3.15) we have

$$P\{L_1 \| \sum_{i=1}^n x_{in}\psi(e_i) \| \ge c_0 L_1^2 \varepsilon_1/8\}$$

$$\le 2^6(c_0 L_1 \varepsilon_1)^{-2} E \| \sum_{i=1}^n x_{in}\psi(e_i) \|^2$$

(3.28) $\qquad\qquad \le 2^6 c_1 p(c_0 L_1 \varepsilon_1)^{-2} \quad < \quad \varepsilon/4.$

From (3.20), (3.23), (3.26–3.28), it follows that

(3.29) $\qquad\qquad P\{\inf_{\gamma \in \Lambda} D(\gamma, L_1) \le 0\} < \varepsilon/2.$

Now we consider the case of $\gamma \in \Lambda^c$. By (3.11) and (3.14) we get $\frac{1}{2} > \varepsilon_1 > \gamma' S(\delta) \gamma = \sum_{i \in J^c}(x'_{in}\gamma)^2$, which implies that $\sum_{i \in J}(x'_{in}\gamma)^2 > \frac{1}{2}$. Since $\#(J) \leq m = [p\delta^{-2}]$, by (3.12) and (3.18) we have

$$
\begin{aligned}
\frac{1}{2} &< \sum_{i \in J \cap I_\gamma} (x'_{in}\gamma)^2 + \sum_{i \in J - I_\gamma} (x'_{in}\gamma)^2 \\
&\leq \sum_{i \in J \cap I_\gamma} (x'_{in}\gamma)^2 + \eta^2 m \quad < \sum_{i \in J \cap I_\gamma} |x'_{in}\gamma| + \frac{1}{4},
\end{aligned}
$$

which implies that

$$
(3.30) \qquad \sum_{i \in J \cap I_\gamma} |x'_{in}\gamma| > \frac{1}{4}.
$$

Put

$$(3.31) \quad A_n = \{\psi(e_i + \alpha) > K \text{ and } \psi(e_i - \alpha) < -K, \text{ for } i \in J\},$$

$$(3.32) \quad B_n = \{\alpha \sum_{i \in J} |\psi(e_i)| < L_2 K/16\},$$

and denote by A_n^c and B_n^c the complementary events of A_n and B_n respectively. By (3.17), (3.19) and $\#(J) \leq m$,

$$
\begin{aligned}
P(a_n^c \cup B_n^c) &< \varepsilon/8 + P\{\alpha \sum_{i \in J} |\psi(e_i)| \geq L_2 K/16\} \\
&< \varepsilon/8 + 16\alpha (L_2 K)^{-1} E \sum_{i \in J} |\psi(e_i)| \quad < \varepsilon/8 + 16 m \alpha c_1^{1/2} (L_2 K)^{-1}
\end{aligned}
$$

$$(3.33) \quad < \varepsilon/8 + \varepsilon/8 \;=\; \varepsilon/4.$$

By (3.12) and (3.19), $L_2|x'_{in}\gamma| \geq L_2\eta \geq 2\alpha$ for $i \in I_\gamma$. Noting this, the monotonicity of ψ and (3.30), we have on the event $A_n \cap B_n$

$$
(3.34) \quad \inf_{\gamma \in \Lambda^c} \int_0^{-L_2 x'_{in}\gamma} \psi(e_i + t)\, dt
$$

$$
\geq \inf_{i \in \Lambda^c} \sum_{i \in J \cap I_\gamma} \left\{ \int_0^{\alpha \, \text{sign}\, (-x'_{in}\gamma)} \psi(e_i)\, dt + \int_{\alpha \, \text{sign}\, (-x'_{in}\gamma)}^{-L_2 x'_{in}\gamma} \psi(e_i + t)\, dt \right\}
$$

$$
\geq \inf_{\gamma \in \Lambda^c} \sum_{i \in J \cap I_\gamma} K(L_2|x'_\epsilon\gamma| - \alpha) - \alpha \sum_{i \in J} |\psi(e_i)|
$$

$$
> L_2 K/8 - L_2 K/16 \;=\; L_2 K/16.
$$

By (3.33) and (3.34), we have

$$
(3.35) \quad P\left\{ \inf_{\gamma \in \Lambda^c} \sum_{i \in J \cap I_\gamma} \int_0^{-L_2 x'_{in}\gamma} \psi(e_i + t)\, dt \leq L_2 K/16 \right\} < \varepsilon/4.
$$

By the monotonicity of ψ, using the facts that $\#(J) \le m$ and $|x'_{in}\gamma| < \eta$ for $i \in I^c_\gamma$, we get

$$\int_0^{-L_2 x'_{in}\gamma} \psi(e_i + t)\, dt \ge \int_0^{-L_2 x'_{in}\gamma} \psi(e_i)\, dt \ge -L_2\eta|\psi(e_i)|, \quad i \in I^c_\gamma.$$

By (2.2) and (3.18) we have

$$P\left\{ \inf_{\gamma \in \Lambda^c} \sum_{i \in J - I_\gamma} \int_0^{-L_2 x'_{in}\gamma} \psi(e_i + t)\, dt \le -L_2 K/32 \right\}$$

$$\le P\{L_2\eta \sum_{i \in J} |\psi(e_i)| \ge L_2 K/32\} \le 32K^{-1}\eta E \sum_{i \in J} |\psi(e_i)|$$

$$(3.36) \quad \le 32K^{-1}\eta m c_1^{1/2} < \varepsilon/8.$$

For $\gamma \in \Lambda^c$, by the monotonicity of ψ we have

$$(3.37) \qquad \sum_{i \in J^c} \int_0^{-L_2 x'_{in}\gamma} \psi(e_i + t)\, dt \ge -L_2 \sum_{i \in J^c} \psi(e_i) x'_{in}\gamma.$$

Let the non–empty sets among $\tilde{U} \cap \Lambda^c, \ldots, \tilde{U}_M \cap \Lambda^c$ be $U_1, \ldots, U_{M_1}$. Take $\gamma_j \in U_j$, $j = 1, \ldots, M_1$. By (2.2) and (3.13),

$$P\left\{ \sup_{1 \le j \le M_1} \sup_{\gamma \in U_j} L_2| \sum_{i \in J^c} \psi(e_i) x'_{in}(\gamma - \gamma_j)| \ge L_2 K/64 \right\}$$

$$\le P\{\varepsilon_2 \| \sum_{i \in J^c} x_{in}\psi(e_i) \| \ge K/64\}$$

$$(3.38) \le (2^6\varepsilon_2 K^{-1})^2 E \| \sum_{i \in J^c} x_{in}\psi(e_i) \|^2 \le (2^6\varepsilon_2 K^{-1})^2 c_1 p < \varepsilon/16.$$

Since $\gamma'_j S(\delta)\gamma_j < \varepsilon_1$, by (2.2) and (3.14) we get

$$P\left\{ \sup_{1 \le j \le M_1} L_2| \sum_{i \in J^c} \psi(e_i) x'_{in}\gamma_j| \ge L_2 K/64 \right\}$$

$$\le \sum_{j=1}^{M_1} (2^6 K^{-1})^2 E| \sum_{i \in J^c} \psi(e_i) x'_{in}\gamma_j|^2$$

$$(3.39) \quad \le \sum_{j=1}^{M_1} (2^6 K^{-1})^2 c_1 \gamma'_j S(\delta)\gamma_j < 2^{12} K^{-2} M c_1\varepsilon_1 < \varepsilon/16.$$

By (3.37) – (3.39), we have

$$(3.40) \qquad P\left\{\inf_{\gamma \in \Lambda^c} \sum_{i \in J^c} \int^{-L_2 x'_{in}\gamma} \psi(e_i + t)\, dt \leq -L_2 K/32\right\} < \varepsilon/8.$$

From (3.35), (3.36) and (3.40) it follows that

$$(3.41) \qquad P\left\{\inf_{\gamma \in \Lambda^c} D(\gamma, L_2) \leq 0\right\} < \varepsilon/2.$$

By (3.7), (3.8), (3.29) and (3.41), we obtain

$$(3.42) \qquad \sup_{n \geq n_0} P\{\| \hat{\beta}_n \| \geq L\} \leq \varepsilon \text{ for } L \geq \max(L_1, L_2),$$

establishing Theorem 1.

References

[1] Bai, Z.D., Rao, C.R. and Wu, Y. (1989). Unified theory of *M*–estimation of linear regression parameters. Technical Report No. 89–04, Center for Multivariate Analysis, Penn State University.

[2] Huber, P.J. (1973). Robust regression: Asymptotics, conjectures and Monte Carlo. *Ann. Statist.* **1**, 799–821.

[3] Jurečková, J. (1989). Consistency of *M*–estimators of vector parameters. *Proc. 4th Prague Symp. Asympt. Statist.* (P. Mandel and M. Hušková eds.) Charles Univ. Prague. pp. 305–311.

[4] Pollard, D. (1991). Asymptotics for least absolute deviation regression estimators. *Econ. Theor.* **7**, 186–199.

[5] Rao, C.R. (1988). Methodology based on the L_1–norm in statistical inference. *Sankhyā* A, **50**, 289–313.

[6] Yohai, V.J. and Maronna, R.A. (1979). Asymptotic behavior of *M*–estimators for the linear model. *Ann. Statist.* **7**, 248–268.

Center for Multivariate Analysis
Pennsylvania State University
and
Graduate School, Academia Sinica
Beijing